Singular
Bilinear Integrals

Singular
Bilinear Integrals

Brian Jefferies
University of New South Wales, Australia

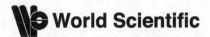

World Scientific

NEW JERSEY · LONDON · SINGAPORE · BEIJING · SHANGHAI · HONG KONG · TAIPEI · CHENNAI · TOKYO

Published by

World Scientific Publishing Co. Pte. Ltd.

5 Toh Tuck Link, Singapore 596224

USA office: 27 Warren Street, Suite 401-402, Hackensack, NJ 07601

UK office: 57 Shelton Street, Covent Garden, London WC2H 9HE

Library of Congress Cataloging-in-Publication Data

Names: Jefferies, Brian, 1956–

Title: Singular bilinear integrals / by Brian Jefferies (University of New South Wales, Australia).

Description: New Jersey : World Scientific, 2017. | Includes bibliographical references.

Identifiers: LCCN 2016048677 | ISBN 9789813207578 (hardcover : alk. paper)

Subjects: LCSH: Vector-valued measures. | Integrals. | Bilinear forms. |
    Ideal spaces. | Vector valued functions.

Classification: LCC QA325 .J44 2017 | DDC 518/.54--dc23

LC record available at https://lccn.loc.gov/2016048677

British Library Cataloguing-in-Publication Data

A catalogue record for this book is available from the British Library.

Printed in Singapore

To the memory of Igor Kluvánek

Preface

The idea for this monograph germinated at the "Vector Measures, Integration and Applications" conference held in Eichstätt (Germany) in September 2008. Three topics concerning bilinear integration inspired by the 1980 survey paper "Applications of Vector Measures" of I. Kluvánek [82] were treated in the conference talk [65]. Bilinear integration treats the problem of integrating a function with values in an infinite dimensional vector space with respect to a measure with values in another infinite dimensional vector space.

The concept of *decoupled* bilinear integrals has proved to be common to diverse applications of vector integration in quantum physics, stochastic analysis, scattering and operator theory. Decoupling is required when the classical theories of bilinear integration of R. Bartle [11] and I. Dobrakov [39,40] cannot be applied, as is the case in the applications just mentioned. The term *singular* is used somewhat loosely in the title to describe the situation where the classical theory of bilinear integration does not work. The study of decoupled bilinear integration affords the opportunity to touch upon the diverse and interesting subjects mentioned that lend themselves to the techniques of functional analysis.

An ingenue to mathematics may be entranced by the idea that Science depends on a reliable notion of the *measurement* of phenomena. Especially in quantum physics, this intuition is proved valid once we are forced to consider the spectral theory of differential operators. In quantum physics the results of physical observations are determined by the values of the self adjoint *spectral measure* associated with the operator determined by an observable quantity, so the study of vector measures and vector integration lies at the foundations of scientific enterprise.

Although the theory of integration of scalar quantities with respect to vector valued measures and the integration of vector valued quantities with respect to scalar valued measures seems to be largely settled, difficulties arise when integrating vector or operator valued functions with respect to vector or operator valued measures. Unfortunately, such problems regularly arise in quantum physics where spectral measures are fundamental in terms of physical observation. This monograph treats the mathematics that evolves from the integration of vector valued functions with respect to spectral measures that features in the representation of solutions in a number of problems in mathematical physics.

Thanks are due to my collaborators L. Garcia-Raffi, S. Okada and P. Rothnie in this work.

Chillingham, 2016　　　　　　　　　　　　　　　　　　　　*Brian Jefferies*

Contents

Chapter 1

Introduction

In classical measure theory, as given in [59] for example, the *variation* of a real or complex valued set function is an essential tool used to estimate the size of its range and space of integrable functions. For a finitely additive set function $\mu : \mathcal{A} \to \mathbb{C}$ defined on an algebra $\mathcal{A}$ of subsets of a nonempty set Σ, the variation $V(\mu) : \mathcal{A} \to [0, \infty]$ of μ is given by

$$V(\mu)(A) = \sup_{\mathcal{P}} \sum_{B \in \mathcal{P}} |\mu(A \cap B)|, \quad A \in \mathcal{P}.$$

The supremum is over all finite partitions $\mathcal{P}$ of Σ be elements of the algebra $\mathcal{A}$. The set function μ has uniformly bounded range on the algebra $\mathcal{A}$ if and only if μ has finite total variation $V(\mu)(\Sigma)$.

Once we move to infinite dimensional spaces and replace the modulus $|\cdot|$ by a norm $\|\cdot\|$, we find that many simple vector valued set functions with uniformly bounded range fail to have finite total variation or its variation may only take the values zero or infinity, see Example 1.1 below.

A simple example appears immediately in a Hilbert space $\mathcal{H}$ with an orthonormal basis $\{e_n\}_{n=1}^{\infty}$. The sequence $\{e_n/n\}_{n=1}^{\infty}$ is summable in $\mathcal{H}$ but not absolutely summable. The total variation $V(m)(\mathbb{N})$ of the countably additive set function $m : A \longmapsto \sum_{n \in A} e_n/n$, $A \subset \mathbb{N}$, is $V(m)(\mathbb{N}) = \sum_{n=1}^{\infty} 1/n = \infty$. By contrast, the *total semivariation*

$$\|m\|(\mathbb{N}) = \sup_{\xi \in \mathcal{H}, \|\xi\| \leq 1} \sum_{n=1}^{\infty} \left| \frac{\xi_n}{n} \right| = \left(\sum_{n=1}^{\infty} \frac{1}{n^2} \right)^{\frac{1}{2}}$$

of m is finite. For a measure m with values in a Banach space X, that is, a countably additive set function $m : \mathcal{E} \to X$ defined on a σ-algebra $\mathcal{E}$ of subsets of a nonempty set Σ, the semivariation $\|m\| : \mathcal{E} \to [0, \infty)$ defined in Section 1.1 below is always finite and features in the consideration of the

1

integration of scalar valued functions with respect to the vector measure m considered in Section 1.2.

The study of vector valued integration received an impetus from the spectral theory of selfadjoint operators on a Hilbert space $\mathcal{H}$ lying at the heart of the mathematical foundations of quantum mechanics [134]. Any selfadjoint operator $T : \mathcal{D}(T) \to \mathcal{H}$ with dense domain $\mathcal{D}(T)$ in $\mathcal{H}$ has a representation

$$T = \int_{\sigma(T)} \lambda \, dP_T(\lambda).$$

The *spectrum* $\sigma(T)$ of T is the set of all real numbers λ for which $\lambda I - T$ is not invertible in the collection $\mathcal{L}(\mathcal{H})$ of bounded linear operators on $\mathcal{H}$ and the *spectral measure* $P_T : \mathcal{B}(\sigma(T)) \to \mathcal{L}(\mathcal{H})$ associated with T is an operator valued measure σ-additive on the Borel σ-algebra $\mathcal{B}(\sigma(T))$ of $\sigma(T) \subset \mathbb{R}$ for the strong operator topology of $\mathcal{L}(\mathcal{H})$, that is, for all pairwise disjoint Borel sets $B_n \in \mathcal{B}(\sigma(T))$, $n = 1, 2, \ldots$, and every $h \in \mathcal{H}$, the sum $\sum_{n=1}^{\infty} P_T(B_n)h$ converges in the norm of $\mathcal{H}$ and

$$P_T\left(\bigcup_{n=1}^{\infty} B_n\right) h = \sum_{n=1}^{\infty} P_T(B_n)h.$$

The operator valued measure P_T has values in the selfadjoint projections so that $P_T(A \cap B) = P_T(A)P_T(B)$ for $A, B \in \mathcal{B}(\sigma(T))$ and $P_T(\sigma(T)) = I$, the identity operator on $\mathcal{H}$ [121, Definition 12.17]. In the case of a *compact* selfadjoint operator T, the integral becomes the sum

$$T = \sum_{\lambda \in \sigma(A)} \lambda . P_T(\{\lambda\})$$

in which $P_T(\{\lambda\})$ is the orthogonal projection onto the finite dimensional subspace of $\mathcal{H}$ generated by eigenvectors of T for the eigenvalue λ.

The strong operator topology of $\mathcal{L}(\mathcal{H})$ is defined by the collection $\{p_h : h \in \mathcal{H}\}$ of seminorms given by

$$p_h(T) = \|Th\|, \quad T \in \mathcal{L}(\mathcal{H}),$$

for each $h \in \mathcal{H}$. Because the strong operator topology is not metrisable unless $\mathcal{H}$ is finite dimensional, we shall consider vector measures taking values in a locally convex Hausdorff topological vector space E over $\mathbb{C}$, briefly a locally convex space or lcs, whose topology is defined by a family of seminorms [123, Section II.4].

The next section is a brief guide to the basic facts concerning measures with values in a locally convex space such as the space $\mathcal{L}(\mathcal{H})$ of operators on a Hilbert space $\mathcal{H}$ endowed with the strong operator topology.

Integration of vector valued functions with respect to scalar measures is considered in Section 1.3. The distinction between variation and semivariation features again with the Bochner integral and the Pettis integral.

Bilinear integration treats the integration of an X-valued function with respect to a Y-valued measure m with a given bilinear map from $X \times Y$ into Z. Typically X, Y and Z are infinite dimensional Banach spaces, but we shall also be concerned with the bilinear maps $(T, x) \longmapsto Tx$ and $(x, T) \longmapsto Tx$ with T a continuous linear map between Banach spaces E and F and $x \in E$. For bilinear integration, the X-semivariation of the Y-valued measure m in Z with respect to the bilinear map $(\cdot, \cdot)$ is a central concept.

In this work, we shall be mainly treating bilinear integration in tensor products, where there exists a tensor product topology τ for which the vector space Z is equal to the complete tensor product $X \widehat{\otimes}_\tau Y$ and $(x, y) \longmapsto x \otimes y$, $x \in X$, $y \in Y$, is the associated bilinear map. Various aspects of the X-semivariation of a Y-valued measure in a tensor product space $X \widehat{\otimes}_\tau Y$ are considered in Section 4.

The fundamental work of R. Bartle [11] treats the integration of an X-valued function with respect to a Y-valued measure m in Z by approximating in the X-semivariation of m. In the case where one of the vector spaces X or Y is finite dimensional, the integrals described in Sections 2 and 3 below are obtained. A further refinement of Bartle's approach due to I. Dobrakov is described in Section 5. The situation where there are sufficiently many sets with finite X-semivariation may be thought of as the case of *regular bilinear integration*.

If $\mathcal{H}$ is an infinite dimensional Hilbert space and P is a spectral measure acting on $\mathcal{H}$, then except in trivial cases, P fails to have finite $\mathcal{H}$-semivariation with respect to the bilinear map $(x, T) \longmapsto Tx$, $x \in \mathcal{H}$, $T \in \mathcal{L}(\mathcal{H})$. The applications of bilinear integration to quantum physics and operator theory considered here give rise to the *singular bilinear integrals* mentioned in the title in which X-semivariation plays no direct role.

By taking our integrals to have values in the tensor products such as $X \widehat{\otimes}_\tau Y$, $\mathcal{L}(\mathcal{H}) \widehat{\otimes}_\tau \mathcal{H}$ or $\mathcal{H} \widehat{\otimes}_\tau \mathcal{L}(\mathcal{H})$ with different choices of the tensor product topology τ, there is a type of *decoupling* between the values of the vector valued function that is being integrated with respect to a vector valued measure. Such a decoupling is already a feature of stochastic integration in probability theory where Brownian paths almost surely have infinite variation on any interval.

A general treatment of bilinear integration in tensor products is given in Chapter 2, based on joint work of the author with S. Okada [70]. The reminder of the monograph is devoted to applications of decoupled bilinear integration to diverse areas of analysis and mathematical physics.

For a bounded linear operator $T : L^2([0,1]) \to L^2([0,1])$ with an integral kernel k, we can try to view the trace $\mathrm{tr}(T)$ of T as the integral $\int_0^1 k(x,x)\,dx$, just as the trace of a matrix $\{a_{ij}\}_{i,j=1}^n$ is the sum $\sum_{i=1}^n a_{ii}$ of the diagonal elements. For a *trace class* operator on $L^2([0,1])$, this is in fact the case for a specific choice of integral kernel k. If we set $\Phi_k(x) = k(x,\cdot) \in L^2([0,1])$, $x \in [0,1]$, then $\mathrm{tr}(T)$ may be viewed as a type of bilinear integral

$$\int_0^1 \langle T_k, dm \rangle = \int_0^1 \langle \Phi_k, dm \rangle$$

with respect to the vector measure $m : B \to \chi_B$, $B \in \mathcal{B}([0,1])$.

Chapter 3 is based on [66] and explores the lattice (order) ideal of traceable operators T on a Banach function space X for which the bilinear integral $\int_\Sigma \langle T, dm \rangle$ exists. By contrast, on a Hilbert space, the collection of trace class operators is an operator ideal. The two classes coincide for hermitian positive operators T, that is, selfadjoint operators T whose spectrum $\sigma(T)$ is a subset of $[0, \infty)$. The ideas here resurface in Chapter 7 in the proof of the CLR inequality in quantum physics.

The topics of this monograph are heavily weighed in the direction of functional analysis and measure theory. A brief guide to probability theory and stochastic processes is given in Chapter 4. The final Section 4.4 is devoted to general stochastic integration which may be viewed as a type of bilinear integration that has been treated in Chapter 2.

The connection between stationary state and time dependent scattering theory is treated in Chapter 5 based on joint work of the author with L. Garcia-Raffi [49]. In order to treat this subject, it is necessary to integrate operator valued functions with respect to spectral measures, so the tools developed in Chapter 2 may be applied.

The treatment of random evolutions in Chapter 6 requires the integration of operator valued functions with respect to operator valued measures to obtain a version of the Feynman-Kac formula. Integrals of this type are of the *decoupled* variety treated in Chapter 2. This material is from [64]. The treatment of progressive measurability in Section 6.3 is from [68].

The Feynman-Kac formula reappears in Chapter 7 in the proof of the CLR inequality for dominated semigroups on a Hilbert space $L^2(\mu)$. As mentioned in Chapter 6, we can associate a σ-additive evolution process with a dominated semigroup on $L^2(\mu)$ and the Feynman-Kac formula with respect to this process is a feature of our proof of the CLR inequality in the semiclassical approximation of quantum physics.

Double operator integrals are a type of singular bilinear integral and the solution of certain operator equations are also expressed as bilinear integrals—the integral of a resolvent with respect to a spectral measure. Chapter 8 touches upon trace class perturbations of selfadjoint operators in scattering theory where we employ double operator integrals. Grothendieck's inequality and our discussion of Lusin filtrations in Chapter 3 are used to prove Peller's characterisation [104] of the class of double operator integrable functions for the space of bounded linear operators. We end with an elementary construction of Krein's spectral shift function as the boundary value of a harmonic function in the upper half plane.

Prerequisites. The subject of 'assumed knowledge' is a fraught consideration for an author of what may be viewed as a research monograph. Topics that may have been routinely discussed in graduate courses in mathematics in some countries 30 years ago may have fallen out of favour, despite lying at the foundations of future scientific discoveries. As should be clear from the preceding discussion, many routine facts from functional analysis are employed in our analysis.

H.H. Schaefer [123] provides a succinct account of topological vector spaces. Basic facts about Banach lattices may be found in [96]. Classical measure theory is exposed in [59] and [16] provides a connection with probability theory. A comprehensive but not exhaustive study of measure theory is given in the volumes of D. Fremlin [45, 46] that complements and is complemented by L. Schwartz [125]. Basic facts about complex analysis are found in [120] and the spectral theorem for selfadjoint operators and distribution theory are treated in [121].

The first few chapters of [38] and [86] contain most of what we need concerning vector valued measures. Other topics that arise relating to operator theory [76], harmonic analysis [54], stochastic processes [28] and scattering theory [136] are treated in the following chapters.

In view of the diverse range of mathematical topics listed, it should be apparent that this monograph promotes the view that the study of integration and measure in infinite dimensions continues to provide insights into mathematical analysis and its applications, just as it did as the mathematical foundations of quantum physics were laid by J. von Neumann in 1932 [134].

1.1   Vector measures

The collection of all continuous linear functionals on a lcs E is denoted by E'. The ordered pair $\langle E, E' \rangle$ forms a *dual pair* for which there is a bilinear map $(x, \xi) \longmapsto \langle x, \xi \rangle$ given by $\langle x, \xi \rangle = \xi(x)$ for $\xi \in E'$, $x \in E$. The notation is convenient because it is analogous to the inner product (h, g) for vectors h, g belonging to a Hilbert space $\mathcal{H}$, which we take to be linear in the first variable and antilinear in the second.

Given an E-valued function Φ defined over some set Σ, or a collection Σ of sets, we use the notation $\langle \Phi, x' \rangle$ to denote the function $\sigma \mapsto \langle \Phi(\sigma), x' \rangle$, $\sigma \in \Sigma$. Similarly, if Φ takes its values in the space $\mathcal{L}(E, F)$ of continuous linear maps from E into the lcs F, then for each $x \in E$ and $y' \in F'$, $\Phi x : \sigma \mapsto \Phi(\sigma)x$ and $\langle \Phi x, y' \rangle : \sigma \mapsto \langle \Phi(\sigma)x, y' \rangle$, for all $\sigma \in \Sigma$.

The Hahn-Banach Theorem establishes that the collection of all seminorms $x \longmapsto |\langle x, \xi \rangle|$, $x \in E$, for $\xi \in E'$ separates the points of E and so defines a locally convex Hausdorff topology $\sigma(E, E')$ on E called the *weak topology*.

Let $(\Sigma, \mathcal{E})$ be a measurable space, that is, $\mathcal{E}$ is a σ-algebra of subsets of a set Σ. The term *measure space* is used for the triple $(\Sigma, \mathcal{E}, \mu)$ with $(\Sigma, \mathcal{E})$ a measurable space and $\mu : \mathcal{E} \to [0, \infty]$ an extended real valued measure, meaning that the equality

$$\mu \left(\bigcup_{n=1}^{\infty} A_n \right) = \sum_{n=1}^{\infty} \mu(A_n)$$

of extended real numbers holds for any pairwise disjoint sets $A_n \in \mathcal{E}, n = 1, 2, \dots$. If a measure μ has infinite values, then the family

$$\{ A \in \mathcal{E} : \mu(A) < \infty \}$$

of subsets of Ω constitutes a δ-*ring*, that is, a *ring* $\mathcal{D}$ of sets closed under the operations of symmetric difference Δ and intersection $\cap$ such that

$$\bigcap_{n=1}^{\infty} A_n \in \mathcal{D}$$

for all $A_n \in \mathcal{D}$, $n = 1, 2, \dots$. By restricting analysis to measures defined on δ-rings, extended real numbers can be avoided.

A *vector measure* $m : \mathcal{E} \to E$ with values in a locally convex space E is a set function that is countably additive (σ-additive) in the locally convex topology of E. By this we mean that for any pairwise disjoint sets $A_n \in \mathcal{E}, n = 1, 2, \dots$, the equality

$$m \left(\bigcup_{n=1}^{\infty} A_n \right) = \sum_{n=1}^{\infty} m(A_n)$$

holds. The sum on the right-hand side converges in the topology of E. In the case that $E = \mathbb{R}$ or $E = \mathbb{C}$, m is called a *scalar measure*.

In the special case of an operator valued measure $M : \mathcal{E} \to \mathcal{L}_s(X)$ acting on a Banach space X, σ-additivity is always assumed to be for the *strong operator topology* of $\mathcal{L}(X)$: for any pairwise disjoint sets $A_n \in \mathcal{E}, n = 1, 2, \ldots$, the equality

$$M \left(\bigcup_{n=1}^{\infty} A_n \right) x = \sum_{n=1}^{\infty} M(A_n) x$$

holds in X for each $x \in X$.

The union $\cup_{n=1}^{\infty} A_n$ is the same however the pairwise disjoint family $\{A_n : n = 1, 2, \ldots\}$ of sets is indexed, so the sums above are also independent of the ordering of the index, giving rise to the notion that a sequence of vectors $x_n, n = 1, 2, \ldots$, in a lcs E is *unconditionally summable* if there exists $x \in E$ with the property that for every neighbourhood U of 0 in E, there is a finite set K of natural numbers such that $x - \sum_{j \in J} x_j \in U$, for any finite set J of natural numbers containing K. The term *weakly unconditionally summable* is used in the case that the sequence is unconditionally summable for the weak topology of E.

A basic result is that as far as unconditional summability is concerned, the weak and norm topologies of a normed space are equivalent.

Theorem 1.1 (Orlicz-Pettis [38, I.4.4]**).** *Let E be a lcs. A sequence of vectors $x_n, n = 1, 2, \ldots$, in E is weakly unconditionally summable if and only if it is unconditionally summable in any topology consistent with the duality between E and E'.*

If m is countably additive in the weak topology of E, then by the Orlicz-Pettis lemma, it is automatically countably additive in the original topology of E. Terms such as 'm-a.e.' have the same meaning as for scalar measures: off an m-null set, by which we mean a set $N \in \mathcal{E}$ such that $m(A) = 0$ for all $A \in \mathcal{E}$ contained in N.

A scalar measure $\mu : \mathcal{E} \to \mathbb{C}$ has bounded range on $\mathcal{E}$ [38, I.1.19], so the *variation* $|\mu|$ of μ is a finite nonnegative measure on $\mathcal{E}$ defined by

$$|\mu|(A) = \sup \left\{ \sum_j |\mu(A_j \cap A)| \right\}$$

for all $A \in \mathcal{E}$. The supremum is taken over the family $\Pi_{\mathcal{E}}$ of all pairwise disjoint subsets $A_1, \ldots, A_k$ of Σ belonging to $\mathcal{E}$, and all $k = 1, 2, \ldots$. The same definition is adopted in the case that $\mathcal{E}$ is an algebra of subsets of Σ and $m : \mathcal{E} \to \mathbb{C}$ is an additive set function with bounded range.

Suppose that $m : \mathcal{E} \to E$ is a vector measure and p is a continuous seminorm on E. The p-*semivariation* of m is the set function $p(m) : \mathcal{E} \to [0, \infty)$ defined by $p(m)(A) = \sup\{|\langle m, \xi \rangle|(A)\}$ for all $A \in \mathcal{E}$. The supremum is taken over all elements ξ belonging to the polar

$$U_p^\circ = \{\xi : |\langle x, \xi \rangle| \leq 1, \ \forall x \in E, \ p(x) \leq 1\}$$

of U_p.

An application of the uniform boundedness principle to the associated family of scalar measures shows that for every continuous seminorm p on E, the p-semivariation takes finite values, see Proposition 1.1 below. However, the p-*variation* of m defined by $V_p(m)(A) = \sup\{\sum_j p(m(A_j \cap A))\}$ may take the value infinity. The supremum here is again taken over $\Pi_{\mathcal{E}}$.

Example 1.1. Let $1 < p < \infty$. Let $\mathcal{B}([0,1])$ be the Borel subsets of $[0,1]$ and let $m : \mathcal{B}([0,1]) \to L^p([0,1])$ be the vector measure defined by $m(A) = \chi_A$, for every $A \in \mathcal{B}([0,1])$. Denote the Lebesgue measure on $\mathcal{B}([0,1])$ by λ.

For any Borel set A contained in $[0,1]$ such that $\lambda(A) > 0$ and any $n = 1, 2, \ldots$, we can find disjoint Borel subsets $A_1, \ldots, A_n$ of A such that $\lambda(A_i) = \lambda(A)/n$ for all $i = 1, \ldots, n$. Then

$$\sum_{i=1}^n \|m(A_i)\|_p = \sum_{i=1}^n (\lambda(A)/n)^{1/p} = n^{1-1/p} \lambda(A)^{1/p}.$$

Plainly this means that the $\| \cdot \|_p$-variation of m is infinite on A.

We may also describe $V_p(m)$ as the smallest positive measure μ such that $p(m(A)) \leq \mu(A)$, for every $A \in \mathcal{E}$. In the case that E is a Banach space with norm $\| \cdot \|$, the semivariation of m is written as $\|m\|$ and the variation of m as $V(m)$. The normed space E is finite dimensional if and only if the norms $m \mapsto \|m\|(\Sigma)$ and $m \mapsto V(m)(\Sigma)$ are equivalent. More accurately, a result of Dvoretsky-Rogers shows that if $\mathcal{E}$ is infinite and $V(m)(\Sigma) < \infty$ for every E-valued measure $m : \mathcal{E} \to E$, then E is finite dimensional [123, IV.10.7, Corollary 3].

The same notation is adopted in the case that $\mathcal{E}$ is an algebra of subsets of Σ and $m : \mathcal{E} \to Y$ is an additive set function with bounded range.

For two vector measures $m : \mathcal{E} \to X$ and $n : \mathcal{E} \to Y$ with values in locally convex spaces X and Y, we write $n \ll m$ if every m-null set is an n-null set. If X, Y are normed spaces, then $\lim_{\|m\|(A) \to 0^+} \|n\|(A) = 0$. It is clear that a set $A \in \mathcal{E}$ is m-null if and only if $p(m)(A) = 0$ for every continuous seminorm p on X.

The p-semivariation of a vector measure is related to its range by the following estimates.

Proposition 1.1 ([38, Proposition I.1.11]). *Let* $m : \mathcal{E} \to E$ *be a vector measure and let* p *be a continuous seminorm on* E. *Then for every* $A \in \mathcal{E}$, *we have*

$$\sup\{p(m(B)) : B \subseteq A, \ B \in \mathcal{E}\} \leq p(m)(A)$$
$$\leq 4 \sup\{p(m(B)) : B \subseteq A, B \in \mathcal{E}\}.$$

Let E be a lcs. A family Λ of E-valued measures on $\mathcal{E}$ is called *uniformly countably additive* if for any sequence of sets A_n, $n = 1, 2, \ldots$, in $\mathcal{E}$ decreasing to the empty set and every continuous seminorm p on E, we have $\sup_{\mu \in \Lambda} p(\mu)(A_n) \to 0$ as $n \to \infty$. The family Λ is *bounded* if $\sup_{\mu \in \Lambda} p(\mu)(\Omega) < \infty$ for every continuous seminorm p on E. For a vector measure m and a continuous seminorm p, the uniform countable additivity and boundedness of $|\langle m, \xi \rangle|$, $\xi \in U_p^\circ$, implies the following result [86, Theorem II.1.1].

Theorem 1.2 (Bartle-Dunford-Schwartz). *Let* Λ *be a bounded and uniformly countably additive family of scalar measures* $\mu : \mathcal{E} \to \mathbb{C}$. *Then there exists a finite positive measure* λ *on* $\mathcal{E}$ *such that* $\lambda(A) \leq \sup_{\mu \in \Lambda} |\mu|(A)$ *for all* $A \in \mathcal{E}$ *and* $\lambda(A) \to 0$, $A \in \mathcal{E}$, *implies that* $\sup_{\mu \in \Lambda} |\mu|(A) \to 0$.

Corollary 1.1. *Let* $m : \mathcal{E} \to E$ *be a vector measure. For every continuous seminorm* p *on* E, *there exists a finite positive measure* λ_p *on* $\mathcal{E}$ *such that* $\lambda_p(A) \leq p(m)(A)$, *for every* $A \in \mathcal{E}$, *and* $\lambda_p(A) \to 0$, $A \in \mathcal{E}$, *implies that* $p(m)(A) \to 0$.

Theorem 1.3 (Nikodym Boundedness Theorem, [38, I.3.1]). *Let* $\langle m_\iota \rangle_{\iota \in I}$ *be a family of vector measures* $m_\iota : \mathcal{E} \to E, \iota \in I$, *such that* $\langle m_\iota(A) \rangle_{\iota \in I}$ *is a bounded subset of* E *for every* $A \in \mathcal{E}$. *Then for every continuous seminorm* p *on* E, $\sup_{\iota \in I} p(m_\iota)(\Sigma) < \infty$.

Theorem 1.4 (Vitali-Hahn-Saks, [38, I.4.8]). *Let* $m_n, n = 1, 2, \ldots,$ *be* E-*valued measures on* $\mathcal{E}$ *such that* $m_n(A), n = 1, 2, \ldots,$ *converges in* E *for every* $A \in \mathcal{E}$. *Then* $\langle m_n \rangle_{n \in \mathbb{N}}$ *is a bounded and uniformly countably additive family of vector measures. In particular, the set function* $m : \mathcal{E} \to E$ *defined by* $m(A) = \lim_{n \to \infty} m_n(A)$ *for each* $A \in \mathcal{E}$ *is countably additive.*

Corollary 1.2. *Let* $m_n, n = 1, 2, \ldots,$ *be* E-*valued measures on* $\mathcal{E}$ *such that the vectors* $m_n(A), n = 1, 2, \ldots,$ *converge in* E *for every* $A \in \mathcal{E}$. *For every* *continuous seminorm p on E, there exists a finite positive measure λ_p on $\mathcal{E}$ such that $\lambda_p(A) \to 0$, $A \in \mathcal{E}$, implies that $\sup_{n \in \mathbb{N}} p(m_n)(A) \to 0$.*

Further important properties of vector measures result from studying their ranges [86, Theorem IV.6.1]. A lcs is said to be *complete* if every Cauchy net converges. It is *quasicomplete* if every bounded Cauchy net converges. A Cauchy sequence $\{x_n\}_{n=1}^{\infty}$ in E is necessarily bounded, so it converges in a quasicomplete space E.

Theorem 1.5. *Let E be a quasicomplete lcs. If $m : \mathcal{E} \to E$ is a vector measure, then the range $m(\mathcal{E})$ of m is relatively weakly compact.*

Of course, set functions are usually defined by an intuitive procedure on an elementary family of sets and then extended, by some method, to a more complicated family. The following extension theorem of Carathéodory-Hahn-Kluvánek [38, I.V.2] does the job for vector measures. For σ-additivity on an *algebra* $\mathcal{A}$, attention is restricted to a countable family of pairwise disjoint sets from $\mathcal{A}$ whose union also belongs to $\mathcal{A}$.

Theorem 1.6. *Let E be a quasicomplete lcs, let $\mathcal{A}$ be an algebra of subsets of Σ and let $\mathcal{E}$ be the σ-algebra generated by $\mathcal{A}$. A σ-additive set function $m : \mathcal{A} \to E$ is the restriction to $\mathcal{A}$ of a unique vector measure $\tilde{m}$ defined on $\mathcal{E}$ if and only if the range $m(\mathcal{A})$ of m is relatively weakly compact, and in this case, $\tilde{m}(\mathcal{E}) \subseteq \overline{m(\mathcal{A})}$.*

It follows from this theorem that if $\mu : \mathcal{E} \to \mathbb{C}$ is a scalar measure defined on the σ-algebra $\mathcal{E}$ of Σ and $m : \mathcal{S} \to E$ is a vector measure with values in the quasicomplete lcs E, then there exists a unique vector measure $\mu \otimes m : \mathcal{E} \otimes \mathcal{S} \to E$ defined on the σ-algebra $\mathcal{E} \otimes \mathcal{S}$ generated by $\mathcal{E}$ and $\mathcal{S}$ such that $(\mu \otimes m)(E \times S) = \mu(E)m(S)$ for every $E \in \mathcal{E}$ and $S \in \mathcal{S}$.

1.2   Integration of scalar functions with respect to a vector valued measure

Let $\mathcal{E}$ be a σ-algebra of subsets of a set Σ. Let $m : \mathcal{E} \to E$ be a vector measure with values in a lcs E. A scalar function $f : \Sigma \to \mathbb{C}$ is *m-integrable* in E if it is integrable with respect to the scalar measure $\langle m, y' \rangle$ for every $y' \in E'$, and for each $A \in \mathcal{E}$, there exists a vector $fm(A) \in E$ such that

$$\langle fm(A), y' \rangle = \int_A f \, d\langle m, y' \rangle, \quad \text{for all } y' \in E'.$$

The set function $fm : \mathcal{E} \to E$ is σ-additive in the original topology of E by the Orlicz-Pettis lemma. We shall often adopt the more conventional notations

$$\int_A f \, dm \quad \text{and} \quad \int_A f(\gamma) \, dm(\gamma)$$

for $fm(A)$, $A \in \mathcal{E}$, and write $m(f) = fm(\Sigma)$ for the definite integral of f with respect to the vector measure m.

It is often convenient and natural to regard integration as forming the product $f.m$ of a function f with respect to a measure m. When we consider the bilinear integral of an X-valued function f with respect to a Y-valued measure m, using analogous notation, the integral $f \otimes m$ has values in the tensor product $X \widehat{\otimes} Y$ completed with respect to a suitable topology.

The case of integration of a scalar function $f : \Sigma \to \mathbb{C}$ with respect to a measure $m : \mathcal{E} \to \mathcal{L}_s(X)$ taking values in the space of operators acting on a Banach space X is of special significance. Then, the dual space of $\mathcal{L}_s(X)$ with the strong operator topology may be identified with $X \otimes X'$ [123, Corollary IV.3.4] and for each $x \in X$ and $x' \in X'$, the function f is $\langle mx, x' \rangle$-integrable and for each $A \in \mathcal{E}$, there exists $fm(A) \in \mathcal{L}(X)$ such that $\langle fm(A)x, x' \rangle = \int_A f \, d\langle mx, x' \rangle$, for all $x \in X$ and $x' \in X'$.

The indefinite integral fm of an m-integrable f with respect to an E-valued measure m is a vector measure, so the p-semivariation of fm with respect to the continuous seminorm p defines the seminorm $p(m)$ on the space $\mathcal{L}^1(m)$ of m-integrable functions, that is, $p(m)(f) = p(fm)(\Omega)$ for every $f \in \mathcal{L}^1(m)$. According to [86, Lemma II.2.2], the equality $p(m)(f) = \sup\{|\langle m, \xi \rangle|(|f|) : \xi \in U_p^\circ\}$, $f \in \mathcal{L}^1(m)$, is valid. Note that this implies that $p(m)(\Omega) = \sup\{p(m(f))\}$, where the supremum is taken over all $\mathcal{S}$-simple functions f with $\|f\|_\infty \leq 1$.

The following convergence theorem for vector measures together with the analogues of the Beppo Levi convergence theorem and the monotone convergence theorem are proved in [86, II.4].

Theorem 1.7. (Dominated convergence) *Let E be a quasicomplete lcs and let $m : \mathcal{S} \to E$ be a vector measure. If $\{f_n\}_{n=1}^\infty$ is a sequence of m-integrable functions converging m-a.e. to a function f, and if there is an m-integrable function g with $|f_n| \leq g$ m-a.e., for all $n = 1, 2, \ldots$, then f is m-integrable, $m(f_n) \to m(f)$ and $p(m)(f - f_n) \to 0$ as $n \to \infty$, for every continuous seminorm p on E.*

It follows that every bounded $\mathcal{S}$-measurable function is m-integrable [86, II.3 Lemma 1] and the space $\mathcal{L}^1(m)$ of functions integrable with respect to a vector measure $m : S \to E$ is a vector lattice.

An m-integrable function f is said to be m-*null* if its indefinite integral is (identically) the zero vector measure. Two m-integrable functions f, g are m-*equivalent* if the function $|f - g|$ is m-null. We also say that f and g are equal m-almost everywhere (m-a.e.). The class of all m-integrable functions m-equivalent to an m-integrable function f is denoted by $[f]_m$.

Denote by $\tau(m)$ the topology on $\mathcal{L}^1(m)$ determined by the family of seminorms $f \mapsto p(m)(f), f \in \mathcal{L}^1(m)$, for every continuous seminorm p on E.

The quotient space of $\mathcal{L}^1(m)$ modulo the subspace of all m-null functions is denoted by $L^1(m)$. The resulting topology, denoted again by $\tau(m)$, turns $L^1(m)$ into a lcs under the corresponding notions of pointwise addition and scalar multiplication almost everywhere. Now $L^1(m) = \{[f]_m : f \in \mathcal{L}^1(m)\}$, so if we put $p(m)([f]_m) = p(m)(f)$, for every $f \in \mathcal{L}^1(m)$ and every continuous seminorm p on E, the resulting system of seminorms defines the topology $\tau(m)$ on $L^1(m)$. It is clear that the topology $\tau(m)$ is analogous to the usual L^1-norm topology of the classical Lebesgue space. In the case where $E = \mathbb{C}$, the space $L^1(m)$ is the standard Lebesgue space with the topology defined by its norm.

As in the case of a scalar measure, for $1 \leq r < \infty$, we may define

$$L^r(m) = \{[f]_m : f|f|^{r-1} \in \mathcal{L}^1(m)\},$$
$$p_r(m)(f) = p(m)(|f|^r), \quad [f]_m \in L^r(m),$$

and give $L^r(m)$ the topology defined by the family of seminorms

$$[f]_m \longmapsto p_r(m)(f), \quad [f]_m \in L^r(m)$$

for every continuous seminorm p on E. For $r = \infty$,

$$\|[f]_m\|_\infty = \inf\{M : |f| \leq M \ m\text{-a.e.}\}$$

and $L^\infty(m)$ is the Banach space of classes $[f]_m$ of S-measurable functions f for which $\|[f]_m\|_\infty < \infty$ equipped with the norm $\| \cdot \|_\infty$.

1.3 Integration of vector valued functions with respect to a scalar measure

In infinite dimensional vector spaces such as linear spaces of functions or operators, there is an unavoidable distinction between 'strong' integrals and 'weak' integrals. It is fair to say that, mostly, estimates that facilitate the convergence of strong integrals are routine whereas, say, proving that certain classes of integrals converge weakly may be relevant to the foundations of mathematics itself.

1.3.1 *The Pettis integral*

A definition similar to integration with respect to vector measures may be adopted for the integration with respect to a scalar measure $\mu : \mathcal{E} \to \mathbb{C}$, of functions $f : \Sigma \to E$ with values in a locally convex space E.

The function f is said to be *scalarly μ-integrable* if for all $y' \in E'$, the scalar function $\langle f, y' \rangle$ is μ-integrable. Then f is *μ-integrable* if it is scalarly integrable, and for each $A \in \mathcal{E}$, there exists a vector $f\mu(A) \in E$ such that

$$\langle f\mu(A), y' \rangle = \int_A \langle f, y' \rangle \, d\mu, \quad \text{for all } y' \in E'.$$

As before, $f\mu : \mathcal{E} \to E$ is σ-additive, and we write $\mu(f) = f\mu(\Sigma)$ and $\int_A f \, d\mu = f\mu(A)$, for every $A \in \mathcal{E}$. The special case of the locally convex space space $E = \mathcal{L}(X)$ of operators acting on a Banach space X has a direct translation, which goes as follows.

A function $f : \Sigma \to \mathcal{L}(X)$ is said to be *scalarly μ-integrable* if for all $x \in X$ and $x' \in X'$, the scalar function $\langle fx, x' \rangle$ is μ-integrable. Then f is *μ-integrable* if it is scalarly integrable, and for each $A \in \mathcal{E}$, there exists a vector $f\mu(A) \in \mathcal{L}(X)$ such that

$$\langle f\mu(A)x, x' \rangle = \int_A \langle fx, x' \rangle \, d\mu, \quad \text{for all } x \in X, \ x' \in X'.$$

We shall sometimes need to distinguish integrability in the weak sense from stronger forms of integrability, which we shall describe shortly; a vector valued function integrable in the above sense is said to be *Pettis μ-integrable* [38, II.3.2].

A survey of the Pettis integral and a list of references up to 2002 has been given by K. Musial [98] and M. Talagrand gives an earlier treatment [130] of deep measure theoretic aspects of the Pettis integral.

The definition of the Pettis integral is natural because it is analogous to the integration of a scalar valued function with respect to a vector measure. However, it is not innocuous—for example, the statement "for every Banach space X, every bounded and scalarly Lebesgue measurable function $f : [0,1] \to X$ is Pettis integrable with respect to Lebesgue measure" is independent of ZFC, the usual axioms of set theory with the Axiom of Choice [98, Theorem 8.2]. Even for a separable Hilbert space $\mathcal{H}$, in which scalar integrablity and Pettis integrability are equivalent, the existence of Pettis integrals like $\int_0^\infty f(t) \frac{dt}{t}$ for certain $\mathcal{H}$-valued functions f is related to fundamental problems in harmonic analysis and operator theory [69]; for these examples, it is clear that $\int_0^\infty \|f(t)\|_\mathcal{H} \frac{dt}{t} = \infty$.

1.3.2   The Bochner integral

Now suppose that $(X, \|\cdot\|_X)$ is a Banach space and μ is a scalar measure on $\mathcal{E}$. A function $f : \Sigma \to X$ is said to be *strongly μ-measurable* if it is the limit μ-almost everywhere of X-valued $\mathcal{E}$-simple functions, that is, functions that are finite sums of functions $x\chi_E$ for any $x \in X$ and $E \in \mathcal{E}$. The integral of X-valued $\mathcal{E}$-simple functions is defined in the obvious way. For strongly μ-measurable functions, $\|f\|_X : \gamma \mapsto \|f(\gamma)\|_X, \gamma \in \Sigma$, is μ-measurable.

We say that a strongly μ-measurable function f is *Bochner μ-integrable* if the integral $\int_\Sigma \|f\|_X \, d|\mu|$ is finite. It turns out that for such functions, there exist X-valued $\mathcal{E}$-simple functions f_k, $k = 1, 2, \ldots$, with the property that $f_k \to f$ μ-a.e. as $k \to \infty$ and the integrals $\int_\Sigma \|f_k - f_j\|_X \, d|\mu|$ converge to zero as $k, j \to \infty$ [38, II.2.2]. Hence, f is Pettis μ-integrable and $\int_A f_k \, d\mu \to f\mu(A)$ in X as $k \to \infty$, uniformly for all $A \in \mathcal{E}$.

Let $1 \le p < \infty$. Then $L^p(\Sigma, \mathcal{E}, \mu; X)$ denotes the vector space of μ-equivalence classes $[f]_\mu$ of μ-strongly measurable functions $f : \Sigma \to X$ such that $\|f\|_X^p$ is μ-integrable. It is a Banach space under the norm

$$\|[f]_\mu\|_p = \left(\int_\Sigma \|f\|_X^p \, d|\mu| \right)^{1/p}.$$

In most circumstances, we write f instead of $[f]_\mu$. The subscript is omitted in the case that μ is Lebesgue measure. For $p = \infty$, the space $L^\infty(\Sigma, \mathcal{E}, \mu; X)$ denotes the Banach space of (equivalence classes of) strongly μ-measurable functions $f : \Sigma \to X$ for which $\|f\|_X$ is μ-essentially bounded. The norm is the μ-essential bound $\|f\|_\infty$ of $\|f\|_X$, that is,

$$\|f\|_\infty = \inf\{c : \|f\|_X \le c \ \mu\text{-a.e.}\}.$$

The same terminology and notation is adopted in the case that μ is a nonnegative measure.

If Σ is a nonempty rectangle in Euclidean space $\mathbb{R}^n$, $\mathcal{E}$ is the Borel σ-algebra $\mathcal{B}\Sigma$ of Σ and μ is the n-dimensional Lebesgue measure, we use the notation $L^p(\Sigma)$ in place of $L^p(\Sigma, \mathcal{E}, \mu)$, and $L^p(\Sigma; X)$ in place of $L^p(\Sigma, \mathcal{E}, \mu; X)$.

1.4 Tensor products

Our brief discussion of topological tensor products is taken from [123, III.6, IV.9]. A lively treatment of the historical background may be found in [38, VIII.5]. We start with the algebraic definition of a tensor product.

Given two vector spaces E and F over the same set of scalars $\mathbb{R}$ or $\mathbb{C}$, the vector space of all bilinear forms on $E \times F$ is denoted by $B(E, F)$. For each $x \in E$ and $y \in F$, the evaluation map $u_{x,y} : f \longmapsto f(x, y)$, $f \in B(E, F)$, is a linear form on $B(E, F)$ and so an element of the algebraic dual $B(E, F)^*$ of the space $B(E, F)$ of bilinear forms. The mapping $\chi : (x, y) \longmapsto u_{x,y}$ of $E \times F$ into $B(E, F)^*$ is itself bilinear. The vector space generated by its range $\chi(E \times F)$ is denoted by $E \otimes F$. The map χ is called the *canonical bilinear map* of $E \times F$ into the *tensor product* $E \otimes F$ of E and F. Denoting the element $u_{x,y}$ of $E \otimes F$ by $x \otimes y$ for each $x \in E$ and $y \in F$, each element of $E \otimes F$ is a finite sum $\sum_j \lambda_j (x_j \otimes y_j)$ with the sum over the empty set being 0.

If G is another vector space over the same set of scalars, the mapping $u \longmapsto u \circ \chi$ is an isomorphism of the space $L(E \otimes F, G)$ of all linear maps from the vector space $E \otimes F$ into G onto the space $B(E, F; G)$ of all bilinear maps from $E \times F$ into G [123, III.6.1]. Consequently, the algebraic dual $(E \otimes F)^*$ of the tensor product $E \otimes F$ of E and F may be identified with the vector space of all bilinear forms on $E \times F$.

Now suppose that E and F are locally convex spaces. The tensor product equipped with a topology τ is written as $E \otimes_\tau F$. A locally convex topology τ on $E \otimes F$ is *compatible with the tensor product* if the following conditions hold:

a) the canonical map $\chi : E \times F \to E \otimes_\tau F$ is separately continuous;
b) for every $\xi \in E'$ and $\eta \in F'$, the linear functional defined by

$$\xi \otimes \eta : x \otimes y \longmapsto \langle x, \xi \rangle . \langle y, \eta \rangle, \quad x \in E, \ y \in F,$$

belongs to $(E \otimes_\tau F)'$;
c) if $G_1 \subset E'$ is equicontinuous on E and $G_2 \subset F'$ is equicontinuous on F, then $G_1 \otimes G_2 = \{\xi \otimes \eta : \xi \in G_1, \ \eta \in G_2\}$ is an equicontinuous family of linear functionals on $E \otimes_\tau F$.

In the context of normed vector spaces X and Y, a *tensor product norm* $\| \cdot \|_\tau$ on $X \otimes Y$ has the property that there exists $C > 0$ such that

(T1) $\|x \otimes y\|_\tau \le C \|x\| \, \|y\|$ for all $x \in X$ and $y \in Y$, and

(T2) $X' \otimes Y'$ may be identified with a linear subspace of the continuous
    dual $(X \otimes_\tau Y)' = (X \widehat{\otimes}_\tau Y)'$ of $X \otimes_\tau Y$ and $\|x' \otimes y'\| \leq C\|x'\| \, \|y'\|$ for
    all $x' \in X'$ and $y \in Y'$.

The topology defined by a tensor product norm is clearly compatible with
the tensor product as defined above. The tensor product $X \otimes Y$ equipped
with the topology defined by the norm $\| \cdot \|_\tau$ is written here as $X \otimes_\tau Y$ and
its completion is $X \widehat{\otimes}_\tau Y$. The term *cross-norm* is used if $C = 1$.

Example 1.2. Let $(\Sigma, \mathcal{E}, \mu)$ be a measure space, $1 \leq p < \infty$ and let X be
a Banach space. For an element $f \in L^p(\Sigma, \mathcal{E}, \mu)$ and a vector $x \in X$, it
is natural to identify $f \otimes x$ with the μ-equivalence class of the X-valued
function $\sigma \longmapsto f(\sigma)x$ defined for μ-almost all σ. Then
$$\|f \otimes x\|_{L^p(\Sigma, \mathcal{E}, \mu; X)} = \|f\|_p \|x\|.$$
Moreover, if $1 < q \leq \infty$ is the dual index satisfying $1/p + 1/q = 1$ and
$g \in L^q(\Sigma, \mathcal{E}, \mu)$ and $x' \in X'$, then
$$\|g \otimes x'\|_{L^q(\Sigma, \mathcal{E}, \mu; X')} = \|g\|_q \|x'\|.$$
For every $f \in L^p(\Sigma, \mathcal{E}, \mu; X)$, including elements of the tensor product
$L^p(\Sigma, \mathcal{E}, \mu) \otimes X \subset L^p(\Sigma, \mathcal{E}, \mu; X)$, we can write
$$\langle f, g \otimes x' \rangle = \int_\Sigma \langle f, x' \rangle . g(\sigma) \, d\mu(\sigma)$$
and because
$$\int_\Sigma |\langle f, x' \rangle| . |g(\sigma)| \, d\mu(\sigma) \leq \|f\|_{L^p(\Sigma, \mathcal{E}, \mu; X)} \|g\|_q \|x'\|,$$
the tensor product $g \otimes x'$ defines a continuous linear function on
$L^p(\Sigma, \mathcal{E}, \mu) \otimes X$ so that the restriction of the norm $\| \cdot \|_{L^p(\Sigma, \mathcal{E}, \mu; X)}$ to
$L^p(\Sigma, \mathcal{E}, \mu) \otimes X$ is a tensor product norm. Each element of $L^p(\Sigma, \mathcal{E}, \mu; X)$
can be approximated in norm by X-valued $\mathcal{E}$-simple functions, so
$L^p(\Sigma, \mathcal{E}, \mu) \otimes X$ is actually dense in the Banach space $L^p(\Sigma, \mathcal{E}, \mu; X)$. Hence
the complete tensor product $L^p(\Sigma, \mathcal{E}, \mu) \widehat{\otimes}_{L^p(\Sigma, \mathcal{E}, \mu; X)} X$ may be identified
with $L^p(\Sigma, \mathcal{E}, \mu; X)$. It is convenient to abbreviate to the simpler notation
$L^p(\Sigma, \mathcal{E}, \mu) \widehat{\otimes}_p X$ so that we can write
$$L^p(\Sigma, \mathcal{E}, \mu; X) = L^p(\Sigma, \mathcal{E}, \mu) \widehat{\otimes}_p X.$$
For the case $p = 2$ and a Hilbert space $\mathcal{H}$, the identification
$$L^2(\Sigma, \mathcal{E}, \mu; \mathcal{H}) = L^2(\Sigma, \mathcal{E}, \mu) \widehat{\otimes}_2 \mathcal{H}$$
can be realised as a Hilbert space tensor product.

    The tensor product norm $\| \cdot \|_{L^p(\Sigma, \mathcal{E}, \mu; X)}$ clearly depends on the special
nature of the L^p-space of pth-Bochner integrable functions. In the next sec-
tion we consider two important examples of tensor product norms defined
for any normed vector spaces X and Y.

1.4.1 *Injective and projective tensor products*

The projective tensor product arises in many applications of bilinear integration in tensor products. Suppose that E and F are locally convex spaces. The finest locally convex topology of $E \otimes F$ for which the canonical bilinear map $\chi : E \times F \to E \otimes F$ is continuous is called the *projective tensor product topology* on $E \otimes F$. We use $E \otimes_\pi F$ to denote the tensor product $E \otimes F$ equipped with the projective topology.

If $\mathcal{U}$ is a neighbourhood base of zero in E and $\mathcal{V}$ is a neighbourhood base of zero in F, then the family of balanced, convex hulls

$$\{\mathrm{bco}(U \otimes V) : U \in \mathcal{U}, \ V \in \mathcal{V}\}$$

is a neighbourhood base of zero for the projective topology. Here the notation $U \otimes V$ means the set of all $x \otimes y$ with $x \in U$, $y \in V$.

The dual space $(E \otimes_\pi F)'$ can be identified with the space $\mathcal{B}(E, F)$ of all continuous bilinear forms on $E \times F$, so that the equicontinuous subsets of $(E \otimes_\pi F)'$ are mapped to the equicontinuous subsets of bilinear forms on $E \times F$ [123, III.6.2].

Suppose that p is a continuous seminorm on E and q is a continuous seminorm on F such that

$$U = \{x \in E : p(x) < 1\}$$

and

$$V = \{y \in F : q(y) < 1\}$$

are the neighbourhoods of zero associated with the seminorms p and q respectively. Then gauge r of $\mathrm{bco}(U \otimes V)$ is given by

$$r(u) = \inf \left\{ \sum_j p(x_j) q(y_j) : u = \sum_j x_j \otimes y_j \right\}$$

with the infimum taken over such finite representations of $u \in E \otimes F$. Moreover, $r(x \otimes y) = p(x)q(y)$ for all $x \in E$ and $y \in F$ [123, III.6.3]. The seminorm $r = p \otimes q$ is called the *tensor product* of the seminorms p and q. For normed vector spaces X and Y, the tensor product of their respective norms is a norm defining the projective topology of $X \otimes_\pi Y$.

The completion of the projective tensor product $E \otimes_\pi F$ is written as $E \widehat{\otimes}_\pi F$. Occasionally we shall employ the following representation of elements of $E \widehat{\otimes}_\pi F$ in the case that E and F are metrisable locally convex spaces or normed vector spaces.

Theorem 1.8 ([123, III.6.4]). *Let E and F be metrisable locally convex spaces and let $u \in E \widehat{\otimes}_\pi F$. Then there exists an absolutely summable sequence $\{\lambda_j\}_{j=1}^\infty$ of scalars and null sequences $\{x_j\}_{j=1}^\infty \subset E$, $\{y_j\}_{j=1}^\infty \subset F$ such that*

$$u = \sum_{j=1}^\infty \lambda_j x_j \otimes y_j. \tag{1.1}$$

The sum converges absolutely in $E \widehat{\otimes}_\pi F$. If E and F are normed vector spaces, then

$$\|u\|_\pi = \inf \left\{ \sum_{j=1}^\infty |\lambda_j| \|x_j\|_E \|y_j\|_F \right\}$$

where the infimum is over all representations (1.1).

Example 1.3. Let $(\Sigma, \mathcal{E}, \mu)$ be a measure space and let X be a Banach space. In Example 1.2 we saw how the identification

$$L^p(\Sigma, \mathcal{E}, \mu; X) = L^p(\Sigma, \mathcal{E}, \mu) \widehat{\otimes}_p X$$

is valid for a tensor product norm in the case $1 \le p < \infty$. For the case $p = 1$, we have $L^1(\Sigma, \mathcal{E}, \mu; X) = L^1(\Sigma, \mathcal{E}, \mu) \widehat{\otimes}_\pi X$ [123, III.6.5]. If $X = L^1(\Omega, \mathcal{S}, \nu)$ for another measure space $(\Omega, \mathcal{S}, \nu)$, then the identities

$$L^1(\Sigma, \mathcal{E}, \mu) \widehat{\otimes}_\pi L^1(\Omega, \mathcal{S}, \nu) = L^1(\Sigma, \mathcal{E}, \mu; L^1(\Omega, \mathcal{S}, \nu))$$
$$= L^1(\Omega, \mathcal{S}, \nu; L^1(\Sigma, \mathcal{E}, \mu))$$

obtain and in case μ and ν are σ-finite measures

$$L^1(\Sigma, \mathcal{E}, \mu) \widehat{\otimes}_\pi L^1(\Omega, \mathcal{S}, \nu) = L^1(\Sigma \times \Omega, \mathcal{E} \otimes \mathcal{S}, \mu \otimes \nu).$$

The case of general measure spaces is worked out in [45, 253Yi]—some care needs to be exercised defining the 'product measure'.

In the case $1 < p < \infty$, there is no tensor product norm $\| \cdot \|_\tau$ defined by a method that works for *all* normed vector spaces E and F such that $L^p(\Sigma, \mathcal{E}, \mu; X) = L^p(\Sigma, \mathcal{E}, \mu) \widehat{\otimes}_\tau X$ for every Banach space X [38, p. 253]. The tensor product norm given in Example 1.2 is specific to the space $L^p(\Sigma, \mathcal{E}, \mu)$.

The projective tensor product $L^2(\Sigma, \mathcal{E}, \mu) \widehat{\otimes}_\pi L^2(\Omega, \mathcal{S}, \nu)$ has special significance in operator theory. We examine it more closely in Proposition 2.4 in Chapter 2 and Section 3.1 of Chapter 3.

Let E and F be nonzero locally convex spaces over the scalars $\mathbb{C}$. An element $u = \sum_j x_j \otimes y_j$ of the tensor product $E \otimes F$ may be viewed as a separately continuous bilinear form $f_u : E'_\sigma \times F'_\sigma \to \mathbb{C}$ given by

$$f_u(x', y') = \sum_j \langle x_j, x' \rangle \langle y_j, y' \rangle, \quad x' \in E', \ y' \in Y', \tag{1.2}$$

that is, $x' \longmapsto f_u(x', y')$, $x' \in E'$, is continuous for the weak topology $\sigma(E', E)$ for each fixed $y' \in F'$ and $y' \longmapsto f_u(x', y')$, $y' \in F'$, is continuous for the weak topology $\sigma(F', F)$ for each fixed $x' \in E'$.

The collection $\mathfrak{B}_e(E'_\sigma, F'_\sigma)$ of all separately continuous bilinear forms on $E'_\sigma \times F'_\sigma$ is equipped with the topology e of uniform convergence on all sets $S \times T$ with S equicontinuous in E' and T equicontinuous in F'. The relative topology of $\mathfrak{B}_e(E'_\sigma, F'_\sigma)$ on the subspace $E \otimes F$ is called the *injective tensor product topology* ϵ. Its completion $E \widehat{\otimes}_\epsilon F$, is sometimes written as $E \check{\otimes} F$ whereas $E \widehat{\otimes}_\pi F$ is abbreviated to $E \hat{\otimes} F$. If E and F are complete locally convex spaces, then $\mathfrak{B}_e(E'_\sigma, F'_\sigma)$ is itself complete, so $E \widehat{\otimes}_\epsilon F$ may be identified with the closure of $E \otimes F$ in the space $\mathfrak{B}_e(E'_\sigma, F'_\sigma)$ of separately continuous bilinear forms [123, IV.9.1].

In the case that E and F are normed vector spaces and $u = \sum_j x_j \otimes y_j \in E \otimes F$, the norm

$$\|u\|_\epsilon = \sup \left\{ \left| \sum_j \langle x_j, x' \rangle \langle y_j, y' \rangle \right| : x' \in B_{E'}, \ y' \in B_{F'} \right\}$$

defines the injective tensor product topology ϵ with respect to the closed unit balls $B_{E'}$ and $B_{F'}$ of the spaces E', F' dual to E and F, respectively.

As mentioned earlier, the dual space $(E \otimes_\pi F)'$ of the projective tensor product $E \otimes_\pi F$ of two nonzero locally convex spaces E and F can be identified with the space $\mathcal{B}(E, F)$ of all continuous bilinear forms on $E \times F$. The characterisation of $(E \otimes_\epsilon F)'$ leads to an important operator ideal that arises in the theory of double operator integrals we consider in Chapter 8.

The dual $(E \otimes_\epsilon F)'$ may be identified with a subspace of $\mathcal{B}(E, F) = (E \otimes_\pi F)'$ which we now identify.

Theorem 1.9 ([123, IV.9.2]). *The continuous bilinear form* $v \in \mathcal{B}(E, F)$ *represents an element of the dual space* $(E \otimes_\epsilon F)'$ *if and only if there exist closed, equicontinuous sets* $S \subset E'_\sigma$, $T \subset F'_\sigma$ *and a Radon measure* $\gamma : \mathcal{B}(S \times T) \to [0, \infty)$ *such that*

$$\langle u, v \rangle = \int_{S \times T} (ju)(x', y') \, d\gamma(x', y'), \quad u \in E \otimes F,$$

for the embedding $j : E \otimes F \to \mathfrak{B}(E'_\sigma, F'_\sigma)$ *defined by formula (1.2).*

Bilinear forms $v \in \mathcal{B}(E, F)$ representing an element of $(E \otimes_\epsilon F)'$ are called *integral* and a linear map $u \in \mathcal{L}(E, F'_\sigma)$ is called 1-*integral* if there exists an integral bilinear form v such that $\langle y, ux \rangle = v(x, y)$ for all $x \in E$, $y \in F$. The term integral operator is used in a different sense in Chapter 3, so the prefix '1' serves to distinguish between the two concepts. A 1-integral map $u \in \mathcal{L}(E, F'_\sigma)$ has the representation

$$ux = \int_{S \times T} \langle x, x' \rangle y' \, d\gamma(x', y') \tag{1.3}$$

with $\gamma : \mathcal{B}(S \times T) \to [0, \infty)$ a Radon measure, as described above. The integral is a Pettis integral because the strong measurability of the integrand may not be valid.

Example 1.4. Let $(\Sigma, \mathcal{E}, \mu)$ and $(\Omega, \mathcal{S}, \nu)$ be σ-finite measure spaces and $E = L^1(\Sigma, \mathcal{E}, \mu)$, $F = L^1(\Omega, \mathcal{S}, \nu)$. Suppose that the continuous linear map $u : L^1(\Sigma, \mathcal{E}, \mu) \to L^\infty(\Omega, \mathcal{S}, \nu)$ is 1-integral. The space $L^\infty(\Omega, \mathcal{S}, \nu)$ is the space dual to $L^1(\Omega, \mathcal{S}, \nu)$ with the duality

$$\langle f, g \rangle = \int_\Omega f(\omega) g(\omega) \, d\nu(\omega), \quad f \in L^1(\Omega, \mathcal{S}, \nu), \; g \in L^\infty(\Omega, \mathcal{S}, \nu).$$

Then the linear map u has the representation (1.3) for some Radon measure γ on the compact product set $S \times T$ in $E'_\sigma \times F'_\sigma$. Let $\eta : \mathcal{B}(S) \to [0, \infty)$ be the Radon measure defined by the marginal measure $A \longmapsto \gamma(A, T)$, $A \in \mathcal{B}(S)$.

Each element x of $E = L^1(\Sigma, \mathcal{E}, \mu)$ uniquely defines a continuous function $x' \longmapsto \langle x, x' \rangle$, $x' \in S$. Let $v_1 : E \to L^\infty(S, \mathcal{B}(S), \eta)$ be the corresponding embedding and $j : L^\infty(S, \mathcal{B}S, \eta) \to L^1(S, \mathcal{B}S, \eta)$ the natural inclusion. Then $u = v_2 \circ j \circ v_1$ for the bounded linear map $v_2 : L^1(S, \mathcal{B}(S), \eta) \to L^\infty(\Omega, \mathcal{S}, \nu)$ defined by

$$v_2 f = \int_{S \times T} f(x') y' \, d\gamma(x', y'), \quad f \in L^1(S, \mathcal{B}(S), \eta).$$

The integral converges as a Pettis integral in F'_σ because T is a compact subset of $L^\infty(\Omega, \mathcal{S}, \nu)$ for the weak*-topology, or more simply, the Radon-Nikodym Theorem ensures that the measure

$$B \longmapsto \int_{S \times T} f(x') \langle \chi_B, y' \rangle \, d\gamma(x', y'), \quad B \in \mathcal{B}(T),$$

has a unique density $v_2 f$. Consequently, u admits the factorisation

$$\begin{array}{ccc} L^1(\mu) & \overset{u}{\longrightarrow} & L^\infty(\nu) \\ v_1 \downarrow & & \uparrow v_2 \\ L^\infty(\eta) & \underset{j}{\longrightarrow} & L^1(\eta) \end{array}$$

which means that u is strictly 1-integral according to the definition given in [36, p. 95].

The injective norm of the tensor product

$$E \otimes_\epsilon F = L^1(\Sigma, \mathcal{E}, \mu) \otimes_\epsilon L^1(\Omega, \mathcal{S}, \nu)$$

is the relative norm of the space $\mathfrak{B}_e(E'_\sigma, F'_\sigma)$ of separately continuous bilinear forms on $E'_\sigma \times F'_\sigma$. A bounded linear map $u : L^1(\mu) \to L^\infty(\nu)$ defines a linear functional

$$w \longmapsto \sum_j \langle y_j, ux_j \rangle, \quad w = \sum_j x_j \otimes y_j \in L^1(\mu) \otimes L^1(\nu)$$

on $L^1(\mu) \otimes L^1(\nu)$. The number $\sum_j \langle y_j, ux_j \rangle$ can be expressed as $\mathrm{tr}(uv)$ for the finite rank linear map $v : L^\infty(\nu) \to L^1(\mu)$ given by

$$vf = \sum_j x_j \langle y_j, f \rangle, \quad f \in L^\infty(\nu).$$

The trace $\mathrm{tr}(uv)$ is well-defined because uv is a finite rank operator on $L^\infty(\nu)$ and $\mathrm{tr}(uv)$ is the trace in any matrix representation of the action of the continuous linear map uv, see [36, p. 125] for example. Because

$$\|w\|_{L^1(\mu) \otimes_\epsilon L^1(\nu)} = \|w\|_{\mathfrak{B}_e(E'_\sigma, F'_\sigma)} = \|v\|_{\mathcal{L}(L^\infty(\nu), L^1(\mu))}$$

for $w = \sum_j x_j \otimes y_j \in L^1(\mu) \otimes L^1(\nu)$, it follows that the bounded linear map $u : L^1(\mu) \to L^\infty(\nu)$ is 1-integral if and only if there exists $M > 0$ such that

$$|\mathrm{tr}(uv)| \leq M\|v\|_{\mathcal{L}(L^\infty(\nu), L^1(\mu))} \tag{1.4}$$

for every finite rank operator $v : L^\infty(\nu) \to L^1(\mu)$ [36, Theorem 6.16 (a)]. The equivalence of the factorisation given above and the bound (1.4) is an ingredient of Theorem 8.9 in Chapter 8.

1.4.2 *Grothendieck's inequality*

A discussion of Grothendieck's inequality and its many applications is given in the survey [112]. In this section, attention is limited to some consequences needed to investigate double operator integrals in Chapter 8.

The projective tensor product $\ell^\infty \widehat{\otimes}_\pi \ell^\infty$ is the completion of the tensor product $\ell^\infty \otimes \ell^\infty$ with respect to the norm

$$\|u\|_\pi = \inf \left\{ \sum_{j=1}^n \|x_j\|_\infty \|y_j\|_\infty : u = \sum_{j=1}^n x_j \otimes y_j, \ x_j, y_j \in \ell^\infty \right\}.$$

Another distinguished norm on $\ell^\infty \otimes \ell^\infty$ is given by

$$\gamma_2(u) = \inf \left\{ \sup_{\xi \in \ell^1, \|\xi\|_1 \le 1} \left(\sum_{j=1}^n |\langle x_j, \xi \rangle|^2 \right)^{\frac{1}{2}} \cdot \sup_{\eta \in \ell^1, \|\eta\|_1 \le 1} \left(\sum_{j=1}^n |\langle y_j, \eta \rangle|^2 \right)^{\frac{1}{2}} \right\}$$

where the infimun runs over all possible representations $u = \sum_{j=1}^n x_j \otimes y_j$ for $x_j, y_j \in \ell^\infty$, $j = 1, \ldots, n$ and $n = 1, 2, \ldots$. Then γ_2 may also be viewed as the norm of factorisation through a Hilbert space:

$$\gamma_2(u) = \inf \{ \sup_i \|x_i\| \cdot \sup_j \|y_j\| \}$$

where the infimum runs over all Hilbert spaces $\mathcal{H}$ and all $x_j, y_j \in \mathcal{H}$ for which $u \in \ell^\infty \otimes \ell^\infty$ has the finite representation $u = \sum_{i,j} (x_i, x_j) e_i \otimes e_j$ with respect to the standard basis $\{e_j\}_j$ of ℓ^∞. Another way of viewing $\gamma_2(u)$ is

$$\gamma_2(u) = \inf \left\{ \left\| \left(\sum_{j=1}^n |x_j|^2 \right)^{\frac{1}{2}} \right\|_\infty \cdot \left\| \left(\sum_{j=1}^n |y_j|^2 \right)^{\frac{1}{2}} \right\|_\infty \right\}$$

over representations $u = \sum_{j=1}^n x_j \otimes y_j$, $x_j, y_j \in \ell^\infty$, because

$$\sup_{\xi \in \ell^1, \|\xi\|_1 \le 1} \left(\sum_{j=1}^n |\langle x_j, \xi \rangle|^2 \right)^{\frac{1}{2}} = \sup_{\sum_j |\alpha_j|^2 \le 1} \left\| \sum_{j=1}^n \alpha_j x_j \right\|_\infty$$

$$= \sup_k \left(\sum_{j=1}^n |x_j(k)|^2 \right)^{\frac{1}{2}}$$

$$= \left\| \left(\sum_{j=1}^n |x_j|^2 \right)^{\frac{1}{2}} \right\|_\infty .$$

Proposition 1.2. *Let $\varphi : \mathbb{N} \times \mathbb{N} \to \mathbb{C}$ be a function that defines a Schur multiplier $M_\varphi : \mathcal{L}(\ell^2) \to \mathcal{L}(\ell^2)$, that is, in matrix notation $M_\varphi(\{a_{ij}\}_{i,j \in \mathbb{N}}) = \{\varphi(i,j) a_{ij}\}_{i,j \in \mathbb{N}}$. The following conditions are equivalent.*

(i) $\|M_\varphi\|_{\mathcal{L}(\mathcal{L}(\ell^2))} \le 1$.

(ii) *There exist a Hilbert space $\mathcal{H}$ and functions $x : \mathbb{N} \to B_1(\mathcal{H})$, $y : \mathbb{N} \to B_1(\mathcal{H})$ with values in the closed unit ball $B_1(\mathcal{H})$ of $\mathcal{H}$ such that $\varphi(n,m) = (x(n), x(m))$, $n, m \in \mathbb{N}$.*

(iii) *For all finite subsets* E, F *of* $\mathbb{N}$, *the bound*

$$\left\| \sum_{i \in E, j \in F} \varphi(i,j) e_i \otimes e_j \right\|_{\ell^\infty \otimes_{\gamma_2} \ell^\infty} \leq 1$$

holds.

Proof. Suppose first that φ is zero off a finite set $E \times F$. Then the bound (i) is equivalent to the condition that

$$\left| \sum_{i \in E, j \in F} \varphi(i,j) a_{ij} \alpha(i) \beta(j) \right| \leq 1$$

for all linear maps $a : \ell^2(E) \to \ell^2(F)$ with norm $\|a\| \leq 1$ and matrix $\{a_{ij}\}$ with respect to the standard basis and all $\alpha \in B_1(\ell^2(E))$, $\beta \in B_1(\ell^2(F))$, that is, φ belongs to the polar C_1° of the set C_1 of all matrices $\{\alpha(i) a_{ij} \beta(j)\}_{(i,j) \in E \times F}$ with a, α, β as described. According to [112, Remark 23.4], the set C_1 is itself the polar C_2° of the set C_2 of all matrices $\{\psi_{ij}\}_{(i,j) \in E \times F}$ with $\left\| \sum_{i \in E, m \in F} \psi(i,j) e_i \otimes e_j \right\|_{\ell^\infty \otimes_{\gamma_2} \ell^\infty} \leq 1$. Then (i) holds if and only if φ belongs to $C_2^{\circ\circ} = C_2$, which is exactly condition (iii). Conditions (ii) and (iii) are equivalent by the definition of the norm γ_2. The passage to all of $\mathbb{N} \times \mathbb{N}$ follows from a compactness argument. $\square$

Remark 1.1. a) The argument above uses the factorisation of the norm γ_2^* dual to γ_2 described in [112, Proposition 3.3] and [112, Remark 23.4]—this only relies on the Hahn-Banach Theorem.

b) The representation (8.14) below is the measure space version of the implication (ii) $\implies$ (i) above. The necessity of the condition (8.14) in the general measure space setting is proved using complete boundedness arguments in [129, Theorem 3.3], see also [78, 131].

One version of Grothendieck's inequality from [112] is that the norm γ_2 and the projective tensor product norm are equivalent on $\ell^\infty \otimes \ell^\infty$ with

$$\gamma_2(u) \leq \|u\|_\pi \leq K_G \gamma_2(u), \quad u \in \ell^\infty \otimes \ell^\infty.$$

The constant K_G is Grothendieck's constant. The projective tensor product version of Proposition 1.2 follows, with the same notation.

Proposition 1.3. *Let* E, F *be finite subsets of* $\mathbb{N}$ *and let* $\varphi : \mathbb{N} \times \mathbb{N} \to \mathbb{C}$ *be a function vanishing off* $E \times F$. *Then*

$$\frac{1}{K_G} \left\| \sum_{i \in E, j \in F} \varphi(i,j) e_i \otimes e_j \right\|_{\ell^\infty \otimes_\pi \ell^\infty} \leq \|M_\varphi\|_{\mathcal{L}(\mathcal{L}(\ell^2))}$$

$$= \left\| \sum_{i \in E, j \in F} \varphi(i,j) e_i \otimes e_j \right\|_{\ell^\infty \otimes_{\gamma_2} \ell^\infty}.$$

Passing to infinite sets, a bounded function $\varphi : \mathbb{N} \times \mathbb{N} \to \mathbb{C}$ with $\|M_\varphi\|_{\mathcal{L}(\mathcal{L}(\ell^2))} < \infty$ necessarily has a representation

$$\varphi(i,j) = \sum_{k=1}^{\infty} a(i,k)\beta(j,k), \quad i, j \in \mathbb{N},$$

with $\sum_{k=1}^{\infty} \|a(\cdot, k)\|_\infty \|\beta(\cdot, k)\|_\infty < \infty$, as in Peller's representation (8.12).

The following formulation may be viewed as the dual version of Proposition 1.3, see [112, Equation (3.11)].

Proposition 1.4 ([112, Theorem 2.4]). *Let* $n = 1, 2, \ldots$ *and let* $\{a_{ij}\}_{i,j=1}^n$ *be scalars such that*

$$\left| \sum_{i,j=1}^n a_{ij} \alpha_i \beta_j \right| \leq \sup_i |\alpha_i| \sup_j |\beta_j|$$

for all scalars α_i, β_j, $i, j = 1, \ldots, n$. *Then there exists* $K > 0$ *independent of* n, *such that for any Hilbert space* $\mathcal{H}$ *and any* $x_i \in \mathcal{H}$, $y_j \in \mathcal{H}$, $i, j = 1, \ldots, n$, *the bound*

$$\left| \sum_{i,j=1}^n a_{ij}(x_i, y_j) \right| \leq K \sup_i \|x_i\|_\mathcal{H} \sup_j \|y_j\|_\mathcal{H}$$

holds. The smallest such constant K *valid for all* $\mathcal{H}$ *and* $n = 1, 2, \ldots$ *is Grothendieck's constant* K_G.

1.5 Semivariation

In the context of bilinear integration, Bartle [11] worked with a concept related to semivariation originally introduced in [50]; it is needed in the proof of the bounded convergence theorem for bilinear integrals.

Let $m : S \to Y$ be a vector measure defined on the σ-algebra S of subsets of a set Ω. The X-*semivariation* $\beta_X : S \to [0, \infty]$ of m in $X \otimes_\tau Y$ is defined by

$$\beta_X(m)(A) = \sup \left\{ \left\| \sum_{j=1}^k x_j \otimes m(A_j) \right\|_\tau \right\}. \tag{1.5}$$

The supremum is taken over all pairwise disjoint sets $A_1, \ldots, A_k$ from S contained in $A \in S$ and vectors $x_1, \ldots, x_k$ from X, such that $\|x_j\| \leq 1$ for all $j = 1, \ldots, k$ and $k = 1, 2, \ldots$. A similar notion applies if the canonical bilinear map $(x, y) \mapsto x \otimes y$ from $X \times Y$ into $X \otimes Y$ is replaced by some continuous bilinear map $(x, y) \mapsto xy$ into a locally convex space Z; see [25]. If $X = \mathbb{C}$, then the $\mathbb{C}$-semivariation of m in Y coincides with the usual notion of semivariation of a vector valued measure mentioned in Section 1.2.

Two variations of this theme are relevant to the applications considered in this work. If E and F are Banach spaces and $m : S \to \mathcal{L}(E, F)$ is an additive set function, then

$$\beta_E(m)(A) = \sup \left\{ \left\| \sum_{j=1}^k m(A_j) x_j \right\|_F \right\} \tag{1.6}$$

is the E-*semivariation* of m in F, with the A_j as above and $\|x_j\|_E \leq 1$ for $j = 1, \ldots, k$. The relevant bilinear map from $E \times \mathcal{L}(E, F)$ into F is $(x, T) \mapsto Tx$. Similarly, the $\mathcal{L}(E, F)$-*semivariation* $\beta_{\mathcal{L}(E,F)}(m)$ of an E-valued additive set function m in F is associated with the bilinear map $(T, x) \mapsto Tx$, $T \in \mathcal{L}(E, F)$, $x \in E$. For $F = \mathbb{C}$, the Hahn-Banach Theorem ensures that E'-semivariation $\beta_{E'}(m)$ is equal to the variation $V(m)$ of m with respect to the norm of E.

Unlike the scalar semivariation for a vector measure, there is no guarantee that these bilinear semivariations have values other than 0 or ∞, as happens in some simple examples. Suppose that E is a Banach space and $F = \mathbb{C}$. Then an operator valued measure $\nu : S \to \mathcal{L}(E, \mathbb{C})$ is a finitely additive set function with values in the dual Banach space E'. Denote by $V(\nu)$ the variation of the E'-valued set function ν. It then follows from the Hahn-Banach Theorem that $V(\nu) = \beta_E(\nu)$ on S.

Example 1.5. Let $1 < p < \infty$ and $1/p + 1/q = 1$. Let $S = \mathcal{B}([0, 1])$. The $L^p([0, 1])$-valued measure $m : A \longmapsto \chi_A$ on S defines a measure $\nu : S \to \mathcal{L}(L^q([0, 1]), \mathbb{C})$. Since $V(\nu)(A) = \infty$ for every set A with positive Lebesgue measure [38, Example I.1.16], it follows that the $L^q([0,1])$-semivariation $\beta_{L^q([0,1])}(\nu)$ of ν in $\mathbb{C}$ has only values 0 or ∞.

If for all sets $A_k \in \mathcal{S}$ decreasing to the empty set, we have $\beta_X(m)(A_k) \to 0$ as $k \to \infty$, then we say that the X-semivariation $\beta_X(m)$ is *continuous*. A study of continuity for semivariation has been conducted by Dobrakov [39–41]. If $\beta_X(m)$ is continuous, then $\beta_X(m)(\Omega) < \infty$. In fact, an equivalent formulation for the continuity of $\beta_X(m)$ is that the set of $X \otimes Y$-valued measures $\phi \otimes m$ as ϕ ranges over all X-valued $\mathcal{S}$-simple functions with values in the closed unit ball of X, is bounded and uniformly countably additive for the norm $\| \cdot \|_\tau$. The result of Bartle-Dunford-Schwartz, Theorem 1.2, ensures that there exists a finite nonnegative measure λ on $\mathcal{S}$ such that $\lambda \leq \beta_X(m)$ and $\lim_{\lambda(E) \to 0^+} \beta_X(m)(E) = 0$; see [41, Lemma 2]. In the paper [11], continuity of the semivariation is called, unhelpfully, the *-property*.

Another of Dobrakov's results [39, *-Theorem] implies that if X, Y and $X \widehat{\otimes}_\tau Y$ are Banach spaces for which $X \otimes_\tau Y$ contains no subspace isomorphic to c_0, then the X-semivariation $\beta(m)$ of m in $X \otimes_\tau Y$ is continuous once it is finite.

If X, Y and $X \widehat{\otimes}_\tau Y$ are normed vector spaces, then a Y-valued measure m with finite variation $V(m) : \mathcal{S} \to [0, \infty)$ necessarily has finite X-semivariation in $X \otimes_\tau Y$, by virtue of the separate continuity (T1) of the map $X \times Y \to X \otimes Y$. Moreover, the X-semivariation of m in $X \otimes_\tau Y$ is continuous.

1.5.1   *Semivariation in L^p-spaces*

Let $(\Gamma, \mathcal{E}, \mu)$ be a σ-finite measure space. In this section we examine conditions guaranteeing the finiteness of semivariation for L^p-space valued measures. Here, attention is limited to the case in which X is a Banach space, Y is the Banach space $L^p(\Gamma, \mathcal{E}, \mu)$ and τ is the relative topology of $L^p(\Gamma, \mathcal{E}, \mu; X)$ on $L^p(\Gamma, \mathcal{E}, \mu) \otimes X$. In this case, if $\| \cdot \|_p$ is the norm of $L^p(\Gamma, \mathcal{E}, \mu; X)$ and B_X is the closed unit ball of X, then for each $E \in \mathcal{S}$, the X-semivariation $\beta_X(m)(E)$ of $m : \mathcal{S} \to L^p(\Gamma, \mathcal{E}, \mu)$ in $L^p(\Gamma, \mathcal{E}, \mu; X)$ is denoted by $S \mapsto \widehat{m}_{p,X}(S)$, $S \in \mathcal{S}$. Thus, $\widehat{m}_{p,\mathbb{C}}$ is actually the usual semivariation $\|m\|$ of m as an $L^p(\Gamma, \mathcal{E}, \mu)$-valued measure.

As mentioned in [11], every vector measure with finite variation has finite X-semivariation. It turns out that for L^p-spaces, there is a simple sufficient condition which guarantees finite semivariation, which is not as restrictive as the condition of finite variation.

We first introduce some terminology from vector lattices in the context of L^p-spaces.

The order relation $f \geq_\mu 0$, $f \in L^p(\mu)$, also written as $0 \leq_\mu f$, is taken to mean that f is the equivalence class of a function greater than or equal to zero μ-almost everywhere. The set of elements $f \geq_\mu 0$ of $L^p(\mu)$ is written $L^p(\mu)_+$. We shall drop the subscript μ when it is clear from the context below.

The spaces $L^p(\mu), 1 \leq p \leq \infty$ are complex vector lattices [124, Definition II.11.1] with respect to these order relations. A subset A of $L^p(\mu)$ is said to be *order bounded* if there exists $u \geq_\mu 0$ such that $u \geq_\mu |f|$ for all $f \in A$. Every order bounded subset A of $L^p(\mu)$ has a supremum, which is to say that $L^p(\mu; \mathbb{R})$ is said to be *Dedekind complete* [124, Proposition II.8.3, Exercise II.23].

Definition 1.1. Let $1 \leq p \leq \infty$. A bounded linear operator T from $L^p(\mu)$ to $L^p(\mu)$ is said to be *positive* if $Tf \geq_\mu 0$ for every $f \in L^p(\mu; \mathbb{R})$ such that $f \geq_\mu 0$. A bounded linear operator $T : L^p(\mu) \to L^p(\mu)$ is said to be a *regular operator* if it can be written as a linear combination of positive operators.

The restriction $T_\mathbb{R}$ of a bounded linear operator $T : L^p(\mu) \to L^p(\mu)$ to $L^p(\mu; \mathbb{R})$ may be written as $T_\mathbb{R} = T_1 + iT_2$ for bounded real linear operators T_1, T_2 acting on $L^p(\mu; \mathbb{R})$ and T is regular if and only if both T_1 and T_2 map order bounded intervals in $L^p(\mu; \mathbb{R})$ into order bounded intervals, in which case the positive linear operators $T_1^\pm, T_2^\pm$ on $L^p(\mu; \mathbb{R})$ are defined by setting, for each $j = 1, 2$,

$$T_j^+ f = \sup\{T_j u : 0 \leq_\mu u \leq_\mu f\}, \qquad T_j^- f = \sup\{-T_j u : 0 \leq_\mu u \leq_\mu f\},$$

for every $f \in L^p(\mu; \mathbb{R})$ with $f \geq_\mu 0$. Then $T_\mathbb{R} = T_1^+ - T_1^- + i(T_2^+ - T_2^-)$. The *modulus* $|T| : L^p(\mu) \to L^p(\mu)$ of a regular operator T is defined by the formula $|T|u = \sup_{0 \leq_\mu |f| \leq_\mu u} |Tf|$, for all $u \geq_\mu 0$ [124, Proposition IV.1.6].

The following result is formulated in terms of Banach lattices [138, Section 83] such as $L^p(\mu)$ with μ a σ-finite measure.

Lemma 1.1. *Let X be a Dedekind complete Banach lattice, $\mathcal{T}$ an algebra of subsets of a set Γ and $m : \mathcal{T} \to X$ an additive set function. Then m has order bounded range if and only if there exists a nonnegative additive set function $\nu : \mathcal{T} \to X$ such that $|m(A)| \leq \nu(A)$ for each $A \in \mathcal{T}$, in the order of X. If m is σ-additive and the norm of X is order continuous, then ν may be chosen σ-additive.*

Proof. We look in the direction in which m is assumed to have order bounded range and X is a Dedekind complete real Banach lattice. The implication in the other direction is clear. Let $L = \mathrm{sim}(\mathcal{T})$ be the set of all real valued $\mathcal{T}$-simple functions. Then L is a Riesz space and the map $I_m : L \to X$ defined by $I_m h = \int_\Gamma h\,dm$, $h \in L$, is linear. Moreover, I_m is order bounded because m has order bounded range. By [138, Theorem 83.3], there exist positive linear operators I_m^+ and I_m^- from L into M such that $I_m = I_m^+ - I_m^-$. Then the additive map $\nu : \mathcal{T} \to X$ defined by $\nu(A) = I_m^+(\chi_A) + I_m^-(\chi_A)$, for all $A \in \mathcal{T}$ has the required properties. In the case that m is σ-additive and the norm of X is order continuous, then the map I_m is sequentially order continuous, so [138, Lemma 84.1] shows that ν is σ-additive. The complex case is straightforward. $\qquad\square$

The measure ν constructed above is denoted by $|m|$ as it is the smallest measure dominating m.

Proposition 1.5. *Suppose that $1 \le p < \infty$ and $m : \mathcal{S} \to L^p(\Gamma, \mathcal{E}, \mu)$ is an order bounded measure. Then for every Banach space X, the semivariation $\widehat{m}_{p,X}(\Gamma)$ of m on Γ is finite and continuous.*

Proof. Let $\nu : \mathcal{S} \to L^p(\Gamma, \mathcal{E}, \mu)$ be a pointwise positive vector measure for which $|m(A)| \le \nu(A)$ a.e. for all $A \in \mathcal{S}$; such a vector measure exists by Lemma 1.1. Let $k = 1, 2, \ldots$, and suppose that $x_j \in X$ satisfy $\|x_j\| \le 1$ for all $j = 1, \ldots, k$. Let $E_j, j = 1, \ldots, k$, be pairwise disjoint sets belonging to $\mathcal{S}$. Then

$$\int_\Gamma \left\| \sum_{j=1}^k x_j \big(m(E_j)\big)(\gamma) \right\|^p d\mu(\gamma) \le \int_\Gamma \left(\sum_{j=1}^k \|x_j\|\, \big|(m(E_j))(\gamma)\big| \right)^p d\mu(\gamma)$$

$$\le \int_\Gamma \left(\sum_{j=1}^k \nu(E_j)(\gamma) \right)^p d\mu(\gamma) \le \int_\Gamma (\nu(\Gamma)(\gamma))^p\, d\mu(\gamma) = \|\nu(\Gamma)\|_p^p.$$

$$\square$$

A typical way to produce measures which are not order bounded follows; see also Example 1.8.

Example 1.6. Let $1 \le p \le 2$, $1/p + 1/q = 1$ and let $\phi : [0,1] \to L^p[0,1]$ be defined for all $x, y \in [0,1]$ by

$$\phi(x)(y) = \sum_{k=1}^\infty k \chi_{[1/(k+1), 1/k)}(x) e^{iky}.$$

Then ϕ is Pettis integrable in $L^p[0,1]$ because $\int_0^1 |\langle \phi(x), g \rangle| \, dx < \infty$ for every $g \in L^q[0,1]$ [38, II.3.7]. Let $m : \mathcal{S} \to L^p(\Gamma, \mathcal{E}, \mu)$ be the indefinite Pettis integral of ϕ. If there existed a nonnegative measure $\nu : \mathcal{S} \to L^p(\Gamma, \mathcal{E}, \mu)$ such that $|m(A)| \leq \nu(A)$ for all $A \in \mathcal{S}$, then for each $j = 1, 2, \ldots$,

$$\sum_{k=1}^j |m\big([1/(k+1), 1/k)\big)| = \sum_{k=1}^j \frac{1}{k+1} 1 \leq \nu(\Gamma),$$

which is impossible. Thus, by Lemma 1.1, m cannot be order bounded in $L^p(\Gamma, \mathcal{E}, \mu)$.

The situation for Hilbert spaces is accommodated by Grothendieck's inequality.

Proposition 1.6. *Let $\mathcal{H}$ be a Hilbert space and $m : \mathcal{S} \to L^2(\Gamma, \mathcal{E}, \mu)$ a measure. Let $\|m\| : \mathcal{S} \to [0, \infty)$ be the semivariation of m in $L^2(\Gamma, \mathcal{E}, \mu)$. Then the measure m has finite $\mathcal{H}$-semivariation $\widehat{m}_{2,\mathcal{H}}$ in $L^2(\Gamma, \mathcal{E}, \mu; \mathcal{H})$. Moreover, there exists a constant $C > 0$ independent of $\mathcal{H}$ and m and a finite measure $0 \leq \nu \leq \|m\|$ such that $\lim_{\nu(E) \to 0} \|m\|(S) = 0$ and $\widehat{m}_{2,\mathcal{H}}(S) \leq C\|m\|(S)$, for all $S \in \mathcal{S}$.*

Proof. Let $n = 1, 2, \ldots$ and suppose that $x_1, \ldots, x_n$ belong to the closed unit ball of $\mathcal{H}$. Suppose that $E_1, \ldots, E_n$ are pairwise disjoint sets from $\mathcal{S}$. Then,

$$\int_\Omega \left\| \sum_{j=1}^n x_j \big(m(E_j)\big)(\omega) \right\|_{\mathcal{H}}^2 d\mu(\omega) = \sum_{k,j=1}^n (x_j, x_k)_{\mathcal{H}} \big(m(E_j), m(E_k)\big)_{L^2(\Gamma, \mathcal{E}, \mu)}.$$

Let $a_{j,k}$ be the complex number $\big(m(E_j), m(E_k)\big)_{L^2(\Gamma, \mathcal{E}, \mu)}$ for each $j, k = 1, \ldots, n$. By Grothendieck's inequality Proposition 1.4 or [93, Theorem 2.b.5],

$$\left| \sum_{k,j=1}^n a_{j,k}(x_j, x_k)_{\mathcal{H}} \right| \leq K_G \sup \left| \sum_{k,j=1}^n a_{j,k} s_j t_k \right|, \qquad (1.7)$$

where the supremum on the right is over all complex numbers $s_j, t_k, j, k = 1, \ldots, n$ such that $|s_j| \leq 1$ and $|t_k| \leq 1$ for all $j, k = 1, \ldots m$, and K_G is Grothendieck's constant. But the sum $\sum_{k,j=1}^n a_{j,k} s_j t_k$ is equal to

$$\sum_{k,j=1}^n \big(m(E_j), m(E_k)\big)_{L^2(\Gamma, \mathcal{E}, \mu)} s_j t_k = \left(\sum_{j=1}^n m(E_j) s_j, \sum_{k=1}^n m(E_k) t_k \right)_{L^2(\Gamma, \mathcal{E}, \mu)}$$

$$= \big(m(f), m(g)\big)_{L^2(\Gamma, \mathcal{E}, \mu)},$$

for the scalar $\mathcal{S}$-simple functions $f = \sum_{j=1}^{n} s_j \chi_{E_j}$ and $g = \sum_{k=1}^{n} t_k \chi_{E_k}$. By the Cauchy-Schwarz inequality, $\left| (m(f), m(g)) \right| \leq \|m(f)\|_2 \|m(g)\|_2$.

Since $\sup_{\|u\|_\infty \leq 1} \|m(u)\|_2 \leq M = 2 \sup\{\|x\|_2 : x \in \mathrm{bco}(m(\mathcal{S}))\}$ [86, Lemma IV.6.1], the right-hand side of equation (1.7) is bounded by $K_G M^2$. Here we have appealed to the fact that the vector measure m is bounded on $\mathcal{S}$, so $M < \infty$. It follows that $\int_\Omega \left\| \sum_{j=1}^{n} x_j (m(E_j))(\omega) \right\|_{\mathcal{H}}^2 d\mu(\omega)$ is bounded by $K_G M^2$, that is, the $\mathcal{H}$-semivariation of m is bounded by $\sqrt{K_G} M$. Any measure equivalent to m is also equivalent to $\mathcal{H}$-semivariation $\widehat{m}_{2,\mathcal{H}}$ in $L^2(\Gamma, \mathcal{E}, \mu; \mathcal{H})$. $\square$

Remark 1.2. Let $\mathcal{H}$ be a Hilbert space. If E is any $\mathcal{L}(\mathcal{H})$-valued spectral measure and $h \in \mathcal{H}$, the identity

$$\sum_{n=1}^{\infty} \|E(f_n)h\|_{\mathcal{H}}^2 = \left(E\left(\sum_{n=1}^{\infty} |f_n|^2 \right) h, h \right)$$

ensures that the $\mathcal{H}$-valued measure Eh has bounded ℓ^2-semivariation in $\ell^2(\mathcal{H})$—the Hilbert space tensor product $\mathcal{H} \widehat{\otimes} \ell^2 = \oplus_{j=1}^{\infty} \mathcal{H}$ with norm $\|u\|_{\ell^2(\mathcal{H})}^2 = \sum_{j=1}^{\infty} \|u_j\|_{\mathcal{H}}^2$. The remarkable consequence of Grothendieck's inequality in Proposition 1.6 is that the same holds true when the measure Eh is repaced by *any* $\mathcal{H}$-valued measure.

In the case of Hilbert spaces, the following related result gives a convenient condition for which the operator semivariation of an operator valued measure is finite. In this case, the bilinear map $(u, v) \mapsto u \otimes v$, $u \in \mathcal{L}(\mathcal{H})$, $v \in \mathcal{L}(L^2(\Gamma, \mathcal{E}, \mu))$, has its values in $\mathcal{L}(L^2(\Gamma, \mathcal{E}, \mu; \mathcal{H}))$.

Proposition 1.7. *There exists a positive number C such that for every Hilbert space $\mathcal{H}$, every σ-finite measure space $(\Gamma, \mathcal{E}, \mu)$ and every operator valued measure $M : \mathcal{S} \to \mathcal{L}(L^2(\Gamma, \mathcal{E}, \mu))$, the following bound holds for the $\mathcal{L}(\mathcal{H})$-semivariation $\beta_{\mathcal{L}(\mathcal{H})}(M)$ of M in $\mathcal{L}(L^2(\Gamma, \mathcal{E}, \mu; \mathcal{H}))$:*

$$\beta_{\mathcal{L}(\mathcal{H})}(M)(B) \leq C \sup\{V(M'h)(B) : h \in \mathcal{H}, \|h\|_{\mathcal{H}} \leq 1\}, \qquad \text{for all } B \in \mathcal{S}.$$

Of course, the supremum on the right-hand side of the inequality may be infinite.

Proof. The number $\beta_{\mathcal{L}(H)}(M)(B)$ is the supremum of all numbers $\|\sum_k (A_k \otimes [M(B_k)]) \phi\|_2$ for $A_k \in \mathcal{L}(H)$ with $\|A_k\| \leq 1$, $\{B_k\}$ pairwise disjoint subsets of $B \in \mathcal{S}$ and $\phi \in L^2(\Gamma, \mathcal{E}, \mu) \otimes H$ with $\|\phi\|_2 \leq 1$. Suppose that $\phi = \sum_j h_j \chi_{E_j}$ for finitely many $h_j \in H$ and $\{E_j\}$ pairwise disjoint elements of $\mathcal{E}$ satisfying $\|\phi\|_2^2 = \sum_j \|h_j\|_H^2 \mu(E_j) \leq 1$. Then,

$$\Big\| \sum_k (A_k \otimes [M(B_k)]) \phi \Big\|_2^2 = \sum_{k,k',j,j'} \langle A_k h_j, A_{k'} h_{j'} \rangle \langle M(B_k) \chi_{E_j}, M(B_{k'}) \chi_{E_{j'}} \rangle$$

$$\leq K_G \sup \Big| \sum_{k,k',j,j'} \langle \|h_j\|_H M(B_k) \chi_{E_j}, \|h'_j\|_H M(B_{k'}) \chi_{E_{j'}} \rangle s_{kj} t_{k'j'} \Big|,$$

by Grothendieck's inequality [93, Theorem 2.b.5]. The supremum is over all scalars $|s_{kj}| \leq 1$ and $|t_{k'j'}| \leq 1$.

Let $u_k = \sum_j s_{kj} \|h_j\| \chi_{E_j}$ and $v_{k'} = \sum_{j'} t_{k'j'} \|h_j\| \chi_{E_{j'}}$ be elements of $L^2(\Gamma, \mathcal{E}, \mu)$. Then $\|u_k\|_2 \leq 1$ and $\|v_{k'}\|_2 \leq 1$ and so we have

$$\Big\| \sum_k (A_k \otimes [M(B_k)]) \phi \Big\|_2^2 \leq K_G \sup_{\{u_k\}, \{v_k\}} \Big| \sum_{k,k'} \langle M(B_k) u_k, M(B_{k'}) v_{k'} \rangle \Big|$$

$$\leq 4 K_G \sup_{\{u_k\}} \Big| \sum_{k,k'} \langle M(B_k) u_k, M(B_{k'}) u_{k'} \rangle \Big|$$

$$= 4 K_G \sup_{\{u_k\}} \Big\| \sum_k M(B_k) u_k \Big\|_2^2,$$

by the polarisation identity. The norm $\| \sum_k M(B_k) u_k \|_2$ is given by

$$\Big\| \sum_k M(B_k) u_k \Big\|_2 = \sup_{\|h\|_2 \leq 1} \Big| \langle \sum_k M(B_k) u_k, h \rangle \Big|$$

$$= \sup_{\|h\|_2 \leq 1} \Big| \sum_k \langle u_k, M(B_k)' h \rangle \Big| \leq \sup_{\|h\|_2 \leq 1} V(M'h)(B).$$

$$\square$$

Let Y be a Banach space and $1 \leq p < \infty$. A vector measure $m : \mathcal{S} \to Y$ is said to have finite *p-variation* if there exists $C > 0$ such that for every $n = 1, 2, \ldots$ and every finite family of pairwise disjoint sets $E_j, j = 1, \ldots, n$, the inequality $\sum_{j=1}^n \|m(E_j)\|^p \leq C$ holds.

Proposition 1.8. *Let $1 \leq p < \infty$ and let $m : \mathcal{S} \to L^p(\Gamma, \mathcal{E}, \mu)$ be a measure. Let $\mathcal{U}$ be a σ-algebra of subsets of a set Λ and $\nu : \mathcal{U} \to [0, \infty)$ a finite measure. If the measure m has finite $L^p(\nu)$-semivariation $\widehat{m}_{p,L^p(\nu)}$ in $L^p(\mu \otimes \nu)$, then m has finite p-variation.*

Proof. We may suppose that $\mathcal{U}$ contains infinitely many disjoint non-ν-null sets, otherwise m necessarily has finite p-variation and finite $L^p(\nu)$-semivariation.

Let $E_1, \ldots, E_m$ be pairwise disjoint sets with positive ν-measure. For each $j = 1, \ldots, n$, set $f_j = \chi_{E_j} / \nu(E_j)^{1/p}$. Let $F_1, \ldots, F_m$ be pairwise disjoint sets belonging to $\mathcal{S}$. Then $\|f_j\|_p = 1$ and

$$\widehat{m}_{p,L^p(\nu)}(\Gamma) \geq \left(\int_\Gamma \int_\Lambda \left| \sum_{j=1}^n f_j(\lambda) m(F_j)(\gamma) \right|^p d\nu(\lambda) \, d\mu(\gamma) \right)^{1/p}$$

$$= \left(\int_\Gamma \sum_{j=1}^n |m(F_j)(\gamma)|^p \left(\int_\Lambda |f_j(\lambda)|^p \, d\nu(\lambda) \right) d\mu(\gamma) \right)^{1/p}$$

$$= \left(\int_\Gamma \sum_{j=1}^n |m(F_j)(\gamma)|^p \, d\mu(\gamma) \right)^{1/p} = \left(\sum_{j=1}^n \|m(F_j)\|_p^p \right)^{1/p}.$$

$$\square$$

For the case $p = 2$, we know from Proposition 1.6 that m has finite 2-variation, as follows from the fact that the inclusion of ℓ^1 in ℓ^2 is an absolutely summing operator [38, p. 255].

The following observation is used in Example 1.8 below.

Example 1.7. Let X be an infinite dimensional Banach space. If $\{\lambda_j\}_{j=1}^\infty$ is a sequence of positive numbers such that $\sum_{j=1}^\infty \lambda_j^2 < \infty$, then there exists an unconditionally summable sequence $\{x_j\}_{j=1}^\infty$ in X such that $\|x_j\| = \lambda_j$ [93, Theorem 1.c.2 p. 16].

Now let $1 \leq p < 2$. We can choose $\{\lambda_j\}_{j=1}^\infty$ such that $\sum_{j=1}^\infty \lambda_j^2 < \infty$ and $\sum_{j=1}^\infty \lambda_j^p = \infty$. It follows that there exists an unconditionally summable sequence $\{x_j\}_{j=1}^\infty$ in X such that $\sum_{j=1}^\infty \|x_j\|^p = \infty$. For $X = L^p(\Omega, \mathcal{S}, \mu)$, the vector measure $m(E) = \sum_{j \in E} x_j$ therefore has infinite p-variation, and so it has infinite $L^p(\nu)$-semivariation in $L^p(\nu \otimes \mu)$, by Proposition 1.8.

For the case $p > 2$, an $L^p(\mu)$-valued vector measure with infinite $L^p(\nu)$-semivariation in $L^p(\nu \otimes \mu)$ is constructed in [72].

1.5.2 *Semivariation of positive operator valued measures*

In this section, we see that operator valued measures taking their values in the Banach lattice of positive operators on an L^p-space have bounded $\mathcal{L}(X)$-semivariation for any Banach space X and the same holds true for operator valued measures dominated by a positive measure. The bounded (S, Q)-processes on L^p-spaces considered in Section 6.5 have this property, so Theorems 1.10 and 6.6 below see use in Chapter 6 on random evolutions.

Let $(\Gamma, \mathcal{E}, \mu)$ be a σ-finite measure space and $1 \le p \le \infty$. The strongly closed subspace of $\mathcal{L}(L^p(\Gamma, \mathcal{E}, \mu))$ consisting of all positive operators T is written as $\mathcal{L}_+(L^p(\Gamma, \mathcal{E}, \mu))$, that is, $Tf \ge 0$ for all $f \in L^p(\Gamma, \mathcal{E}, \mu)$ with $f \ge 0$ μ-a.e.. Where it is convenient, the space $L^p(\Gamma, \mathcal{E}, \mu)$ is abbreviated to $L^p(\mu)$ and $L^p_+(\mu)$ denotes the nonnegative elements of $L^p(\mu)$.

Let $(\Omega, \mathcal{S})$ be a measurable space. An operator valued measure $M : \mathcal{S} \to \mathcal{L}_+(L^p(\mu))$ is called a *positive operator valued measure* and we write $M \ge 0$ in this case. For $p = \infty$, we suppose that Mf is weak* σ-additive in $L^\infty(\mu)$ for each $f \in L^\infty(\mu)$. An operator valued measure $N : \mathcal{S} \to \mathcal{L}(L^p(\mu))$ is *dominated* by a positive operator valued measure M if for every $S \in \mathcal{S}$, the inequality $|N(S)| \le M(S)$ holds in the Banach lattice of regular operators on $L^p(\mu)$, that is,

$$|N(S)f| \le M(S)f, \quad \mu\text{-a.e. for every } f \in L^p_+(\mu).$$

Theorem 1.10. *Let X be a Banach space and suppose that the operator valued measure $N : \mathcal{S} \to \mathcal{L}(L^p(\mu))$ is dominated by an operator valued measure $M \ge 0$. Then N has finite $\mathcal{L}(X)$-semivariation $\beta_{\mathcal{L}(X)}(N)$ and*

$$\beta_{\mathcal{L}(X)}(N)(S) \le \|M(S)\|, \quad S \in \mathcal{S}.$$

If $1 \le p < \infty$ and $f \in L^p(\mu)$, then the vector measure Nf has continuous X-semivariation in $L^p(\Gamma, \mathcal{E}, \mu; X)$.

Proof. We first establish the result for the special case $N = M \ge 0$. For $l \in \mathbb{N}$ and $1 \le i \le l$ let $A_i \in \mathcal{L}(X)$ with $\|A_i\|_{\mathcal{L}(X)} \le 1$. For the same finite set of i's let $\{B_i\}_{i=1}^l$ be pairwise disjoint subsets of $B \in \mathcal{S}$. For $k \in \mathbb{N}$ let $g = \sum_{j=1}^k x_j \chi_{G_j}$ be an X-valued $\mathcal{E}$-simple function with $x_j \in X$ and $G_j \in \mathcal{E}$ pairwise disjoint for $j = 1, \ldots, k$. Also assume that $\|g\|_{L^p(\mu; X)} \le 1$.

Then taking pointwise estimates, we have

$$\left\| \sum_i (A_i \otimes [M(B_i)])g \right\|_{\mathcal{L}(L^p(\mu; X))}^p = \int_\Gamma \left\| \sum_{i,j} (A_i x_j)[M(B_i)\chi_{G_j}](\gamma) \right\|_X^p d\mu(\gamma)$$

$$\le \int_\Gamma \left(\sum_{i,j} \|A_i x_j\|_X [M(B_i)\chi_{G_j}](\gamma) \right)^p d\mu(\gamma)$$

$$\le \int_\Gamma \left(\sum_{i,j} \|x_j\|_X [M(B_i)\chi_{G_j}](\gamma) \right)^p d\mu(\gamma).$$

The right-hand side is equal to

$$\left\| \sum_i M(B_i) \left(\sum_j \|x_j\|_X \chi_{G_j} \right) \right\|_{L^p(\mu)}^p$$

$$\leq \left\| \sum_i M(B_i) \right\|_{\mathcal{L}(L^p(\mu))}^p \left\| \sum_j \|x_j\|_X \chi_{G_j}' \right\|_{L^p(\mu)}^p$$

$$= \left\| \sum_i M(B_i) \right\|_{\mathcal{L}(L^p(\mu))}^p \|g\|_{L^p(\mu;X)}^p$$

$$\leq \|M(B)\|_{\mathcal{L}(L^p(\mu))}^p.$$

The last equality follows from the observation that if $A, B \in \mathcal{S}$ with $A \subseteq B$ then $\|M(A)\|_{\mathcal{L}(L^p(\mu))} \leq \|M(B)\|_{\mathcal{L}(L^p(\mu))}$. This is easily seen by noting that $M(B)f - M(A)f = M(B\backslash A)f \geq 0$ holds for all $f \geq 0$. Since X-valued $\mathcal{E}$-simple functions are dense in $L^p(\mu; X)$, by Definition 6.2 this implies $\beta_{\mathcal{L}(X)}(M)(B) \leq \|M(B)\|_{\mathcal{L}(L^p(\mu;X))}$.

Next we show that for an arbitrary $y \in L^p(\mu)$, the $L^p(\mu)$-valued measure My has continuous X-semivariation. This follows immediately from Proposition 1.5 if we can show that My is dominated by a positive vector measure. However, decomposing y first into its real and imaginary parts and then into its positive and negative parts gives us $|My| \leq 2M|y|$. This is easily seen since, for all $A \in \mathcal{S}$,

$$|M(A)y| = |M(A)y_1 + iM(A)y_2|$$
$$= |M(A)y_1^+ - M(A)y_1^- + iM(A)y_2^+ - iM(A)y_2^-|$$
$$\leq M(A)(y_1^+ + y_1^-) + M(A)(y_2^+ + y_2^-)$$
$$= M(A)(|y_1| + |y_2|)$$
$$\leq 2M(A)|y|.$$

Since M is a positive operator valued measure it follows that $M|y| \geq 0$. This completes the proof for the positive case.

To prove that the dominated measure N has finite $\mathcal{L}(X)$-semivariation it suffices to note that $|N(A)\chi_G| \leq M(A)\chi_G$ for all $A \in \mathcal{S}$ and $G \in \mathcal{E}$ and repeat the argument for the positive case. To prove Ny has continuous X-semivariation for each $y \in L^p(\mu)$ it suffices to note that $|Ny| \leq 2M|y|$, that is, the vector measure Ny is dominated by a positive measure. The result then follows from Proposition 1.5. □

1.6 Bilinear integration after Bartle and Dobrakov

Suppose that $\mathcal{S}$ is a σ-algebra of subsets of a set Σ and $m : \mathcal{S} \to E$ is a vector measure with values in a lcs E. If $f_n : \Sigma \to \mathbb{C}$, $n = 1, 2, \ldots$, are m-integrable functions converging pointwise to a function $f : \Sigma \to \mathbb{C}$ m-almost everywhere and $\lim_{n \to \infty} \int_A f_n \, dm$ converges in E for each set $A \in \mathcal{S}$, then the Vitali-Hahn-Saks Theorem ensures that f is m-integrable and

$$\lim_{n \to \infty} \int_A f_n \, dm = \int_A f \, dm, \quad A \in \mathcal{S}. \tag{1.8}$$

Moreover, the indefinite integrals $f_n m$, $n = 1, 2, \ldots$, converge to fm in semivariation, that is,

$$p(f_n m - fm)(\Sigma) = p(m)(f - f_n) \to 0$$

as $n \to \infty$ for every continuous seminorm p on E, see [86, II.5 Theorem 2] and [43, Theorem IV.10.9].

For a set $A \in \mathcal{S}$, the p-semivariation $p(m)(A)$ of m on A is given by

$$
\begin{aligned}
p(m)(A) &= \sup\{|\langle m, \xi \rangle|(A) : \xi \in U_p^\circ\} \\
&= \sup\{\|m(\varphi \chi_A)\| : \varphi \ \mathcal{S}\text{-simple}, \ \|\varphi\|_\infty \le 1\} \\
&= \sup\left\{\left\|\sum_j c_j m(A_j)\right\|\right\},
\end{aligned}
$$

where the last supremum is over all pairwise disjoint subsets $A_1, \ldots, A_k$ of A and all scalars $|c_j| \le 1$ for $j = 1, \ldots, k$ and $k = 1, 2, \ldots$.

In the case of *bilinear* integration, these convergence results suggest that semivariation may also be used to control the convergence of integrals of a vector valued function with respect to a vector valued measure.

Let X and Y be Banach spaces and suppose that $\| \cdot \|_\tau$ is a norm tensor product on $X \otimes Y$. The X-semivariation $\beta_X(m)$ in $X \otimes_\tau Y$ of a Y-valued measure $m : \mathcal{S} \to Y$ is defined by equation (1.5). Following R. Bartle [11] and I. Dobrakov [39], we shall employ the extended real valued set function $\beta_X(m) : \mathcal{S} \to [0, \infty]$ to control the integration of X-valued functions with respect to m in the completed tensor product space $X \widehat{\otimes}_\tau Y$.

Similar remarks apply to other cases of bilinear integration. If E, F are Banach spaces and $m : \mathcal{S} \to \mathcal{L}(E, F)$ is σ-additive for the strong operator topology, then the E-semivariation $\beta_E(m) : \mathcal{S} \to [0, \infty]$ of m in F is defined by formula (1.6). In the preceding case, we may take $F = X \widehat{\otimes}_\tau Y$, $E = X$ and $\tilde{m}(A)x = x \otimes m(A)$ for $x \in X$ and $A \in \mathcal{S}$, so that $\beta_X(\tilde{m}) = \beta_X(m)$.

In order to integrate $\mathcal{L}(E, F)$-valued functions with respect to an E-valued measure and the bilinear map $(T, x) \mapsto Tx$, $T \in \mathcal{L}(E, F)$, $x \in E$, then the $\mathcal{L}(E, F)$-semivariation $\beta_{\mathcal{L}(E,F)}(m) : \mathcal{S} \to [0, \infty]$ of m in F can be employed. The salutary Example 1.5 shows that these set functions may not be helpful for controlling the integration process.

On the other hand, suppose that the X-semivariation $\beta_X(m)$ in $X \otimes_\tau Y$ of a Y-valued measure $m : \mathcal{S} \to Y$ is not only finite on the σ-algebra $\mathcal{S}$, but also continuous as defined in Section 1.5.

For any X-valued $\mathcal{S}$-simple function $\varphi : \Sigma \to X$ given by

$$\varphi = \sum_{j=1}^{n} x_j \chi_{A_j}$$

for pairwise disjoint sets $A_j \in \mathcal{S}$ and $x_j \in X$, $j = 1, \ldots, n$ and $n = 1, 2, \ldots$, the integral $\varphi \otimes m$ is defined by linearity in the usual way as

$$(\varphi \otimes m)(A) = \sum_{j=1}^{n} x_j \otimes m(A_j \cap A), \quad A \in \mathcal{S},$$

independently of the representation of φ. Then $\varphi \otimes m : \mathcal{S} \to X \otimes Y$ is a finitely additive set function. It is sometimes convenient to write $(\varphi \otimes m)(A)$ as $\int_A \varphi \otimes dm$ or $\int_A \varphi(\sigma) \otimes m(d\sigma)$ for $A \in \mathcal{S}$.

Under the assumption that $\beta_X(m)$ is continuous, Theorem 9 of [11] may be adopted as a definition of integrability.

Definition 1.2 (Integration in the sense of Bartle). Let $\mathcal{S}$ be a σ-algebra of subsets of a nonempty set Σ. Let X and Y be Banach spaces and suppose that $\| \cdot \|_\tau$ is a norm tensor product on $X \otimes Y$. Suppose that the X-semivariation $\beta_X(m)$ in $X \otimes_\tau Y$ of the Y-valued measure $m : \mathcal{S} \to Y$ is continuous.

A function $f : \Sigma \to X$ is said to be *(Bartle) m-integrable* in $X \widehat{\otimes}_\tau Y$ if there exist X-valued simple $\mathcal{S}$-functions φ_n, $n = 1, 2, \ldots$, converging to f m-almost everywhere such that $(\varphi_n \otimes m)(A)$ converges in $X \widehat{\otimes}_\tau Y$ as $n \to \infty$ for each $A \in \mathcal{S}$. The notation $(f \otimes m)(A)$ and $\int_A f \otimes dm$ is used to denote the limit for $A \in \mathcal{S}$.

According to [11, Theorem 10], there is an analogue of the convergence result (1.8) for Bartle bilinear integrals.

Theorem 1.11. *Let X, Y and $m : \mathcal{S} \to Y$ be as in Definition 1.2.*

If $f_n : \Sigma \to X$, $n = 1, 2, \ldots$, are functions that are (Bartle) m-integrable in $X \widehat{\otimes}_\tau Y$ converging m-a.e. to $f : \Sigma \to X$ and for every $\epsilon > 0$ there exists $\delta > 0$ such that $\left\| \int_A f_n \otimes dm \right\|_\tau < \epsilon$ for all $A \in \mathcal{S}$ such that $\beta_X(m)(A) < \delta$, *then f is (Bartle) m-integrable and*

$$\lim_{n \to \infty} \int_A f_n \otimes dm = \int_A f \otimes dm$$

in $X \widehat{\otimes}_\tau Y$ *uniformly for* $A \in \mathcal{S}$*, so that* $\lim_{n \to \infty} \|f \otimes m - f_n \otimes m\|_\tau(\Sigma) = 0$.

The Vitali Convergence Theorem is proved in [11, Theorem 10] without the assumption that $\beta_X(m)$ is continuous or that m is countably additive. There almost everywhere convergence is replaced by the analogue of 'convergence in measure' with respect to the subadditive set function $\beta_X(m)$.

In the case of Banach spaces E and F and an operator valued measure $m : \mathcal{S} \to \mathcal{L}(E, F)$ that is σ-additive for the strong operator topology, the E-semivariation $\beta_E(m)$ of m defined by formula (1.6) has the property that $\|m\|_{\mathcal{L}(E,F)} \leq \beta_E(m)$ for the *scalar* semivariation $\|m\|_{\mathcal{L}(E,F)}$ with respect to the uniform operator norm of $\mathcal{L}(E, F)$, for we can choose $x_j = c_j x$ with $\|x\|_E \leq 1$ and $|c_j| \leq 1$ in (1.6). Hence, if the E-semivariation $\beta_E(m)$ of m is continuous, then m is necessarily σ-additive for the *uniform* operator topology—a condition that is rarely satisfied in applications.

Unless the E-semivariation $\beta_E(m)$ of the operator valued measure m is controlled by a measure and so continuous, the Vitali Convergence Theorem of [11, Theorem 10] is difficult to apply. In a series of papers [39–41], I. Dobrakov circumvented this obstruction by the elementary device of approximating by simple functions based on sets where $\beta_E(m)$ is finite. Further expositions of Dobrakov's approach have appeared in [42, 103].

The papers [39–41] are formulated in terms of the δ-rings $\mathcal{S}_1$ of sets $A \in \mathcal{S}$ for which $\beta_E(m)(A) < \infty$ and the σ-ring it generates. For the basic Example 1.5, $\mathcal{S}_1$ consists of subsets of $[0, 1]$ with Lebesgue measure zero. In order to avoid examples with a trivial set of integrable functions, we shall suppose that the operator valued measure m has σ-*finite* E-semivariation $\beta_E(m)$ in F, that is, there exist pairwise disjoint sets $\Sigma_j \in \mathcal{S}$, $j = 1, 2, \ldots$, such that $\Sigma = \cup_{j=1}^{\infty} \Sigma_j$ and $\beta_E(m)(\Sigma_j) < \infty$ for each $j = 1, 2, \ldots$. The same terminology is used when the vector measure $m : \mathcal{S} \to Y$ is said to have σ-*finite* X-semivariation $\beta_X(m)$ in $X \otimes_\tau Y$. Then the equality $\sigma(\mathcal{S}_1) = \mathcal{S}$ holds.

Definition 1.3 (Integration in the sense of Dobrakov). Let $\mathcal{S}$ be a σ-algebra of subsets of a nonempty set Σ. Let X and Y be Banach spaces and suppose that $\| \cdot \|_\tau$ is a norm tensor product on $X \otimes Y$. Suppose that the X-semivariation $\beta_X(m)$ in $X \otimes_\tau Y$ of the Y-valued measure $m : \mathcal{S} \to Y$ is σ-finite. Let $\mathcal{S}_1 = \{A \in \mathcal{S} : \beta_X(m)(A) < \infty\}$.

A function $f : \Sigma \to X$ is said to be *(Dobrakov) m-integrable* in $X \widehat{\otimes}_\tau Y$ if there exist X-valued $\mathcal{S}_1$-simple functions φ_n, $n = 1, 2, \ldots$, converging to f m-almost everywhere such that $(\varphi_n \otimes m)(A)$ converges in $X \widehat{\otimes}_\tau Y$ as $n \to \infty$ for each $A \in \mathcal{S}$. The notation $(f \otimes m)(A)$ and $\int_A f \otimes dm$ is used to denote the limit for $A \in \mathcal{S}$.

That the definite integral $f \otimes m$ is independent of the approximating sequence of simple functions is proved in [39, Theorem 2], with a clarification in [42, Theorem 7].

According to [39, Theorem 2], there is an analogue of the convergence result (1.8) for Dobrakov bilinear integrals.

Theorem 1.12. *Let X, Y and $m : \mathcal{S} \to Y$ be as in Definition 1.3.*

If $f_n : \Sigma \to X$, $n = 1, 2, \ldots$, are functions that are (Dobrakov) m-integrable in $X \widehat{\otimes}_\tau Y$ converging m-a.e. to $f : \Sigma \to X$ and $\{ f_n \otimes m : n \in \mathbb{N} \}$ is a uniformly countably additive collection of $X \widehat{\otimes}_\tau Y$-valued measures, then f is (Dobrakov) m-integrable and

$$\lim_{n \to \infty} \int_A f_n \otimes dm = \int_A f \otimes dm$$

in $X \widehat{\otimes}_\tau Y$ uniformly for $A \in \mathcal{S}$, so that $\lim_{n \to \infty} \| f \otimes m - f_n \otimes m \|_\tau (\Sigma) = 0$.

As mentioned in Section 1.5, if $\beta_X(m)$ is continuous, then $\mathcal{S}_1 = \mathcal{S}$, so Theorem 1.12 is a genuine improvement of Theorem 1.11 and the two notions of bilinear integration coincide [39, Example 4, p. 535].

In [39, Example 7″], an example is given of a bounded and Dobrakov integrable function $f : \Sigma \to E$ and a uniformly countably additive operator valued measure $m : \mathcal{S} \to \mathcal{L}(E, F)$ such that $\beta_E(m)$ is not continuous and f is not even $\beta_E(m)$-measurable in the sense of [11].

For the remainder of this section, we use the term m-integrable instead of (Dobrakov) m-integrable.

The connection with the Bochner integral is noted in [39, Theorem 6] from the obervation that the inequality $\beta_X(m) \leq V(m)$ holds between X-semivariation $\beta_X(m)$ in $X \otimes_\tau Y$ of the Y-valued measure $m : \mathcal{S} \to Y$ and the variation $V(m) : \mathcal{S} \to [0, \infty]$ of m.

Proposition 1.9. *Let X, Y and $X \widehat{\otimes}_\tau Y$ be Banach spaces, with τ a tensor product topology norm on $X \otimes Y$. Suppose that $m : \mathcal{S} \to Y$ is a measure for which there exist sets $\Sigma_k \in \mathcal{S}$, $k = 1, 2, \ldots$, increasing to Σ such that the total variation $V(m)(\Sigma_k)$ of m on Σ_k is finite for each $k = 1, 2, \ldots$, that is, m has σ-finite variation in Y.*

If $f : \Sigma \to X$ is Bochner integrable with respect to the σ-finite measure $V(m)$, then f is m-integrable in $X \widehat{\otimes}_\tau Y$ and

$$V(f \otimes m) \leq \int_\Sigma \|f\| \, dV(m).$$

The nature of bilinear integration is unlike the cases in which either the measure or function is scalar; for example, *bounded* vector valued functions need not be integrable with respect to a vector valued measure [39, Example 7'].

Example 1.8. A measure $m : S \to Y$ can have σ-finite variation without having continuous X-semivariation, or even finite X-semivariation.

Let $1 \leq p < 2$, and $X = Y = L^p([0,1])$. We give $X \otimes Y$ the relative topology τ of $L^p([0,1]^2)$, so that $X \widehat{\otimes}_\tau Y = L^p([0,1]^2)$.

Suppose that $\{y_j\}_{j=1}^\infty$ is an unconditionally summable sequence in $L^p([0,1])$ such that $\sum_{j=1}^\infty \|y_j\|^p = \infty$ [93, Theorem 1.c.2 p. 16]. Let $\Omega = \mathbb{N}$ and let S be the family of all subsets of Ω. Define the Y-valued measure $m : S \to Y$ by $m(A) = \sum_{k \in A} y_k$ for every $A \in S$.

Let $E_j, j = 1, 2, \ldots$, be pairwise disjoint subsets of $[0,1]$ with positive Lebesgue measure $|E_j|$. Set $f_j = \chi_{E_j}/|E_j|^{1/p}$ for each $j = 1, 2, \ldots$. Then $\|f_j\|_p = 1, j = 1, 2, \ldots$, and as $k \to \infty$, we have

$$\int_0^1 \Big\| \sum_{j=1}^k f_j y_j(t) \Big\|_{L^p([0,1])}^p \, dt = \sum_{j=1}^k \int_0^1 |y_j(t)|^p \, dt = \sum_{j=1}^k \|y_j\|^p \to \infty.$$

The vector measure m has infinite X-semivariation but σ-finite variation.

It is easy to check from Definition 1.3 that a function $G : \Omega \to L^p([0,1])$ is m-integrable in $L^p([0,1]^2)$ if and only if $\{G(j) \otimes y_j\}_{j=1}^\infty$ is unconditionally summable in $L^p([0,1]^2)$, because $G\chi_{\{1 \leq j \leq n\}}$ are S_1-simple functions converging pointwise to G in $L^p([0,1])$ as $n \to \infty$. This is obviously guaranteed by the condition $\sum_{j=1}^\infty \|G(j)\|_p \|f_j\|_p < \infty$ of Proposition 1.9. However, the bounded $L^p([0,1])$-valued function $j \mapsto f_j$ is not m-integrable.

Example 1.9. Let $1 \leq p < 2$ and suppose that $\{y_j\}_{j=1}^\infty$ is an unconditionally summable sequence in $L^p([0,1])$ such that $\sum_j \|y_j\|^p = \infty$. Then, as above, there exist positive functions f_j on $[0,1]$ such that $\int_0^1 f_j(t)^p dt = 1$ and $\sum_j y_j f_j$ does not converge in $L^p([0,1]^2)$.

Now let $\{a_j\}$ be an absolutely summable sequence of positive scalars and set $g_j = a_j f_j$. The $L^p([0,1])$-valued measure $m : A \mapsto \sum_{j \in A} g_j$ has finite variation $V(m) : A \mapsto \sum_{j \in A} a_j$. Set $x_j = y_j/a_j$. Then $\{a_j x_j\}_{j=1}^\infty$ is unconditionally summable in $L^p([0,1])$, but $\sum_j x_j \otimes g_j$ does not converge in $L^p([0,1]^2)$. In other words, the function $j \mapsto x_j$ is Pettis integrable with respect to $V(m)$ in $L^p([0,1])$, but it is *not* m-integrable in $L^p([0,1]^2)$. Of course, a sequence which is *Bochner* integrable with respect to $V(m)$ is necessarily m-integrable.

The following example is a version of [39, Example 7''] adapted to tensor products.

Example 1.10. Let $1 \le p < 2$ and $0 < a < 1/p - 1/2$. According to Orlicz's Lemma [93, Theorem 1.c.2 p. 16], there exists an unconditionally summable sequence $\{z_j\}_{j=1}^\infty$ in $L^p([0,1])$ such that $\|z_j\|_p = 1/(j^{1/2} \ln(j+1))$. For each $j = 1, 2, \ldots$, set $y_j = j^{-a} z_j$. Then

$$\sum_{j=1}^\infty \|y_j\|_p^p = \sum_{j=1}^\infty \frac{1}{j^{ap} j^{p/2} \ln(j+1)^p} = \infty.$$

Neverthless, $\{j^a y_j\}_{j=1}^\infty = \{z_j\}_{j=1}^\infty$ is unconditionally summable in $L^p([0,1])$. Let $m(A) = \sum_{j \in A} y_j$ for every subset A of $\mathbb{N}$.

Claim 1.1. *The function $f : j \mapsto j^a 1$ with values in $L^p([0,1])$, is m-integrable in $L^p([0,1]^2)$, but it is not (Bartle) m-integrable in $L^p([0,1]^2)$.*

Proof. We already know that f is m-integrable, because $\{f(j)y_j\}_{j=1}^\infty$ is unconditionally summable in $L^p([0,1])$. Let ϕ be an $L^p[0,1]$-valued simple function. Then for each $j \in \mathbb{N}$, we have $\|f(j) - \phi(j)\|_p \ge |\|f(j)\|_p - \|\phi(j)\|_p| = |j^a - \|\phi(j)\|_p|$. Let J be an integer greater than $(\max_j \|\phi(j)\|_p + 1)^{1/a}$. Then for all $j \ge J$, the inequality $\|f(j) - \phi(j)\|_p \ge 1$ holds.

Let $A = \{j : \|f(j) - \phi(j)\|_p \ge 1\}$. The argument used in Example 1.8 now shows that the $L^p([0,1])$-semivariation of m in $L^p([0,1]^2)$ of the set A is greater than $\sum_{j \ge J} \|y_j\|_p^p = \infty$. Consequently, f cannot be approximated in $L^p([0,1])$-semivariation of m by simple functions, so it cannot be (Bartle) m-integrable in $L^p([0,1]^2)$. $\qquad\square$

Both the Bartle and Dobrakov approaches to bilinear integration utilise semivariation to control the convergence of the integrals—we call these *regular* bilinear integrals. In the next chapter, we see how the use of semivariation can be avoided for bilinear integrals with values in tensor products of vector spaces as is required for bilinear integration with respect to spectral measures employed in subsequent chapters.

Chapter 2

Decoupled bilinear integration

The approach considered in Section 1.6 to bilinear integration in $X\widehat{\otimes}_\tau Y$ of an X-valued function $f : \Sigma \to X$ with respect to a Y-valued measure $m : \mathcal{S} \to Y$ is adequate for Banach spaces X and Y once we know that m has σ-finite X-semivariation in $X\widehat{\otimes}_\tau Y$.

Example 1.5 gives a particularly simple X'-valued measure m for which the X-semivariation $\beta_X(m) = V(m)$ of m in $\mathbb{C}$ has only the values 0 or ∞. It is worthwhile looking at this example in greater detail because it is representative of many vector measures arising in applications.

For $p = 2$ in Example 1.5, the $L^2([0,1])$-valued measure m is equal to $Q\mathbf{1}$ for the spectral measure $Q : \mathcal{B}(\mathbb{R}) \to \mathcal{L}(L^2([0,1]))$ such that $Q(B)$ is the selfadjoint projection operator $h \longmapsto \chi_B.h$, $h \in L^2([0,1])$, for each $B \in \mathcal{B}(\mathbb{R})$ and $\mathbf{1}$ is the constant function equal to one on the interval $[0,1]$. The spectral measure Q is associated with the position operator of a quantum particle on the line.

According to the Spectral Theorem for selfadjoint operators, any spectral measure $P : \mathcal{B}(\mathbb{R}) \to \mathcal{L}(\mathcal{H})$ in a separable Hilbert space $\mathcal{H}$ is unitarily equivalent to a spectral measure similar to Q except for a discrete part, so like the vector measure m of Example 1.5, the variation $V(Ph)$ will have values that are either zero or infinity for $h \in \mathcal{H}$, except in trivial cases. On the other hand, for any vector $h \in \mathcal{H}$ and pairwise disjoint Borel subsets

$B_1, B_2, \ldots$ of $\mathbb{R}$, with respect to the inner product $(\cdot, \cdot)$ of $\mathcal{H}$ we have

$$\|P(\cup_{n=1}^\infty B_n)h\|_{\mathcal{H}}^2 = (P(\cup_{n=1}^\infty B_n)h, P(\cup_{n=1}^\infty B_n)h)$$

$$= \sum_{n,m=1}^\infty (P(B_n)h, P(B_m)h)$$

$$= \sum_{n,m=1}^\infty (P(B_n \cap B_m)h, h)$$

$$= \sum_{n=1}^\infty \|P(B_n)h\|_{\mathcal{H}}^2.$$

Hence, the $\mathcal{H}$-valued measure $Ph : B \longmapsto P(B)h$, $B \in \mathcal{B}(\mathbb{R})$, necessarily has *finite 2-variation*. More generally, *any* Hilbert space valued measure has finite 2-variation by Propositions 1.6 and 1.8.

Given an operator valued function $f : \Sigma \to \mathcal{L}(X, Y)$ and a vector measure $m : \mathcal{S} \to X$, leaving measurability conditions aside for the moment, we wish to consider the integral $\int_\Sigma f \, dm \in Y$ in generality sufficient to treat scattering theory in Chapter 5.

For an element $y' \in Y'$ of the space Y' dual to Y, the X'-valued function $y' \circ f : \Omega \to X'$ is defined by

$$\langle x, (y' \circ f)(\omega) \rangle = \langle f(\omega)x, y' \rangle, \quad x \in X, \ \omega \in \Omega.$$

It is reasonable to expect that the identity

$$\left\langle \int_\Sigma f \, dm, y' \right\rangle = \int_\Sigma \langle y' \circ f, dm \rangle$$

ought to hold for each $y' \in Y'$, in which the right-hand side is the integral of the X'-valued function $y' \circ f$ acting on the range of the X-valued measure m. However, the total variation $V(m)(\Omega)$ of m satisfies the equation

$$V(m)(\Omega) = \sup_{\|s\|_\infty \leq 1} \int_\Sigma \langle s, dm \rangle$$

by the Hahn-Banach Theorem. The supremum is taken over all X'-valued $\mathcal{S}$-simple functions $s = \sum_{k=1}^n x_k' \chi_{B_k}$ with $\|x_k'\| \leq 1$ and pairwise disjoint $B_k \in \mathcal{S}$ for $k = 1, \ldots, n$ and $n = 1, 2, \ldots$, where

$$\int_\Sigma \langle s, dm \rangle = \sum_{k=1}^n \langle m(B_k), x_k' \rangle.$$

As mentioned above, in the case that $m = Ph$ for a spectral measure P, the variation $V(m)$ may have only the values 0 and ∞. Because there may be so few sets on which m has finite variation, the approaches considered in Section 1.6 are unsuited to vector measures m of this type. ·

In the special case of the integration of X-valued functions with respect to Y-valued measures in the tensor product $X\widehat{\otimes}_\tau Y$, we can exploit the product structure of the tensor product to avoid consideration of X-semivariation in $X\widehat{\otimes}_\tau Y$ of the measure.

Expanding on this approach, suppose that an $\mathcal{L}(X,Y)$-valued function has an integral with respect to an X-valued measure that takes its values in $\mathcal{L}(X,Y)\widehat{\otimes}_\tau X$. A careful choice of the tensor product topology τ facilitates a continuous bilinear extension J_1 of the product map $T \otimes x \mapsto Tx$ to $\mathcal{L}(X,Y)\widehat{\otimes}_\tau X$ so that the indefinite integral fm may be realised as $J_1 \circ (f \otimes m)$. Similarly, the integral mf of an X-valued function f with respect to an $\mathcal{L}(X,Y)$-valued measure may be realised as $J_2 \circ (f \otimes m)$ for the continuous bilinear extension J_2 of the product map $x \otimes T \mapsto Tx$ to $X\widehat{\otimes}_\tau \mathcal{L}(X,Y)$.

The intermediary *decoupled* integral $f \otimes m$ may exist even if m has diminished variational bounds such as the case where X is equal to a Hilbert space $\mathcal{H}$ and $m = Ph$ with respect to a spectral measure P and $h \in \mathcal{H}$.

A basic example arises in the Hilbert space $\mathcal{H} = L^2(\mu)$ for a σ-finite measure μ and the vector measure $m : A \mapsto \chi_A$. For a suitable choice of the tensor product $\mathcal{H}\widehat{\otimes}_\tau \mathcal{H}$, functions k for which $\Phi_k : x \mapsto k(x,\cdot)$, $x \in \Sigma$, is m-integrable $\mathcal{H}\widehat{\otimes}_\tau \mathcal{H}$ determine a bounded integral operator $T_k : \mathcal{H} \to \mathcal{H}$ with integral kernel k and the scalar

$$\int_\Sigma \langle T_k, dm \rangle = J_1((\Phi_k \otimes m)(\Sigma))$$

is a generalised trace of the operator T_k. The choice $\tau = \pi$ for the projective tensor product topology π is associated with the operator ideal of trace class operators and $\int_\Sigma \langle T_k, dm \rangle$ is the usual trace of T_k. This example from operator theory is expanded upon in Chapter 3.

Like the measure Ph, a Gaussian random measure $m : \mathcal{B}([0,T]) \to L^2(P)$ is orthogonally scattered [95] with finite 2-variation. For an adapted E-valued simple process X, the decoupled integral $X \otimes m$ has values in $L^2(P \otimes P, E)$. For a suitable Banach space E, the *stochastic integral* $X.m$ can be written as $J \circ (X \otimes m)$ for a continuous extension of the product map $J : f \otimes g \mapsto f.g$ of an E-valued random function f and a random scalar function g. The bilinear naure of stochastic integration is more fully explored in Chapter 4.

In the next section, decoupled bilinear integrals $f \otimes m$ are studied in more detail.

2.1   Bilinear integration in tensor products

Let X, Y and $X \widehat{\otimes}_\tau Y$ be Banach spaces, with τ a norm tensor product topology on $X \otimes Y$ as in Section 1.5. Let $m : \mathcal{S} \to Y$ be a vector measure defined on the σ-algebra $\mathcal{S}$ of subsets of a set Ω. As usual, the integral of X-valued functions with respect to m is first defined for elementary functions.

An X-*valued* $\mathcal{S}$-*simple function* is a function ϕ for which there exist $k = 1, 2, \ldots$, sets $E_j \in \mathcal{S}$ and vectors $c_j \in X, j = 1, \ldots, k$, such that $\phi = \sum_{j=1}^{k} c_j \chi_{E_j}$. The integral $\phi \otimes m$ of ϕ with respect to the Y-valued measure m is defined by $(\phi \otimes m)(A) = \sum_{j=1}^{k} c_j \otimes [m(A \cap E_j)]$, for all $A \in \mathcal{S}$. Then $(\phi \otimes m)(A) \in X \otimes Y$.

We shall make some restrictive assumptions concerning the spaces X and Y and their tensor product $X \otimes Y$ allowing us to integrate a class of functions more general than the simple functions. As mentioned in Section 1.5, property (T2) of a norm tensor product topology τ ensures that $X' \otimes Y'$ may be identified with a linear subspace of the continuous dual $(X \otimes_\tau Y)' = (X \widehat{\otimes}_\tau Y)'$ of $X \otimes_\tau Y$. With suitable modifications of the definitions, the results of this section work also for the tensor product $X \otimes_\tau Y$ of two locally convex spaces X, Y.

Definition 2.1. The topology τ determined by a tensor product norm $\| \cdot \|_\tau$ is said to be *completely separated* if the subspace $X' \otimes Y'$ of $(X \otimes_\tau Y)'$ separates the *completion* $X \widehat{\otimes}_\tau Y$ of the normed space $X \otimes_\tau Y$, that is, if $u \in X \widehat{\otimes}_\tau Y$ and $\langle u, x' \otimes y' \rangle = 0$ for all $x' \in X'$ and $y' \in Y'$, then $u = 0$.

An equivalent formulation of the condition that τ is completely separated, is that if $\{s_n\}_{n \in \mathbb{N}}$, is any τ-Cauchy sequence in $X \otimes Y$ for which

$$\lim_{n \to \infty} \langle s_n, x' \otimes y' \rangle = 0, \quad \text{for all } x' \in X' \text{ and } y' \in Y',$$

then $\lim_{n \to \infty} \|s_n\|_\tau = 0$. For a completely separated tensor product topology τ, the completion $X \widehat{\otimes}_\tau Y$ of $X \otimes_\tau Y$ is naturally identified with a subspace of the completion of $X \otimes Y$ in the topology $\sigma(X \otimes Y, X' \otimes Y')$.

If one of the Banach spaces X and Y has the approximation property, then the projective tensor product topology on $X \otimes Y$ is completely separated [88, 43.2 (7)]. Because the injective tensor product $X \widehat{\otimes}_\epsilon Y$ may be identified with the closure of $X \otimes Y$ in the space of separately continuous bilinear forms $\mathfrak{B}_e(X'_\sigma, Y'_\sigma)$ [123, IV.9.1] and $X' \otimes Y'$ clearly separates $\mathfrak{B}_e(X'_\sigma, Y'_\sigma)$, it follows that ϵ is a completely separated tensor product topology.

We begin by stating an elementary but useful condition [87, 18.4 (7)].

Proposition 2.1. *Let τ be a norm tensor product topology on $X \otimes Y$. If the closed unit ball of $X \widehat{\otimes}_\tau Y$ in the norm $\| \cdot \|_\tau$ is closed for the topology $\sigma(X \otimes Y, X' \otimes Y')$, then τ is completely separated.*

For example, it follows from the Hahn-Banach theorem and Proposition 2.1 that for $1 \leq p \leq \infty$, the $L^p(\Gamma, \mathcal{E}, \mu; X)$-topology on $L^p(\Gamma, \mathcal{E}, \mu) \otimes X$ is completely separated.

The following ubiquitous convergence result is a variant of Vitali's convergence theorem [43, III.3.6].

Lemma 2.1. *Let $(\Omega, \mathcal{S})$ be a measurable space and $\mu : \mathcal{S} \to \mathbb{C}$, a scalar measure. Suppose that f_k, $k = 1, 2, \ldots$, are μ-integrable scalar functions converging μ-almost everywhere to a scalar function f, with the property that the sequence $\{f_k \mu(A)\}_{k=1}^\infty$ converges for each $A \in \mathcal{S}$. Then f is μ-integrable and $f_k \mu(A) \to f \mu(A)$ uniformly for $A \in \mathcal{S}$, as $k \to \infty$.*

Proof. The measures $f_k \mu$, $k = 1, 2, \ldots$, are uniformly countably additive on $\mathcal{S}$, by the Vitali-Hahn-Saks theorem [38, I.5.6]. An appeal to Egorov's measurability theorem [43, III.5.12], ensures that there exists an increasing family of sets $\Omega_j \in \mathcal{S}$ such that $\cup_j \Omega_j$ is a set of full μ-measure, and for each $j = 1, 2, \ldots$, the functions f_k converge to f uniformly on Ω_j as $k \to \infty$.

Let $\epsilon > 0$ and choose j so large that $|f_k \mu|(\Omega \backslash \Omega_j) < \epsilon$ for all $k = 1, 2, \ldots$. Then for K large enough, $|f_k - f_l| < \epsilon$ on Ω_j, for all $k, l \geq K$. Hence, for all $A \in \mathcal{S}$ and $k, l \geq K$,

$$|f_k \mu(A) - f_l \mu(A)| \leq |f_k \mu|(\Omega \setminus \Omega_j) + |f_l \mu|(\Omega \setminus \Omega_j) + (|f_k - f_l|.|\mu|)(\Omega_j)$$
$$< 2\epsilon + \epsilon |\mu|(\Omega).$$

Because ϵ is any positive number, $\lim_{k \to \infty} f_k \mu(A)$ converges uniformly for all $A \in \mathcal{S}$, the function f is integrable, and $\lim_{k \to \infty} f_k \mu(A) = f \mu(A)$, for all $A \in \mathcal{S}$. □

A similar result holds for strongly measurable Pettis integrable with values in a Banach space X.

For an X-valued $\mathcal{S}$-simple function ϕ, the integral $\phi \otimes m$ is σ-additive in $X \otimes_\tau Y$ by property (T1) of a tensor product topology, see Section 1.5. The following lemma is needed for Definition 2.2 to make sense.

Lemma 2.2. *Let τ be a completely separated norm topology on $X \otimes Y$. Suppose that $s_k, k = 1, 2, \ldots$, are X-valued $\mathcal{S}$-simple functions for which*

$\{(s_k \otimes m)(A)\}_{k=1}^{\infty}$ *is τ-Cauchy in $X \otimes Y$ for each $A \in \mathcal{S}$ and $s_k \to 0$ m-a.e. Then $(s_k \otimes m)(A) \to 0$ in $X \otimes_\tau Y$ for each $A \in \mathcal{S}$.*

Proof. For each $x' \in X'$ and $y' \in Y'$, the scalars

$$\langle (s_k \otimes m)(A), x' \otimes y' \rangle = \int_A \langle s_k, x' \rangle \, d\langle m, y' \rangle, \quad k = 1, 2, \ldots,$$

converge as $k \to \infty$ for every $A \in \mathcal{S}$, and $\langle s_k, x' \rangle \to 0$ $\langle m, y' \rangle$-a.e. An appeal to Lemma 2.1, shows that $\lim_{k \to \infty} \langle (s_k \otimes m)(A), x' \otimes y' \rangle = 0$ is true for all $x' \in X'$ and $y' \in Y'$. But we know that $(s_k \otimes m)(A), k = 1, 2, \ldots,$ is already τ-Cauchy in $X \otimes Y$, so the fact that τ is completely separated tells us that $\lim_{k \to \infty} (s_k \otimes m)(A) = 0$ in τ. $\square$

Our bilinear integral is defined by adopting the conclusion of [11, Theorem 9], a translation to the bilinear context of "Dunford's second integral", or in modern parlance, the Pettis integral for strongly measurable functions.

Definition 2.2. Let $(\Omega, \mathcal{S})$ be a measurable space and X, Y Banach spaces. Suppose that τ is a completely separated norm tensor product topology on $X \otimes Y$. Let $m : \mathcal{S} \to Y$ be a Y-valued measure.

A function $\phi : \Omega \to X$ is said to be *m-integrable* in $X \widehat{\otimes}_\tau Y$ if there exist X-valued $\mathcal{S}$-simple functions $\phi_k, k = 1, 2, \ldots,$ such that $\phi_k \to \phi$ m-a.e. as $k \to \infty$, and $\{(\phi_k \otimes m)(A)\}_{k=1}^{\infty}$ converges in $X \widehat{\otimes}_\tau Y$ for each $A \in \mathcal{S}$. Let

$$(\phi \otimes m)(A) = \int_A \phi(\omega) \otimes dm(\omega)$$

denote this limit. Sometimes, we write $m(\phi)$ for the integral $(\phi \otimes m)(\Omega)$.

To check that $\phi \otimes m$ is well-defined, suppose that we have some other X-valued $\mathcal{S}$-simple functions $\phi'_j, j = 1, 2, \ldots,$ such that $\phi'_j \to \phi$ m-a.e. as $j \to \infty$ and the sequence $\{(\phi'_j \otimes m)(A)\}_{j=1}^{\infty}$ converges in $X \widehat{\otimes}_\tau Y$ for each $A \in \mathcal{S}$. Then $[\phi'_j - \phi_j] \to 0$ m-a.e. as $j \to \infty$ and $\{([\phi'_j - \phi_j] \otimes m)(A)\}_{j=1}^{\infty}$ converges in $X \widehat{\otimes}_\tau Y$, for each $A \in \mathcal{S}$, as $j \to \infty$. By Lemma 2.2, we must have $(\phi \otimes m)(A) = \lim_{j \to \infty} (\phi_j \otimes m)(A) = \lim_{j \to \infty} (\phi'_j \otimes m)(A)$, for each set $A \in \mathcal{S}$.

The set function $\phi \otimes m$ is the setwise limit of σ-additive set functions $\phi_k \otimes m, k = 1, 2, \ldots,$ so by the Vitali-Hahn-Saks theorem [38, I.5.6], it is itself σ-additive for the topology τ. It is easy to see that the map $(f, m) \mapsto f \otimes m$ is bilinear, in the obvious sense.

Let $m : \mathcal{S} \to Y$ be a measure. If $A \in \mathcal{S}$ and the restriction of m to the σ-algebra $\mathcal{S} \cap A = \{E \cap A : E \in \mathcal{S}\}$ has σ-finite X-semivariation, then we say that m has σ-finite X-semivariation *on A*.

The following result is a useful consequence of σ-finite X-semivariation.

Theorem 2.1. *Let X, Y and $X\widehat{\otimes}_\tau Y$ be Banach spaces, where τ is a completely separated norm tensor product topology on $X \otimes Y$. Suppose that the vector measure $m : S \to Y$ has σ-finite X-semivariation in $X \otimes_\tau Y$.*

If f_j, $j = 1, 2, \ldots$, are m-integrable functions such that f_j converges m-a.e. to an X-valued function f, and $\{f_j \otimes m(A)\}_{n=1}^\infty$ converges in $X\widehat{\otimes}_\tau Y$ for each $A \in S$, then f is m-integrable in $X\widehat{\otimes}_\tau Y$ and $\{(f_j \otimes m)(A)\}_{n=1}^\infty$ converges in $X\widehat{\otimes}_\tau Y$ to $(f \otimes m)(A)$, uniformly for $A \in S$.

Proof. Let $\| \cdot \|_\tau$ denote the norm of $X\widehat{\otimes}_\tau Y$. As usual, $\|\mu\|_\tau : S \to [0, \infty)$ denotes the semivariation of a measure $\mu : S \to X\widehat{\otimes}_\tau Y$. Because the vector measures $\{f_j \otimes m\}_{j=1}^\infty$ converge setwise in $X\widehat{\otimes}_\tau Y$, they are uniformly bounded by the Nikodym boundedness theorem and uniformly countably additive by the Vitali-Hahn-Saks theorem, so there exists a nonnegative measure $\nu : S \to [0, \infty)$ such that $\lim_{\nu(E) \to 0} \|f_j \otimes m\|_\tau(E) = 0$, uniformly for $j = 1, 2, \ldots$. Moreover, ν may be chosen with the property that the bound $0 \leq \nu(E) \leq \sup_j \|f_j \otimes m\|_\tau(E)$ holds for all $E \in S$ [38, I.2.5].

Let $N = \{\omega \in \Omega : \lim_{j \to \infty} f_j(\omega) \neq f(\omega)\}$. Then N is an m-null set, so by Corollary 2.1, $\|f_j \otimes m\|_\tau(N) = 0$ for all $j = 1, 2, \ldots$. It follows from the inequality above that N is a ν-null set.

Let $\epsilon > 0$ and choose $\delta > 0$ such that for every set $E \in S$ with the property that $\nu(E) < \delta$, the inequality $\|f_j \otimes m\|_\tau(E) < \epsilon/4$ holds for all $j = 1, 2, \ldots$. Let B be the closed unit ball of X. There exist increasing sets $\Omega_k, k = 1, 2, \ldots$, belonging to S on which the $(B, \| \cdot \|_\tau)$-semivariation is finite, and whose union is Ω. The σ-additivity of the measure ν guarantees that for some $K \in \mathbb{N}$, we have $\nu(\Omega_K^c) < \delta/2$.

An appeal to Egorov's theorem [43, III.5.12] ensures that there exists a set B_δ such that $\nu(B_\delta^c) < \delta/2$ and $\|f_k - f\|_X \to 0$ uniformly on B_δ. Let $A_\delta = B_\delta \cap \Omega_K$. Then $\nu(A_\delta^c) < \delta$, $\|f_k - f\|_X \to 0$ uniformly on A_δ as $k \to \infty$ and $\beta_{(B,r)}(A_\delta) < \infty$. Choose $K_\epsilon = 1, 2, \ldots$ such that

$$\sup_{\omega \in A_\delta} \|f(\omega) - f_k(\omega)\|_X < \frac{\epsilon}{4\beta_{(B,r)}(A_\delta) + 1},$$

for all $k \leq K_\epsilon$. It follows from Lemma 2.3 that for every $A \in S$,

$$\|(f_j \otimes m)(A) - (f_k \otimes m)(A)\|_\tau \leq \|([f_j - f_k] \otimes m)(A \cap A_\delta)\|_\tau + \epsilon/2$$
$$\leq \| \|f_j - f_k\|_X \chi_{A_\delta}\|_\infty \beta_{(B,r)}(A_\delta) + \epsilon/2 < \epsilon,$$

for all $j, k \leq K_\epsilon$, so $\{f_k \otimes m(A)\}_{k=1}^\infty$ converges in $X\widehat{\otimes}_\tau Y$, uniformly for all $A \in S$.

It remains to prove that f is integrable. Each function f_k is integrable, so applying the same process to f_k, and choosing a subsequence $\{f_{k_j}\}_{j=1}^{\infty}$ of $\{f_k\}_{k=1}^{\infty}$, if necessary, we obtain X-valued $\mathcal{S}$-simple functions ϕ_j, $j = 1, 2, \ldots$, and an increasing family of sets $D_j \in \mathcal{S}$, $j = 1, 2, \ldots$, such that

(1) $\cup_{j=1}^{\infty} D_j$ is a set of full ν-measure,
(2) $\sup_{\omega \in D_j} \|f(\omega) - f_{k_j}(\omega)\|_X < 1/j$,
(3) $\|(f_{k_l} \otimes m)(A) - (f_{k_j} \otimes m)(A)\|_\tau < 1/j$, for all $l \geq j$ and all $A \in \mathcal{S}$,
(4) $\sup_{\omega \in D_j} \|f_{k_j}(\omega) - \phi_j(\omega)\|_X < 1/j$,
(5) $\|(f_{k_j} \otimes m)(A) - (\phi_j \otimes m)(A)\|_\tau < 1/j$, for all $A \in \mathcal{S}$,

for all $j = 1, 2, \ldots$. Hence, $\phi_j \to f$ m-a.e., and $\{(\phi_j \otimes m)(A)\}_{j=1}^{\infty}$ converges uniformly for $A \in \mathcal{S}$ to $\lim_{k \to \infty}(f_k \otimes m)(A)$. According to Definition 2.2, f is m-integrable and $\lim_{k \to \infty}(f_k \otimes m)(A) = (f \otimes m)(A)$, uniformly for $A \in \mathcal{S}$. $\square$

As a consequence of the proof above, it is evident that if $f : \Omega \to X$ is m-integrable and m has σ-finite X-semivariation on the set $\{f \neq 0\}$, then the measure $f \otimes m$ has σ-finite X-semivariation.

Remark 2.1. Under further technical assumptions involving continuity of semivariation and related concepts, I. Dobrakov [40, Theorem 17] obtains an analogue of the Lebesgue dominated convergence theorem. For our applications in Chapter 5, we need a form of bounded convergence theorem for the integrals of operator valued functions with respect to operator valued measures; this does not readily fit into Dobrakov's scheme of things. We prove our convergence result in Theorem 6.5 in Chapter 6.

In the present context, the notion of integrability of Definition 2.2 differs from (Dobrakov) integrability if and only if m fails to have σ-finite X-semivariation in $X \otimes_\tau Y$ [70, Corollary 3.6]; for an example concerning such a measure, see Examples 2.2 and 2.3.

We state here the bounded convergence theorem of Bartle [11, Theorem 7, Lemma 3] in our setting.

Theorem 2.2. *Suppose that X, Y and $X \otimes_\tau Y$ are as in Theorem 2.1 and the Y-valued measure m has continuous X-semivariation in $X \otimes_\tau Y$. Then every strongly m-measurable, bounded function $f : \Omega \to Y$ is m-integrable. Moreover, if $f_k : \Omega \to Y$, $k = 1, 2, \ldots$, is a uniformly bounded sequence of Y-valued functions converging to f m-almost everywhere, then as $k \to \infty$,* the integrals $\int_A f_k \otimes dm$ converge to $\int_A f \otimes dm$ in $X \widehat{\otimes}_\tau Y$, uniformly for $A \in \mathcal{S}$.

We point out some facts that are easily established. In the case that $X = \mathbb{C}$, a function $f : \Omega \to \mathbb{C}$ is m-integrable in the sense above if and only if it is m-integrable in the sense of vector measures described in Section 1.3. For the case $Y = \mathbb{C}$ and X is a Banach space, a function $f : \Omega \to X$ is m-integrable in the sense above if and only if it is strongly m-measurable in X and Pettis m-integrable. In both cases, the class of functions so obtained coincides with the integral of Bartle [11].

As mentioned previously, if X, Y and $X \widehat{\otimes}_\tau Y$ are Banach spaces and the X-semivariation of m in $X \widehat{\otimes}_\tau Y$ is continuous, then a function $\phi : \Omega \to X$ is m-integrable in $X \widehat{\otimes}_\tau Y$ if and only if it is (Bartle) m-integrable. In this case, both integrals agree. The assumption that τ is a completely separated tensor product topology allows us to avoid using X-semivariation to define integration with respect to m; examples of measures without finite X-semivariation, and so without the continuous X-semivariation are given in Example 2.2 below.

Proposition 2.2. *Let X and Y Banach spaces, and τ a completely separated norm tensor product topology on $X \otimes Y$. Let $m : \mathcal{S} \to Y$ be a Y-valued measure. If $\phi : \Omega \to X$ is m-integrable in $X \widehat{\otimes}_\tau Y$, then for all $x' \in X'$ and $y' \in Y'$, the scalar function $\langle \phi, x' \rangle$ is integrable with respect to the scalar measure $\langle m, y' \rangle$ and the equality*

$$\left\langle \int_A \phi \otimes dm \, , \, x' \otimes y' \right\rangle = \int_A \langle \phi, x' \rangle \, d\langle m, y' \rangle \tag{2.1}$$

is valid.

Moreover, the X-valued function ϕ is integrable with respect to the scalar measure $\langle m, y' \rangle$, the scalar valued function $\langle \phi, x' \rangle$ is integrable with respect to the Y-valued measure m and the following equalities hold for all $A \in \mathcal{S}$:

$$\left\langle \int_A \phi \otimes dm \, , \, x' \otimes y' \right\rangle = \left\langle \int_A \phi \, d\langle m, y' \rangle \, , \, x' \right\rangle$$
$$= \left\langle \int_A \langle \phi, x' \rangle \, dm \, , \, y' \right\rangle. \tag{2.2}$$

Proof. Given X-valued $\mathcal{S}$-simple functions ϕ_k, $k = 1, 2, \ldots$, such that $\phi_k \to \phi$ m-a.e. as $k \to \infty$, and $\{(\phi_k \otimes m)(A)\}_{k=1}^\infty$ converges in $X \widehat{\otimes}_\tau Y$ to $[\phi \otimes m](A)$, for each $A \in \mathcal{S}$, the scalar measures $\langle (\phi_k.\langle m, y' \rangle), x' \rangle = \langle (\langle \phi_k, x' \rangle.m), y' \rangle = \langle \phi_k, x' \rangle.\langle m, y' \rangle$, $k = 1, 2, \ldots$, converge setwise to the scalar measure $\langle \phi \otimes m, x' \otimes y' \rangle$, for each $x' \in X'$ and $y' \in Y'$. Here we have used property (T2) of τ given in Section 1.5.

According to the convergence lemma, Lemma 2.1, the scalar function $\langle \phi, x' \rangle$ is integrable with respect to the scalar measure $\langle m, y' \rangle$ and the equality

$$\int_A \langle \phi, x' \rangle \, d\langle m, y' \rangle = \lim_{k \to \infty} \langle [\phi_k \otimes m](A), x' \otimes y' \rangle = \langle [\phi \otimes m](A), x' \otimes y' \rangle$$

holds by property (T2) of τ. As mentioned in Section 1.5, the maps $x' \otimes I_Y$ and $I_X \otimes y'$ are τ-continuous for every $x' \in X'$ and $y' \in Y'$, so the completeness of the normed spaces X and Y ensures that the sequence

$$\{(x' \otimes I_Y)([\phi_k \otimes m](A))\}_{k=1}^{\infty}$$

converges in Y to, say, $\mu_{x'}(A)$ and $\{(I_X \otimes y')([\phi_k \otimes m](A))\}_{k=1}^{\infty}$ converges in X to, say, $\nu_{y'}(A)$, for each $A \in \mathcal{S}$.

On examination of the definition of vector integration outlined in Section 1.3, we see immediately that ϕ is $\langle m, y' \rangle$-integrable with indefinite integral $\nu_{y'}$, and $\langle \phi, x' \rangle$ is m-integrable with integral $\mu_{x'}$; the equalities (2.1) follow from (2.2). $\qquad\square$

Corollary 2.1. *Let X and Y be Banach spaces, and τ a completely separated norm tensor product topology on $X \otimes Y$. Let $m : \mathcal{S} \to Y$ be a Y-valued measure. If $\phi : \Omega \to X$ is m-integrable in $X \widehat{\otimes}_\tau Y$, then $\phi \otimes m \ll m$.*

Proof. Let $A \in \mathcal{S}$ be a set for which $m(B) = 0$ for all $B \subseteq A$ belonging to $\mathcal{S}$, that is, A is an m-null set. Then for each $y' \in Y'$, we have $|\langle m, y' \rangle|(A) = 0$. An appeal to Proposition 2.2 shows that $\langle \phi, x' \rangle$ is $\langle m, y' \rangle$-integrable for each $x' \in X'$, so we have $\int_A |\langle \phi, x' \rangle| \, d|\langle m, y' \rangle| = 0$. By (2.1), $\langle [\phi \otimes m](B), x' \otimes y' \rangle = 0$ for all subsets $B \in \mathcal{S}$ of A. The result follows from the assumption that τ is a completely separated tensor product topology. $\qquad\square$

Corollary 2.2. *Let X and Y Banach spaces, and τ a completely separated norm tensor product topology on $X \otimes Y$. Suppose that $m : \mathcal{S} \to Y$ is a Y-valued measure. If $\phi : \Omega \to X$ is m-integrable in $X \widehat{\otimes}_\tau Y$, and $f : \Omega \to \mathbb{C}$ is a bounded $\mathcal{S}$-measurable function, then $f\phi$ is m-integrable, ϕ is $f.m$-integrable and the equalities $(f\phi) \otimes m = \phi \otimes (f.m) = f.(\phi \otimes m)$ hold.*

Proof. Bounded scalar valued measurable functions are integrable with respect to a vector measure taking values in a sequentially complete locally convex space [86, Lemma II.3.1], so f is necessarily $\phi \otimes m$-integrable. Then,

$$\langle f(\phi \otimes m), x' \otimes y' \rangle = f.\langle (\phi \otimes m), x' \otimes y' \rangle = f.[\langle \phi, x' \rangle.\langle m, y' \rangle], \quad \text{by } (2.1)$$
$$= [f\langle \phi, x' \rangle].\langle m, y' \rangle = \langle f\phi, x' \rangle.\langle m, y' \rangle$$
$$= \langle \phi, x' \rangle.[f\langle m, y' \rangle] = \langle \phi, x' \rangle.\langle fm, y' \rangle.$$

Once we prove that $f\phi$ is m-integrable and ϕ is $f.m$-integrable, the desired equalities are seen by appealing to Proposition 2.2 and the assumption that τ is completely separated.

Let f_j, $j = 1, 2, \ldots$, be $\mathcal{S}$-simple functions converging uniformly to f on Ω and suppose that ϕ_k, $k = 1, 2, \ldots$, satisfy the assumptions of Definition 2.2. By the Nikodym boundedness theorem [38, I.3.1], the $X \otimes_\tau Y$-valued measures $\phi_k \otimes m$, $k = 1, 2, \ldots$, are uniformly τ-bounded on $\mathcal{S}$, so as $j \to \infty$, the vectors $\big(f_j.[\phi_k \otimes m]\big)(A) \in X \otimes Y$ converge uniformly to $\big(f.[\phi_k \otimes m]\big)(A)$ as the set A ranges over $\mathcal{S}$ and k takes the values $1, 2, \ldots$. In particular, $f_k \phi_k \to f\phi$ m-a.e. and $[(f_k\phi_k) \otimes m](A)$ converges in $X \widehat{\otimes}_\tau Y$ as $k \to \infty$, for each $A \in \mathcal{S}$. Only a glance at Definition 2.2 is needed to see that $f\phi$ is m-integrable.

On the other hand, f is m-integrable because Y is sequentially complete, and

$$\lim_{k \to \infty} \big([\phi_k \otimes (f.m)]\big)(A) = \lim_{k \to \infty} \lim_{j \to \infty} \big([\phi_k \otimes (f_j.m)]\big)(A)$$
$$= \lim_{j \to \infty} \lim_{k \to \infty} \big([\phi_k \otimes (f_j.m)]\big)(A)$$

exists for each $A \in \mathcal{S}$, hence ϕ is fm-integrable. $\qquad\square$

The following bounded convergence result will be useful later.

Lemma 2.3. *Let X and Y Banach spaces, and τ a completely separated norm tensor product topology on $X \otimes Y$. Suppose that $m : \mathcal{S} \to Y$ is a Y-valued measure. If $\phi : \Omega \to X$ is bounded and m-integrable in $X\widehat{\otimes}_\tau Y$, then for every $\epsilon > 0$ and every continuous seminorm p on X, there exist X-valued $\mathcal{S}$-simple functions $\phi_k, k = 1, 2, \ldots$, such that $p(\phi_k(\omega)) \le \|p \circ \phi\|_\infty + \epsilon$ for all $\omega \in \Omega$ and $k = 1, 2, \ldots$, the functions ϕ_k converge to ϕ m-a.e. and $(\phi_k \otimes m)(A) \to (\phi \otimes m)(A)$ as $k \to \infty$ for each $A \in \mathcal{S}$.*

Proof. As ϕ is m-integrable, there exist X-valued $\mathcal{S}$-simple functions $\psi_k, k = 1, 2, \ldots$, such that $\psi_k \to \phi$ m-a.e. and $(\psi_k \otimes m)(A) \to (\phi \otimes m)(A)$ as $m \to \infty$ for each $A \in \mathcal{S}$. Let ϵ be a positive number and p a continuous seminorm on X. For each $k = 1, 2, \ldots$, let

$$\phi_k = \psi_k \chi_{A_k}, \quad A_k = \bigcap_{j=k}^{\infty} \{\omega : \|\psi_j(\omega)\| \le \|\phi\|_\infty + \epsilon\}.$$

By the triangle inequality, $\phi_k(\omega) \to \phi(\omega)$ as $k \to \infty$ at all points $\omega \in \Omega$ at which $\psi_k(\omega) \to \phi(\omega)$ as $k \to \infty$, and $\cup_{k=1}^{\infty} A_k$ is a set of full m-measure. The equality

$$(\phi_k \otimes m)(A) = (\psi_k \otimes m)(A \cap A_k)$$

is valid for all $k = 1, 2, \ldots$ and all $A \in \mathcal{S}$.

By the Vitali-Hahn-Saks theorem, $\{\psi_k \otimes m\}_{k=1}^{\infty}$ is a uniformly countably additive family of $X \otimes_\tau Y$-valued measures, so that for every τ-continuous seminorm r on $X \otimes Y$,

$$\lim_{j \to \infty} \sup_k r\left[(\psi_k \otimes m)(A \cap A_j) - (\psi_k \otimes m)(A)\right] = 0.$$

Hence, for each $A \in \mathcal{S}$, $(\phi_k \otimes m)(A) \to (f \otimes m)(A)$ as $k \to \infty$. $\qquad \square$

Another standard property of vector integrals is that continuous linear maps can be dragged inside the integral to act on the integrand—a property which takes the following form in the present context.

Suppose that X_j, Y_j, $j = 1, 2$ are locally convex spaces and τ_1 is a tensor product topology on $X_1 \otimes Y_1$, and τ_2 is a tensor product topology on $X_2 \otimes Y_2$. The *tensor product* of two linear maps $S : X_1 \to X_2$ and $T : Y_1 \to Y_2$, is the linear map $S \otimes T : X_1 \otimes Y_1 \mapsto X_2 \otimes Y_2$ defined for each $x \otimes y \in X_1 \otimes Y_1$ by $(S \otimes T)(x \otimes y) = (Sx) \otimes (Ty)$. There is no guarantee that $S \otimes T$ is continuous from τ_1 to τ_2 if S and T are continuous. However, if $S \otimes T : X_1 \otimes_{\tau_1} Y_1 \to X_2 \otimes_{\tau_2} Y_2$ *is* continuous, then the same symbol $S \otimes T$ denotes the associated continuous linear map between the completions $X_1 \widehat{\otimes}_{\tau_1} Y_1$ and $X_2 \widehat{\otimes}_{\tau_2} Y_2$.

Proposition 2.3. *Suppose that X_j, Y_j, $j = 1, 2$, are locally convex spaces and τ_1 is a completely separated tensor product topology on $X_1 \otimes Y_1$, and τ_2 is a completely separated tensor product topology on $X_2 \otimes Y_2$. Let $m : \mathcal{S} \to Y_1$ be a measure and suppose that $S : X_1 \to X_2$ and $T : Y_1 \to Y_2$ are continuous linear maps whose tensor product $S \otimes T : X_1 \otimes_{\tau_1} Y_1 \to X_2 \otimes_{\tau_2} Y_2$ is continuous.*

If $\phi : \Omega \to X_1$ is m-integrable in $X_1 \widehat{\otimes}_{\tau_1} Y_1$, then $S\phi$ is Tm-integrable in $X_2 \widehat{\otimes}_{\tau_2} Y_2$ and

$$(S \otimes T) \int_A \phi \otimes dm = \int_A [S\phi] \otimes d[Tm], \qquad \text{for every } A \in \mathcal{S}.$$

Proof. Let ϕ_k, $k = 1, 2, \ldots$, be X_1-valued $\mathcal{S}$-simple functions satisfying the assumptions of Definition 2.2. Then $S\phi_k \to S\phi$ m-a.e. as $k \to \infty$, because S is continuous. The continuity of T guarantees that $Tm := T \circ m$

is a Y_2-valued measure. Since $S \otimes T$ is τ_1-τ_2-continuous, the sequence $\{(S \otimes T)([\phi_k \otimes m](A))\}_{k=1}^\infty$ converges in $X_2 \widehat{\otimes}_{\tau_2} Y_2$ for each $A \in \mathcal{S}$. A glance at Definition 2.2 is enough to complete the proof. $\qquad\square$

2.2 Order bounded measures

We return to the setting of Subsection 1.5.1 where the semivariation of L^p-valued measures was examined. Let $(\Gamma, \mathcal{E}, \mu)$ be a σ-finite measure space. The modulus $|m| : \mathcal{S} \to L_+^p(\mu)$ of an order bounded measure $m : \mathcal{S} \to L^p(\mu)$ was defined in Lemma 1.1.

For any Banach space X, if $X \otimes L^p(\mu)$ is equipped with the relative topology τ of $L^p(\mu; X)$, then $X \widehat{\otimes}_\tau L^p(\mu)$ may be identified with $L^p(\mu; X)$.

According to Proposition 1.5, an order bounded measure $m : \mathcal{S} \to L^p(\mu)$ has finite and continuous X-semivariation $\widehat{m}_{p,X}$ in $L^p(\mu; X)$ for any Banach space X, so integration in $L^p(\mu; X)$ in the sense of Definitions 1.2, 1.3 and 2.2 coincide for the vector measure m.

Theorem 2.3. *Suppose that $1 \le p < \infty$ and $m : \mathcal{S} \to L^p(\Gamma, \mathcal{E}, \mu)$ is an order bounded measure and $f : \Omega \to X$ is strongly m-measurable. If $\|f\|_X$ is $|m|$-integrable, then f is m-integrable in $L^p(\mu; X)$ and the inequality*

$$\left\| \left(\int_A f \otimes dm \right)(\gamma) \right\|_X \le \left(\int_A \|f\|_X \, d|m| \right)(\gamma) \qquad (2.3)$$

holds for all $A \in \mathcal{S}$ and for μ-almost every $\gamma \in \Gamma$.

Proof. First we establish the estimate for simple functions. Let $l \in \mathbb{N}$. Suppose that $h = \sum_{j=1}^l x_j \chi_{E_j}$ is an X-valued $\mathcal{S}$-simple function with $x_j \in X$ and $E_j \in \mathcal{S}$ pairwise disjoint for $j = 1, \ldots, l$. Then for each $A \in \mathcal{S}$ and μ-almost every $\gamma \in \Gamma$,

$$\left\| \left(\int_A h \otimes dm \right)(\gamma) \right\|_X = \left\| \sum_{j=1}^l x_j (m(E_j \cap A))(\gamma) \right\|_X$$

$$\le \sum_{j=1}^l \|x_j\|_X (|m|(E_j \cap A))(\gamma)$$

$$= \left(\int_A \|h\|_X \, d|m| \right)(\gamma).$$

The positivity of the vector measure $|m|$ is crucial here.

Next we prove that f is m-integrable in $L^p(X)$. By assumption, f is strongly m-measurable and so there exists a sequence $\{\psi_j\}_{j=1}^\infty$ of X-valued

S-simple functions such that $\psi_j \to f$ m-almost everywhere as $j \to \infty$. Now let

$$f_j(\omega) = \begin{cases} \psi_j(\omega) & \text{if } \|\psi_j(\omega)\|_X \leq 2\|f(\omega)\|_X \\ 0 & \text{if } \|\psi_j(\omega)\|_X > 2\|f(\omega)\|_X. \end{cases}$$

Then each f_j is an X-valued S-simple function such that $f_j \to f$ m-almost everywhere and further $\|f_j(\omega)\|_X \leq 2\|f(\omega)\|_X$ for all $\omega \in \Omega$. Thus to ensure integrability it suffices to show that the sequence $\{\int_A f_j \otimes dm\}_{j=1}^\infty$ converges in $L^p(\mu; X)$ for each $A \in S$.

Let $j, k \in \mathbb{N}$. By construction, $\|f_j(\cdot) - f(\cdot)\|_X \leq 3\|f(\cdot)\|_X$ and by assumption $\|f\|_X$ is $|m|$-integrable so, making use of inequality (2.3) and dominated convergence for vector measures [86, II.4], we have

$$\left\| \int_A (f_j - f_k) \otimes dm \right\|_{L^p(X)} \leq \int_A \|f_j - f_k\|_X \, d|m|$$

$$\leq \int_A \|f_j - f\|_X \, d|m| + \int_A \|f - f_k\|_X \, d|m|$$

$$\to 0$$

as $j, k \to \infty$. Thus f is m-integrable in $L^p(\mu; X)$.

Finally, we establish that inequality (2.3) holds for the function f. We know that $\lim_{j \to \infty} \int_A f_j \otimes dm = \int_A f \otimes dm$ in $L^p(\mu; X)$ for each $A \in S$. By taking an appropriate subsequence, if necessary, we may assume that

$$\left\| \left(\int_A f_j \otimes dm \right)(\gamma) \right\|_X \to \left\| \left(\int_A f \otimes dm \right)(\gamma) \right\|_X$$

for μ-almost every $\gamma \in \Gamma$ also. Since $\|f\|_X$ is assumed to be $|m|$-integrable, dominated convergence for vector measures again ensures that

$$\int_A \|f_j\|_X \, d|m| \to \int_A \|f\|_X \, d|m|$$

in $L^p(\mu)$ as $j \to \infty$ for all $A \in S$.

By taking a further subsequence, if necessary, we may assume that $\left(\int_A \|f_j\|_X \, d|m| \right)(\gamma) \to \left(\int_A \|f\|_X \, d|m| \right)(\gamma)$ for μ-almost every $\gamma \in \Gamma$ as well. This guarantees that inequality (2.3) holds for the function f. $\qquad \square$

2.3 The bilinear Fubini theorem

A decoupling strategy to integrate an $\mathcal{L}(X, Y)$-valued function with respect to an X-valued measure was outlined at the beginning of the chapter. Although we shall be examining such integrals in greater detail in

Chapter 6 when we consider scattering theory, a simple formulation of Fubini's Theorem is facilitated by the underlying bilinear structure.

Let X, Y be Banach spaces. A locally convex space E is said to be *bilinear admissible* for X, Y if

a) E contains the vector space $\mathcal{L}(X, Y) \otimes X$ as a dense subspace,
b) the composition map $J : \mathcal{L}(X, Y) \times X \to Y$ defined by

$$J(T, x) = Tx, \quad T \in \mathcal{L}(X, Y), \ x \in X,$$

has a continuous linear extension $J_E : E \to Y$ from $\mathcal{L}(X, Y) \otimes X$ to E.
c) for $x \in X$, $x' \in X'$ and $y' \in Y'$, the linear functional defined by

$$x \otimes y' \otimes x' : T \otimes u \longmapsto \langle Tx, y' \rangle \langle u, x' \rangle, \quad T \in \mathcal{L}(X, Y), \ u \in X,$$

is continuous on $\mathcal{L}(X, Y) \otimes X$ for the relative topology of E.
d) the family of all linear functionals $x \otimes y' \otimes x'$ for $x \in X$, $x' \in X'$ and $y' \in Y'$ separates points of E.

If $Y = X$, then we merely say E is *bilinear admissible* for X. If τ is a completely separating tensor product topology on $\mathcal{L}_s(X, Y) \otimes X$ for the space $\mathcal{L}(X, Y)$ endowed with the strong operator topology, then we may take $E = \mathcal{L}(X, Y) \widehat{\otimes}_\tau X$, the completion of the linear space $\mathcal{L}(X, Y) \otimes X$ in the locally convex topology τ. Sometimes the *quasicompletion* [87, 23.1] is taken.

Remark 2.2. If the Banach space X has the *approximation property* [123, III.9], then $X \otimes X'$ separates points of the projective tensor product $X' \widehat{\otimes}_\pi X$ [88, 43.2(12)] and this is precisely the property needed to define the *trace* of a nuclear operator on X [94]. In Example 2.1, the separation property is what we need to define the generalised trace $\int_0^1 \Phi_\varphi \, dm$ when X is a Banach function space such as $L^p([0, 1])$, $1 \le p < \infty$.

Definition 2.3. Suppose that the locally convex space E is bilinear admissible for the Banach spaces X and Y. Let $(\Omega, \mathcal{S})$ be a measurable space. A function $f : \Omega \to \mathcal{L}(X, Y)$ is said to be *m-integrable in E* for a vector measure $m : \mathcal{S} \to X$, if for each $x \in X$, $x' \in X'$, $y' \in Y'$, the scalar function $\langle fx, y' \rangle$ is integrable with respect to the scalar measure $\langle m, x' \rangle$ and for each $S \in \mathcal{S}$, there exists an element $(f \otimes m)(S)$ of E such that

$$\langle (f \otimes m)(S), x \otimes y' \otimes x' \rangle = \int_S \langle fx, y' \rangle \, d\langle m, x' \rangle \tag{2.4}$$

for every $x \in X$, $x' \in X'$ and $y' \in Y'$.

If f is m-integrable in E, then $fm(S) \in Y$ is defined for each $S \in \mathcal{S}$ by

$$fm(S) = J_E\big((f \otimes m)(S)\big).$$

We also denote $fm(S)$ by $\int_S f\, dm$ or $\int_S f(\omega)\, dm(\omega)$.

Because the linear space $X \otimes Y' \otimes X'$ separates points of E, the vector $(f \otimes m)(S) \in E$ is well-defined for each $S \in \mathcal{S}$. The same definition is adopted if $\mathcal{S}$ is generated by the δ-ring $\mathcal{S}_0$ [16, Definition 1.2.13] and $m : \mathcal{S}_0 \to X$ is a vector measure on $\mathcal{S}_0$.

The existence of the E-valued set function $f \otimes m$ is formulated here in the weak sense. In the case that $E = \mathcal{L}(X, Y)\widehat{\otimes}_\tau Y$ for a suitable completely separating tensor product topology τ on $\mathcal{L}(X, Y) \otimes Y$, the existence of the E-valued measure $f \otimes m$ is usually verified along the lines of Definition 2.2.

In the case that X is the set of scalars, fm is the indefinite (Pettis) integral of a Y-valued function with respect to a scalar measure m as in Subsection 1.3.1. We shall use the term *Bochner* integral to distinguish the stronger integration process when f is approximated in the norm of Y, see Subsection 1.3.2.

Definition 2.3 facilitates a simple version of Fubini's Theorem in the operator context. In the following statement, an $\mathcal{L}(X, Y)$-valued function is said to be μ-integrable, if it is Pettis μ-integrable in $\mathcal{L}(X, Y)$ for the strong operator topology, as defined in Subsection 1.3.1.

Theorem 2.4. *Suppose that the locally convex space E is bilinear admissible for the Banach spaces X and Y. Let $(\Omega, \mathcal{S})$ be a measurable space and $(\Gamma, \mathcal{E}, \mu)$ a σ-finite measure space, and $m : \mathcal{S} \to X$ a vector measure. Suppose that $f : \Omega \times \Gamma \to \mathcal{L}(X, Y)$ is $(m \otimes \mu)$-integrable in E. If*

(i) *for m-almost all $\omega \in \Omega$, the $\mathcal{L}(X, Y)$-valued function $f(\omega, \cdot)$ is μ-integrable, and*

(ii) *for μ-almost all $\gamma \in \Gamma$, the $\mathcal{L}(X, Y)$-valued function $f(\cdot, \gamma)$ is m-integrable in E,*

then the function

$$\omega \longmapsto \int_\Gamma f(\omega, \gamma)\, d\mu(\gamma)$$

is m-integrable in E, the function

$$\gamma \longmapsto \int_\Omega f(\omega, \gamma)\, dm(\omega)$$

is integrable in Y with respect to μ, and the equalities

$$\int_{\Omega \times \Gamma} f \, d(m \otimes \mu) = \int_{\Omega} \left(\int_{\Gamma} f(\omega, \gamma) \, d\mu(\gamma) \right) dm(\omega)$$

$$= \int_{\Gamma} \left(\int_{\Omega} f(\omega, \gamma) \, dm(\omega) \right) d\mu(\gamma)$$

hold.

Proof. Let $\Phi(\omega) = \int_{\Gamma} f(\omega, \gamma) \, d\mu(\gamma)$ for all $\omega \in \Omega$ for which $f(\omega, \cdot)$ is μ-integrable and $\Phi(\omega) = 0$ otherwise. For each $x \in X$, $y' \in Y'$, $x' \in X'$ and $S \in \mathcal{S}$, we have

$$\langle \Phi(\omega)x, y' \rangle = \int_{\Gamma} \langle f(\omega, \gamma)x, y' \rangle \, d\mu(\gamma),$$

so that

$$\int_{S} \langle \Phi(\omega)x, y' \rangle \, d\langle m, x' \rangle = \int_{S} \left(\int_{\Gamma} \langle f(\omega, \gamma)x, y' \rangle \, d\mu(\gamma) \right) d\langle m, x' \rangle$$

$$= \langle (f \otimes (m \otimes \mu))(S \times \Gamma), x \otimes y' \otimes x' \rangle$$

by the scalar version of Fubini's Theorem. It follows that Φ is m-integrable in E and

$$\int_{S} \Phi(\omega) \otimes dm(\omega) = (f \otimes (m \otimes \mu))(S \times \Gamma),$$

hence $\int_{\Omega} \Phi(\omega) \, dm(\omega) = J_E((f \otimes (m \otimes \mu))(\Omega \times \Gamma)) = \int_{\Omega \times \Gamma} f \, d(m \otimes \mu)$. A similar appeal to the scalar version of Fubini's Theorem applies to the other iterated integral. $\qquad \square$

Because we are dealing with *vector valued* integrals, the existence of the integrals (i) and (ii) almost everywhere is not ensured by the integrablity of f with respect to the product measure $m \otimes \mu$.

Sometimes we shall need a simple modification of Theorem 2.4 in which condition b) of the definition of a bilinear admissible space E is replaced by

b') there exists a lcs F containing $\mathcal{L}(X, Y) \otimes X$ and embedded in E such that the composition map $J : \mathcal{L}(X, Y) \times X \to Y$ defined by

$$J(T, x) = Tx, \quad T \in \mathcal{L}(X, Y), \; x \in X,$$

has a continuous linear extension $J_F : F \to Y$ from $\mathcal{L}(X, Y) \otimes X$ to F.

Then Definition 2.3 can be modified by writing

$$fm(S) = J_F\big((f \otimes m)(S)\big)$$

if $(f \otimes m)(S) \in F$ for $S \in \mathcal{S}$.

Theorem 2.5. *Suppose that the lcs E, F are as just described. Let $(\Omega, \mathcal{S})$ be a measurable space and $(\Gamma, \mathcal{E}, \mu)$ a σ-finite measure space, and $m : \mathcal{S} \to X$ a vector measure. Suppose that $f : \Omega \times \Gamma \to \mathcal{L}(X, Y)$ is $(m \otimes \mu)$-integrable in E and $(m \otimes \mu)(\Omega \times \Gamma) \in F$. If*

(i) *for m-almost all $\omega \in \Omega$, the $\mathcal{L}(X, Y)$-valued function $f(\omega, \cdot)$ is μ-integrable, and*

(ii) *for μ-almost all $\gamma \in \Gamma$, the $\mathcal{L}(X, Y)$-valued function $f(\cdot, \gamma)$ is m-integrable in E and $\int_{\Omega} f(\omega, \gamma) \otimes dm(\omega) \in F$,*

then the function

$$\omega \longmapsto \int_{\Gamma} f(\omega, \gamma)\, d\mu(\gamma)$$

is m-integrable in E, the function

$$\gamma \longmapsto \int_{\Omega} f(\omega, \gamma)\, dm(\omega)$$

is integrable in Y with respect to μ, and the equalities

$$\int_{\Omega \times \Gamma} f\, d(m \otimes \mu) = \int_{\Omega} \left(\int_{\Gamma} f(\omega, \gamma)\, d\mu(\gamma) \right) dm(\omega)$$

$$= \int_{\Gamma} \left(\int_{\Omega} f(\omega, \gamma)\, dm(\omega) \right) d\mu(\gamma)$$

hold.

Example 2.1. Let $(\Omega, \mathcal{S}, \mu)$ be a finite measure space. The space of all μ-equivalence classes of $\mathcal{S}$-measurable scalar functions is denoted by $L^0(\mu)$. It is equipped with the topology of convergence in μ-measure and vector operations pointwise μ-almost everywhere. Any Banach space X that is a subspace of $L^0(\mu)$ with the properties that

(i) X is an order ideal of $L^0(\mu)$, that is, if $g \in X$, $f \in L^0(\mu)$ and $|f| \leq |g|$ μ-a.e., then $f \in X$, and

(ii) if $f, g \in X$ and $|f| \leq |g|$ μ-a.e., then $\|f\|_X \leq \|g\|_X$, is called a *Banach function space* (based on $(\Omega, \mathcal{S}, \mu)$). We suppose that X contains constant functions and $m : \mathcal{S} \longmapsto \chi_S$, $S \in \mathcal{S}$, is σ-additive in X, for example, X is σ-order continuous, see [99, Corollary 3.6]. If X is reflexive and μ is non-atomic, then it follows from [99, Corollary 3.23] that the values of the variation $V(m)$ of m are either zero or infinity.

Suppose that $\varphi : \Omega \times \Omega \to \mathbb{C}$ is a jointly measurable function and $T_\varphi : X \to X$ is a bounded linear operator such that

$$T_\varphi f = \int_\Omega \varphi(\,\cdot\,, t) f(t)\, d\mu(t)$$

for a dense set of $f \in X$. Suppose also that $\Phi_\varphi(s) = [\varphi(s, \cdot)] \in X'$ for μ-almost all $s \in \Omega$, that is, there exists $K_s > 0$ such that

$$\left| \int_\Omega \varphi(s, t) f(t)\, d\mu(t) \right| \leq K_s \|f\|_X, \quad f \in X.$$

Such operators are called *Carleman operators* in [55].

A bounded linear operator T_φ for which there exists a bilinear $(X, \mathbb{C})$-admissible space E such that Φ_φ is m-integrable in E, is a type of *generalised trace class operator* and $\int_\Omega \Phi_\varphi\, dm = J_E \int_\Omega \Phi_\varphi \otimes dm$ is the trace of T_φ. For example, if X has the approximation property [123, III.9] and $E = X' \hat{\otimes}_\pi X$, then T_φ is a nuclear operator and $\int_\Omega \Phi_\varphi\, dm$ is actually the trace of T_φ [94], see Example 2.3 for the Hilbert space case. There are closed subspaces of ℓ^p, $1 \leq p < \infty$, $p \neq 2$, without the approximation property.

As we shall see in the next chapter, the value $\int_0^1 \Phi_\varphi\, dm = \frac{1}{2}$ is obtained for the Volterra integral operator T_φ on $L^2([0, 1])$ with a judicious choice of the auxiliary spaces E and F. The Volterra integral operator is Hilbert-Schmidt but not trace class on $L^2([0, 1])$.

The integral $\int_0^1 \Phi_\varphi\, dm$ is a type of *singular bilinear integral* referred to in the title of this work, because the diagonal $\{(t, t) : t \in [0, 1]\}$ has measure zero in $[0, 1]^2$, so that the integral $\int_0^1 \varphi(t, t)\, dt$ is not well-defined for a general integral kernel φ associated with the operator T_φ. As we shall see in the next chapter, the auxiliary spaces E, F determine the averaging process of φ around the diagonal and the density of the measure $A \longmapsto \int_A \Phi_\varphi\, dm$, $A \in \mathcal{B}([0, 1])$, with respect to Lebesgue measure represents the average $t \longmapsto \tilde{\varphi}(t, t)$ of φ around the diagonal.

2.4 Examples of bilinear integrals

There are no surprises in this section. We apply the preceding theory to some natural examples to show that the expected class of integrable functions is obtained, and their definite integrals give the expected operators. Nevertheless, the examples also illustrate the difficulty with applying the classical theories of bilinear integration in the context of the integration of operator valued functions with respect to operator valued measures.

The following example shows that the X-semivariation of a Y-valued measure in $X \otimes_\tau Y$ may take only the values zero and infinity.

Example 2.2. Let $m : \mathcal{B}([0,1]) \to L^2([0,1])$ be the vector measure $m(B) = \chi_B$, $B \in \mathcal{B}([0,1])$. Then the $L^1[0,1]$-semivariation of m in $L^1([0,1]; L^2([0,1]))$ is infinite on any Borel set E with positive Lebesgue measure $|E|$.

For, let n be any positive integer and suppose that $E_j, j = 1, \ldots, n$, are pairwise disjoint sets with Lebesgue measure $|E_j| = |E|/n, j = 1, \ldots, n$, — the range of the Lebesgue measure on the Borel σ-algebra $\mathcal{B}(E)$ of E is the interval $[0, |E|]$. Let $f_j = \chi_{E_j}/|E_j|$ for each $j = 1, \ldots, n$. The $L^1([0,1]; L^2([0,1]))$-norm of $\sum_{j=1}^n f_j \otimes m(E_j)$ is

$$\Big\| \sum_{j=1}^n f_j \otimes m(E_j) \Big\|_1 = \int_0^1 \Big\| \sum_{j=1}^n f_j(x) m(E_j) \Big\|_2 dx = |E|^{1/2} n^{1/2}.$$

Because n is any positive integer, the $L^1[0,1]$-semivariation of m in the space $L^1([0,1]; L^2([0,1]))$ is infinite on E. Of course, the $L^2([0,1])$-semivariation of m in $L^1([0,1]; L^2([0,1]))$ and the $L^1[0,1]$-semivariation of m in $L^1([0,1]^2)$ are finite; see Proposition 1.5.

The only $L^1[0,1]$-valued functions which are m-integrable in the space $L^1([0,1]; L^2([0,1]))$ in the sense of Definition 1.2 or Definition 1.3, are the null functions. Nevertheless, it is natural to consider the integration in $L^1([0,1]; L^2([0,1]))$ of functions with values in $L^1[0,1]$, with respect to the $L^2([0,1])$-valued measure m: elements of $L^1([0,1]; L^2([0,1]))$ are associated with the space $\mathcal{H}_{2,1}$ of Hille-Tamarkin operators [124, p. 282], the bounded linear operators $T : L^2([0,1]) \to L^2([0,1])$ with integral kernel $k : [0,1] \times [0,1] \to \mathbb{C}$ such that $Tu(x) = \int_0^1 k(x,y)u(y)\,dy$ for $u \in L^2([0,1])$ and

$$\left(\int_0^1 |k(x,y)|^2\, dy \right)^{\frac{1}{2}} dx < \infty.$$

Because the $L^1[0,1]$-semivariation of m in $L^1([0,1]; L^2([0,1]))$ is infinite, the conditions of Theorem 2.2 do not hold and bounded operator valued functions need not be integrable.

The following example of trace class operators shall assume importance in Chapter 3.

Example 2.3. Let $m : \mathcal{B}([0,1]) \to L^2([0,1])$ be the vector measure defined by $m(B) = \chi_B$ for every $B \in \mathcal{B}([0,1])$. Then the $L^2([0,1])$-semivariation of m in the projective tensor product $L^2([0,1]) \otimes_\pi L^2([0,1])$ (see [88, Section 41.2]) is infinite on any Borel set A with positive Lebesgue measure $|A|$.

To see this, let n be any positive integer and suppose that $A_j, j = 1, \ldots, n$, are pairwise disjoint subsets of A, with Lebesgue measure $|A|/n$. Let $\phi_j = (n/|A|)^{1/2}\chi_{A_j}$ for each $j = 1, \ldots, n$. Then $\Phi : L^2([0,1]) \otimes_\pi L^2([0,1]) \to \mathbb{C}$, defined by $\Phi(f \otimes g) = \langle f|g \rangle$ for every $f \in L^2([0,1])$ and every $g \in L^2([0,1])$, is continuous but

$$\Phi\left(\sum_{j=1}^{n} \phi_j \otimes m(A_j)\right) = \sum_{j=1}^{n}(\phi_j|m(A_j)) = |A|^{1/2}n^{1/2}.$$

Because n is any positive integer, the $L^2([0,1])$-semivariation of m in the space $L^2([0,1]) \otimes_\pi L^2([0,1])$ is infinite on A.

Nevertheless, Proposition 2.4 below demonstrates that it is natural to consider the integration in the vector space $L^2([0,1])\widehat{\otimes}_\pi L^2([0,1])$ of functions with values in $L^2([0,1])$, with respect to the $L^2([0,1])$-valued measure m. Because the $L^2([0,1])$-semivariation of m in $L^2([0,1])\widehat{\otimes}_\pi L^2([0,1])$ is infinite, bounded vector valued functions need not be integrable; see Example 1.8.

Remark 2.3. The only $L^2([0,1])$-valued functions which are (Dobrakov) or (Bartle) m-integrable in the tensor product space $L^2([0,1])\widehat{\otimes}_\pi L^2([0,1])$ are the null functions.

The space $\mathcal{C}_2(L^2([0,1]))$ of Hilbert-Schmidt operators acting on $L^2([0,1])$ is endowed with the Hilbert-Schmidt norm, [88, Section 42.4]. Let $\mathcal{K}$ denote the isometric isomorphism from $L^2([0,1]^2)$ onto $\mathcal{C}_2(L^2([0,1]))$, which sends an element k of $L^2([0,1]^2)$ to the Hilbert-Schmidt operator $T_k : L^2([0,1]) \to L^2([0,1])$ with *kernel* k, [135, Theorem 6.11], that is, $(T_k\phi)(x) = \int_0^1 k(x,y)\phi(y)\,dy$ for almost all $x \in [0,1]$ and for all $\phi \in L^2([0,1])$.

For all $\phi, \psi \in L^2([0,1])$ and $k \in L^2([0,1]^2)$, we have the equality

$$([\mathcal{K}k]\phi, \, \psi) = (k, \, \bar{\phi} \otimes \psi) = \int_{[0,1]^2} k(x,y)\phi(y)\overline{\psi(x)}\,dxdy.$$

By appealing to the representation given in Theorem 1.8, the projective tensor product $L^2([0,1])\widehat{\otimes}_\pi L^2([0,1])$ may be identified with a linear subspace of $L^2([0,1]^2)$ in the obvious way.

Then the image $\mathcal{C}_1(L^2([0,1]))$ of $L^2([0,1])\widehat{\otimes}_\pi L^2([0,1])$ under $\mathcal{K}$ is the space of *nuclear operators* on $L^2([0,1])$. The nuclear and trace class operators on the Hilbert space $L^2([0,1])$ are the same. If we equip $\mathcal{C}_1(L^2([0,1]))$ with the nuclear norm, [88, 42.5.(8)], then $\mathcal{K}$ induces an isometry from $L^2([0,1])\widehat{\otimes}_\pi L^2([0,1])$ onto $\mathcal{C}_1(L^2([0,1]))$. Further properties of trace class operators are mentioned in Section 3.1 below. These observations indicate that for practical reasons the projective tensor product $L^2([0,1])\widehat{\otimes}_\pi L^2([0,1])$ is a worthy object of study.

The following proposition illustrates our claim that, in the present context, we have written down the 'right' definition of integration of vector valued functions with respect to vector valued measures in tensor product spaces. Subsequent chapters lend more evidence for this claim.

In the following proposition, given a function $k \in L^2([0,1]^2)$ and a point $x \in [0,1]$, let $[k(x,\cdot)]$ denote the equivalence class in $L^2([0,1])$ containing $k(x,\cdot)$. Moreover let $Q : \mathcal{B}([0,1]) \to \mathcal{L}(L^2([0,1]))$ denote the measure given by $Q(A)\phi = \chi_A\phi$ for every $A \in \mathcal{B}([0,1])$ and $\phi \in L^2([0,1])$. Then Q is a spectral measure; that is, $Q(A \cap B) = Q(A)Q(B)$ for all $A,\, B \in \mathcal{B}([0,1])$, and $Q([0,1])$ is the identity operator.

Proposition 2.4. *Let* $m : \mathcal{B}([0,1]) \to L^2([0,1])$ *be the vector measure given by* $m(B) = \chi_B, B \in \mathcal{B}([0,1])$. *A function* $f : [0,1] \to L^2([0,1])$ *is m-integrable in* $L^2([0,1])\widehat{\otimes}_\pi L^2([0,1])$ *if and only if there exists a function* $k : [0,1]^2 \to \mathbb{C}$ *such that*

(i) k *is the kernel of a trace class operator; and*
(ii) *the set* $\{x \in [0,1] : f(x) = [k(x,\cdot)]$ *in* $L^2([0,1])\}$ *is a set of full measure.*

If f is m-integrable and $A \in \mathcal{B}([0,1])$, then $[f \otimes m](A)$ is equal to the equivalence class in $L^2([0,1]^2)$ of the function

$$(x,y) \mapsto \chi_A(x)k(x,y), \qquad (x,y) \in [0,1]^2.$$

Moreover, the equality $\mathcal{K}([f \otimes m](A)) = Q(A)\mathcal{K}k$ *is valid for each $A \in \mathcal{B}([0,1])$.*

Proof. Suppose first that the conditions (i) and (ii) are satisfied. According to [88, 42.5.(5)], the operator T_k has a representation

$$T_k\phi = \sum_{j=1}^{\infty} \eta_j(\phi,\, g_j)h_j, \qquad \phi \in L^2([0,1]),$$

where $\{\eta\}_{j=1}^{\infty}$ is an absolutely summable scalar sequence, and $\{g_j\}_{j=1}^{\infty}$ and $\{h_j\}_{j=1}^{\infty}$ are orthonormal sequences in the Hilbert space $L^2([0,1])$. By the Beppo Levi convergence theorem, $\sum_{j=1}^{\infty} |\eta_j| |g_j(x)| |h_j(y)| < \infty$, so that we can define $\xi(x,y) = \sum_{j=1}^{\infty} \eta_j g_j(x) h_j(y)$, for almost all $(x,y) \in [0,1]^2$. Then the function ξ belongs to $L^2([0,1]^2)$. The operators T_k and T_ξ, defined by the kernels k and ξ, respectively, are equal. However, $L^2([0,1]) \otimes L^2([0,1])$ separates elements of $L^2([0,1]^2)$, so $k = \xi$ almost everywhere on $[0,1]^2$.

Given $j = 1, 2, \dots$, the functions g_j and h_j may be expressed as $g_j = \sum_{l=1}^{\infty} \phi_{jl}$ and $h_j = \sum_{n=1}^{\infty} \psi_{jn}$, both almost everywhere in $[0,1]$ and with respect to the norm topology of $L^2([0,1])$, for some scalar valued sequences $\{\phi_{jl}\}_{l=1}^{\infty}$ and $\{\psi_{jn}\}_{n=1}^{\infty}$ of $\mathcal{B}([0,1])$-simple functions with $\sum_{l=1}^{\infty} \|\phi_{jl}\|_2 \le 2$ and $\sum_{n=1}^{\infty} \|\psi_{jn}\|_2 \le 2$, respectively. It then follows from the condition (ii) that for almost all $x \in [0,1]$ we have $\sum_{j,l,n=1}^{\infty} |\lambda_j| |\phi_{jl}(x)| \|\psi_{jn}\|_2 < \infty$, and so the identity $f(x) = \sum_{j,l,n=1}^{\infty} \eta_j \phi_{jl}(x) \psi_{jn}$ holds in the norm topology of $L^2([0,1])$.

Because $\int_A (\phi_{jl}(x)\psi_{jn}) \otimes dm(x) = \chi_A \phi_{jl} \otimes \psi_{jn}$ as elements of the Banach space $L^2([0,1]) \widehat{\otimes}_\pi L^2([0,1])$ for all $j, l, n \in \mathbb{N}$ and because the triple sequence $\{\eta_j [\chi_A \phi_{jl}] \otimes \psi_{jn}\}_{j,l,n=1}^{\infty}$ is summable in the projective tensor topology for every $A \in \mathcal{S}$, the function f is m-integrable according to Definition 2.2 and

$$\int_A f \otimes dm = \sum_{j,l,n=1}^{\infty} \eta_j [\chi_A \phi_{jl}] \otimes \psi_{jn} \,, \qquad A \in \mathcal{B}([0,1]).$$

Conversely suppose that f is m-integrable in $L^2([0,1]) \widehat{\otimes}_\pi L^2([0,1])$. Then the element $(f \otimes m)(A)$ of $L^2([0,1]) \widehat{\otimes}_\pi L^2([0,1])$ is expressed as a function $k \in \mathcal{L}^2([0,1])$ so that $\mathcal{K}(\int_0^1 f \otimes dm)$ is a trace class operator with kernel k. In terms of the inner product $(\cdot\,,\cdot)$ of $L^2([0,1])$, we have

$$\left(\int_0^1 f \otimes dm \,, \overline{\phi} \otimes \psi \right) = \int_0^1 (f\,, \overline{\phi}) \, d(m, \psi)$$

so that

$$\int_0^1 (f\,, \overline{\phi}) \, d(m, \psi) = \left(\mathcal{K}\left[\int_0^1 f \otimes dm \right] \phi, \psi \right) = \int_{[0,1]^2} k(x,y)\phi(y)\overline{\psi(x)} \, dx dy,$$

for all $\phi, \psi \in L^2([0,1])$. On taking ψ to be the characteristic function of a Borel set, we see that for each $\phi \in L^2([0,1])$, the equality $(f(x)\,|\,\overline{\phi}) = \int_0^1 k(x,y)\phi(y) \, dy$ holds for almost all $x \in [0,1]$. The separability of $L^2([0,1])$ ensures that (ii) holds. Finally the formula $\mathcal{K}([f \otimes m](A)) = Q(A)\mathcal{K}k$ is valid for each $A \in \mathcal{B}([0,1])$. $\qquad \square$

Example 2.4. Let m be the vector measure defined in Proposition 2.4. Let

$$k(x,y) = \sum_{n=1}^{\infty} n\chi_{[1/(n+1),1/n)}(x)\,\chi_{[1/(n+1),1/n)}(y)$$

for all x, $y \in [0,1]$. Then $\int_0^1 k(x,y)^2\,dy \le 1$ for all $x \in [0,1]$, but a straightforward calculation shows that k is not the kernel of a trace class operator. The function $f : [0,1] \to L^2([0,1])$ defined by $f(x) = [k(x,\cdot)]$ for all $x \in [0,1]$ is therefore a bounded $L^2([0,1])$-valued function which is *not* m-integrable in $L^2([0,1])\widehat{\otimes}_\pi L^2([0,1])$.

On the other hand, replacing $L^2([0,1])$ by $L^1([0,1])$ in Proposition 2.4, if $k \in L^1([0,1]^2)$, then by Fubini's Theorem the function $x \mapsto f(x,\cdot)$ has values in $L^1([0,1])$ for almost all $x \in [0,1]$. Combined with the observation that $L^1([0,1])\widehat{\otimes}_\pi L^1([0,1]) \equiv L^1([0,1]^2)$ [123, Section 6.5], the following statement follows easily.

Proposition 2.5. *Let* $m : \mathcal{B}([0,1]) \to L^1([0,1])$ *be the vector measure given by* $m(B) = \chi_B, B \in \mathcal{B}([0,1])$. *A function* $f : [0,1] \to L^1([0,1])$ *is m-integrable in* $L^1([0,1])\widehat{\otimes}_\pi L^1([0,1])$ *if and only if there exists a function* $k \in L^1([0,1]^2)$ *such that the set* $\{x \in [0,1] : f(x) = [k(x,\cdot)]\ \text{in}\ L^1([0,1])\}$ *is a set of full measure.*

If f *is m-integrable and* $A \in \mathcal{B}([0,1])$, *then* $[f \otimes m](A)$ *is equal to the equivalence class in* $L^1([0,1]^2)$ *of the function* $(x,y) \mapsto \chi_A(x)k(x,y)$, $(x,y) \in [0,1]^2$.

Remark 2.4. Let $(\Sigma, \mathcal{S})$ be a measurable space and X, Y Banach spaces. Bilinear integration of a function $\phi : \Sigma \to X$ with respect to a measure $m : \mathcal{S} \to Y$ in the projective tensor product $X\widehat{\otimes}_\pi Y$ is more singular than bilinear integration in the injective tensor product $X\widehat{\otimes}_\epsilon Y$ because the bilinear map

$$(x',y') \longmapsto \int_\Sigma \langle \phi, x' \rangle\, d\langle m, y' \rangle, \quad x' \in X',\ y' \in Y',$$

is separately continuous, so m necessarily has finite X-semivariation in $X\widehat{\otimes}_\epsilon Y$, see [47, p. 327]. Consequently, the integral in the sense of Definition 2.2 coincides with Bartle's bilinear integral for the injective tensor product topology. Bilinear integration in injective tensor products is considered in [102] and [47] gives an application to generalised Carleman operators [55].

We shall need the following 'uniform' version of the Lebesgue dominated convergence theorem for vector valued functions.

Lemma 2.4. *Let $(\Gamma, \mathcal{T}, \nu)$ be a finite measure space, $1 \le p < \infty$ and X a Banach space. Suppose that I is an index set and $f_{n,\iota} : \Gamma \to X$, $n = 1, 2, \ldots, \iota \in I$, are strongly ν-measurable functions for which there exists a nonnegative function $g \in L^p(\nu)$ with the property that for each $n = 1, 2, \ldots$ and $\iota \in I$, the bound $\|f_{n,\iota}(\gamma)\| \le g(\gamma)$ holds for ν-almost all $\gamma \in \Gamma$. Suppose that for every $\epsilon > 0$, $\lim_{n \to \infty} [\sup_{\iota \in I} \nu(\{\|f_{n,\iota} - f_\iota\| \ge \epsilon\})] = 0$.*

Then for each $\iota \in I$, the X-valued function f_ι belongs to $L^p(\nu; X)$ and

$$\lim_{n \to \infty} \left[\sup_{\iota \in I} \int_\Gamma \|f_{n,\iota}(\gamma) - f_\iota(\gamma)\|^p \, d\nu(\gamma) \right] = 0.$$

Proof. The usual argument applies: for each $N > 0$,

$$\int_\Gamma \|f_{n,\iota}(\gamma) - f_\iota(\gamma)\|^p \, d\nu(\gamma) \le \int_{\{g \ge N\}} \|f_{n,\iota}(\gamma) - f_\iota(\gamma)\|^p \, d\nu(\gamma)$$
$$+ \epsilon^p \nu(\{g \le N\}) + 2^p N^p \nu(\{\|f_{n,\iota} - f_\iota\| \ge \epsilon\})$$
$$\le 2^p \int_{\{g \ge N\}} g(\gamma)^p \, d\nu(\gamma)$$
$$+ \epsilon^p \nu(\Gamma) + 2^p N^p \nu(\{\|f_{n,\iota} - f_\iota\| \ge \epsilon\}).$$

Choosing ϵ small enough, then N large enough, and then n sufficiently large, we can ensure that $\sup_{\iota \in I} \int_\Gamma \|f_{n,\iota}(\gamma) - f_\iota(\gamma)\|^p \, d\nu(\gamma)$ is as small as we like. $\square$

Proposition 2.6. *Let $(\Gamma, \mathcal{T}, \nu)$ be a measure space, $1 \le p < \infty$ and X a Banach space. Suppose that I is an index set and $f_{n,\iota} : \Gamma \to X$, $n = 1, 2, \ldots, \iota \in I$, are strongly ν-measurable functions for which there exists a nonnegative function $g \in L^p(\nu)$ with the property that for each $n = 1, 2, \ldots$ and $\iota \in I$, the bound $\|f_{n,\iota}(\gamma)\| \le g(\gamma)$ holds for ν-almost all $\gamma \in \Gamma$.*

Suppose that for ν-almost all $\gamma \in \Gamma$,

$$\lim_{n \to \infty} [\sup_{\iota \in I} \{\|f_{n,\iota}(\gamma) - f_\iota(\gamma)\|\}] = 0.$$

Then for each $\iota \in I$, the X-valued function f_ι belongs to $L^p(\nu; X)$ and

$$\lim_{n \to \infty} \left[\sup_{\iota \in I} \int_\Gamma \|f_{n,\iota}(\gamma) - f_\iota(\gamma)\|^p \, d\nu(\gamma) \right] = 0.$$

Proof. Let $\epsilon > 0$ and choose a set Γ_ϵ of finite ν-measure so large that

$$\int_{\Gamma_\epsilon^c} \|f_{n,\iota}(\gamma) - f_\iota(\gamma)\|^p \, d\nu(\gamma) \le 2^p \int_{\Gamma_\epsilon^c} g(\gamma)^p \, d\nu(\gamma) < \epsilon,$$

for all $\iota \in I$ and $n = 1, 2, \ldots$. By the preceding result, it is enough to prove that

$$\lim_{n \to \infty} [\sup_{\iota \in I} \nu(\{\gamma \in \Gamma_\epsilon : \|f_{n,\iota}(\gamma) - f_\iota(\gamma)\| \geq \delta\})] = 0, \qquad (2.5)$$

for every $\epsilon > 0$ and $\delta > 0$.

If for every countable subfamily J of I, (2.5) is true when the index set I is replaced by J, then it is true for the index set I itself, so we may assume from the outset that I is itself countable. The argument is standard: for each $\delta > 0$, the set

$$\bigcup_{m=1}^{\infty} \bigcap_{\iota \in I} \bigcap_{n \geq m} \{\gamma \in \Gamma_\epsilon : \|f_{n,\iota}(\gamma) - f_\iota(\gamma)\|_X < \delta\}$$

has full ν-measure in Γ_ϵ. Let $\alpha > 0$. Then for some $n_0 = 1, 2, \ldots$, we have

$$\nu\left(\bigcup_{\iota \in I} \bigcup_{n = n_0}^{\infty} \{\gamma \in \Gamma_\epsilon : \|f_{n,\iota}(\gamma) - f_\iota(\gamma)\|_X \geq \delta\} \right) < \alpha. \qquad (2.6)$$

For every $\iota \in I$ and $n \geq n_0$, the set

$$S_{n,\iota}(\delta) := \{\gamma \in \Gamma_\epsilon : \|f_{n,\iota}(\gamma) - f_\iota(\gamma)\|_X \geq \delta\}$$

is contained in the set in the argument of the measure ν in (2.6), so $\nu(S_{n,\iota}(\delta)) < \alpha$, for every $\iota \in I$ and $n \geq n_0$. But α is any positive number, so (2.5) follows. $\square$

The following result is an analogue of the Fubini-Tonelli theorem in which the integral with respect to a product measure is replaced by a bilinear integral—the existence of the 'iterated integrals' guarantees the existence of the bilinear integral.

Theorem 2.6. *Let $(\Gamma, \mathcal{T}, \nu)$ be a σ-finite measure space, $1 \leq p < \infty$, $(\Omega, \mathcal{S})$ a measurable space and $m : \mathcal{S} \to X$ a measure with values in a Banach space X. Let $f : \Omega \times \Gamma \to \mathbb{C}$ be an $(\mathcal{S} \otimes \mathcal{T})$-measurable function with the property that*

(i) *$f(\omega, \cdot) \in L^p(\nu)$, for m-almost all $\omega \in \Omega$,*
(ii) *$f(\cdot, \gamma)$ is m-integrable in X for ν-almost all $\gamma \in \Gamma$, and*
(iii) *there exist $g \in L^p(\nu)$ such that for every $A \in \mathcal{S}$, the bound*

$$\left\| \int_A f(\omega, \gamma) \, dm(\omega) \right\|_X \leq g(\gamma)$$

holds for ν-almost all $\gamma \in \Gamma$.

Then $\omega \mapsto f(\omega, \cdot)$, $\omega \in \Omega$, is m-integrable in $L^p(\nu; X)$, and for ν-almost all $\gamma \in \Gamma$, the equality $\left[\int_\Omega f(\omega, \cdot) \otimes dm(\omega) \right](\gamma) = \int_\Omega f(\omega, \gamma) \, dm(\omega)$ holds.

Proof. We prove the result first in the case that ν is a finite measure and $f \geq 0$. Suppose that s_n, $n = 1, 2, \ldots$, are $(\mathcal{S} \otimes \mathcal{T})$-simple functions increasing to f. Each function $s_n(\cdot, \gamma)$, $n = 1, 2, \ldots$, is bounded and so m-integrable, for each $\gamma \in \Gamma$. Then, for every $A \in \mathcal{S}$ and $n = 1, 2, \ldots$, $\left\| \int_A s_n(\omega, \gamma) \, dm(\omega) \right\| \leq 4g(\gamma)$ for ν-almost all $\gamma \in \Gamma$, because the inequality

$$\| \mu(u) \|_X \leq 4 \| u \|_\infty \sup_{A \in \mathcal{S}} \| \mu(A) \|_X$$

is valid for any vector measure $\mu : \mathcal{S} \to X$ with values in a Banach space X, and any bounded μ-measurable function u.

Now for ν-almost all $\gamma \in \Gamma$, $\int_A s_n(\omega, \gamma) \, dm(\omega) \to \int_A f(\omega, \gamma) \, dm(\omega)$, uniformly for all $A \in \mathcal{S}$, by dominated convergence for vector measures, Theorem 1.7. The X-valued functions $\left[\int_A s_n(\omega, \cdot) \, dm(\omega) \right] : \gamma \mapsto \int_A s_n(\omega, \gamma) \, dm(\omega)$, $n = 1, 2, \ldots$, therefore converge in X, uniformly for $A \in \mathcal{S}$, at ν-almost all points $\gamma \in \Gamma$, and are dominated in norm by $4g$. Dominated convergence in $L^p(\nu; X)$, Proposition 2.6, ensures that the X-valued functions $\left[\int_A s_n(\omega, \cdot) \, dm(\omega) \right]$, $n = 1, 2, \ldots$, converge in $L^p(\nu; X)$, uniformly for all $A \in \mathcal{S}$.

An appeal to (i) and dominated convergence shows that $s_n(\omega, \cdot) \to f(\omega, \cdot)$ in $L^p(\nu)$, for m-almost all $\omega \in \Omega$. Unfortunately, the $L^p(\nu)$-valued functions $s_n(\omega, \cdot)$, $n = 1, 2, \ldots$, need not be $L^p(\nu)$-valued $\mathcal{S}$-simple functions, so we cannot apply our definition of bilinear integration without further analysis.

Let $B \in \mathcal{S} \otimes \mathcal{T}$. If B_k belongs to the algebra generated by product sets from $\mathcal{S}$ and $\mathcal{T}$ and $\chi_{B_k \Delta B} \to 0$ $m \otimes \nu$-a.e. as $k \to \infty$, then by Fubini's theorem, for ν-almost all $\gamma \in \Gamma$, $\chi_{B_k \Delta B}(\omega, \gamma) \to 0$ for m-almost all $\omega \in \Omega$. By dominated convergence for vector measures, $\int_A \chi_{B_k}(\omega, \gamma) \, dm(\omega) \to \int_A \chi_B(\omega, \gamma) \, dm(\omega)$ in X, uniformly for all $A \in \mathcal{S}$. The bound $\int_A \chi_{B_k}(\omega, \gamma) \, dm(\omega) \leq \|m\|(A)$, $k = 1, 2, \ldots$, and the finiteness of ν ensures that the X-valued functions $\int_A \chi_{B_k}(\omega, \cdot) \, dm(\omega)$, $k = 1, 2, \ldots$, converge in $L^p(\nu; X)$, uniformly for $A \in \mathcal{S}$ by Proposition 2.6. Moreover, by another appeal to dominated convergence and Fubini's theorem, $\chi_{B_k}(\omega, \cdot) \to \chi_B(\omega, \cdot)$ in $L^p(\nu)$ for m-almost all $\omega \in \Omega$. But χ_{B_k}, $k = 1, 2, \ldots$, are $L^p(\nu)$-valued $\mathcal{S}$-simple functions for which $\left[\int_A \chi_{B_k}(\omega, \cdot) \otimes dm(\omega) \right](\gamma) = \int_A \chi_{B_k}(\omega, \gamma) \, dm(\omega)$, $k = 1, 2, \ldots$, so the function $\omega \mapsto \chi_B(\omega, \cdot)$, $\omega \in \Omega$, is necessarily m-integrable in $L^p(\nu; X)$ and the equality

$$\left[\int_A \chi_B(\omega, \cdot) \otimes dm(\omega)\right](\gamma) = \int_A \chi_B(\omega, \gamma)\, dm(\omega)$$

holds for ν-almost all $\gamma \in \Gamma$.

Now let λ be a scalar measure equivalent to m. The convergence of strongly measurable vector valued functions λ-a.e. implies convergence in measure, so by the argument above, we can find scalar valued simple functions ϕ_n, $n = 1, 2, \ldots$, based on product sets $S \times T$, with $S \in \mathcal{S}$ and $T \in \mathcal{T}$, such that $\lambda(\{\omega \in \Omega : \|s_n(\omega, \cdot) - \phi_n(\omega, \cdot)\|_{L^p(\nu)} \geq 1/n\}) < 1/n$ and

$$\left\|\int_A [s_n(\omega, \cdot) - \phi_n(\omega, \cdot)] \otimes dm(\omega)\right\|_p \leq 2^{-n}, \quad n = 1, 2, \ldots.$$

By passing to subsequences, if necessary, we may suppose that $\|s_n(\omega, \cdot) - \phi_n(\omega, \cdot)\|_{L^p(\nu)} \to 0$ for m-almost all $\omega \in \Omega$, as $n \to \infty$. Then $\phi_n(\omega, \cdot) \to f(\omega, \cdot)$ in $L^p(\nu)$ for m-almost all $\omega \in \Omega$ and $\int_A \phi_n(\omega, \cdot) \otimes dm(\omega)$, $n = 1, 2, \ldots$, converges in $L^p(\nu; X)$, uniformly for $A \in \mathcal{S}$. Because the functions $\omega \mapsto \phi_n(\omega, \cdot)$, $\omega \in \Omega$, are actually $L^p(\nu)$-valued $\mathcal{S}$-simple functions for $n = 1, 2, \ldots$, it follows that the function $\omega \mapsto f(\omega, \cdot)$ is m-integrable and

$$\left[\int_\Omega f(\omega, \cdot) \otimes dm(\omega)\right](\gamma) = \lim_{n \to \infty} \int_\Omega \phi_n(\omega, \gamma)\, dm(\omega) = \int_\Omega f(\omega, \gamma)\, dm(\omega),$$

for ν-almost all $\gamma \in \Gamma$.

We now consider the case of a complex valued function f. If the inequality (iii) holds, then it also holds for the real and imaginary parts of f, and their positive and negative parts, possibly with g replaced by $4g$, so the result follows for all complex valued functions f in the case that $\nu(\Gamma) < \infty$.

Now suppose that ν is σ-finite and let $\Gamma_k \in \mathcal{T}$, $k = 1, 2 \ldots$, be increasing sets with finite ν-measure, whose union is Γ. If conditions (i)-(iii) hold, then $\omega \mapsto Q(\Gamma_k)f(\omega, \cdot)$ is m-integrable in $L^p(\nu)$ for each $k = 1, 2, \ldots$. Here Q is the spectral measure of multiplication by characteristic functions acting on $L^p(\nu)$. By (iii), for ν-almost all $\gamma \in \Gamma$, we have

$$\left\|\left[\int_A [Q(B)f(\omega, \cdot)] \otimes dm(\omega)\right](\gamma)\right\|_X \leq \chi_B(\gamma)g(\gamma), \quad A \in \mathcal{S}, B \in \mathcal{T}.$$

In particular, the functions $\int_A [Q(\Gamma_n)f(\omega, \cdot)] \otimes dm(\omega)$, $n = 1, 2, \ldots$, converge in $L^p(\nu; X)$, uniformly for $A \in \mathcal{S}$, as $n \to \infty$.

Let λ be a measure equivalent to m and choose $L^p(\nu)$-valued $\mathcal{S}$-simple functions ϕ_n such that $\lambda(\{\omega \in \Omega : \|Q(\Gamma_n)f(\omega, \cdot) - \phi_n(\omega, \cdot)\|_{L^p(\nu)} \geq 1/n\}) < 1/n$ and

$$\left\|\int_A [Q(\Gamma_n)f(\omega, \cdot) - \phi_n(\omega, \cdot)] \otimes dm(\omega)\right\|_p \leq 2^{-n}, \quad A \in \mathcal{S}, n = 1, 2, \ldots.$$

Then $\lim_{n\to\infty} \int_A [\phi_n(\omega,\cdot)] \otimes dm(\omega) = \lim_{n\to\infty} \int_A [Q(\Gamma_n)f(\omega,\cdot)] \otimes dm(\omega)$ in $L^p(\nu; X)$, uniformly for $A \in \mathcal{S}$. On passing to a subsequence, if necessary, as $n \to \infty$, the functions $\phi_n(\omega,\cdot)$ converge in $L^p(\nu)$, for m-almost all $\omega \in \Omega$, to $\lim_{n\to\infty} Q(\Gamma_n)f(\omega,\cdot) = f(\omega,\cdot)$. It follows that $f(\omega,\cdot)$ is m-integrable in $L^p(\nu)$ and for ν-almost all $\gamma \in \Gamma$,

$$\left[\int_\Omega f(\omega,\cdot) \otimes dm(\omega) \right](\gamma) = \lim_{n\to\infty} \int_\Omega [\chi_{\Gamma_n}(\gamma)f(\omega,\gamma)]\, dm(\omega)$$

$$= \int_\Omega f(\omega,\gamma)\, dm(\omega).$$

$\square$

Chapter 3

Operator traces

The study of traces and the operator ideal of trace class operators has a venerable tradition with applications to quantum physics and scattering theory [127], algebraic geometry [9] and non-commutative geometry [21].

The trace of a linear map $T : \mathcal{H} \to \mathcal{H}$ on a Hilbert space $\mathcal{H}$ with finite dimension $n = 1, 2, \ldots$, is the number $\sum_{j=1}^{n} a_{jj}$ for any matrix representation $\{a_{jk}\}_{j,k=1}^{n}$ of the linear map T with respect to a basis of $\mathcal{H}$.

By analogy with the finite dimensional case, if $T_k : L^2([0,1]) \to L^2([0,1])$ is a linear operator with an integral kernel k, then one might hope that

$$\text{tr}(T_k) = \int_0^1 k(x,x) \, dx \tag{3.1}$$

if T_k is a *trace class* operator as in Section 3.1 below. However,

$$\{(x,x) : x \in [0,1]\}$$

is a set of measure zero in $[0,1]^2$ and if $k = k_1$ almost everywhere on $[0,1]^2$, then $T_k = T_{k_1}$, so the right-hand side of equation (3.1) is not well-defined. It turns out that formula (3.1) is valid if the kernel k is continuous on $[0,1] \times [0,1]$ or otherwise has a specific representation such as equation (3.3) below.

The right-hand side of formula (3.1) may be viewed as a bilinear integral $\int_0^1 \langle T_k, dm \rangle := \int_0^1 \Phi_k \, dm$ of the type considered in Example 2.1 of the preceding chapter. Unless T_k is a positive operator in the sense of complex Hilbert spaces, that is, $(T_k u, u) \geq 0$ for all $u \in L^2([0,1])$, the convergence of the bilinear integral $\int_0^1 \langle T_k, dm \rangle$ is not sufficient to guarantee that T_k is a trace class operator. As we shall see, the collection of bounded linear operators T_k for which the bilinear integral $\int_0^1 \langle T_k, dm \rangle$ converges constitutes a *lattice ideal* $\mathfrak{C}_1(L^2([0,1]))$ of regular operators on $L^2([0,1])$ that includes the *operator ideal* $\mathcal{C}_1(L^2([0,1]))$ of trace class operators on $L^2([0,1])$. In Chapter 7, the bilinear integral $\int_\Sigma \langle T_k, dm \rangle$ is featured in the proof of the CLR inequality for dominated semigroups on a Hilbert space $L^2(\Sigma, \mathcal{E}, \mu)$.

3.1   Trace class operators

The *singular values* $\{\lambda_j\}_{j=1}^\infty$ of a compact linear operator $T : \mathcal{H} \to \mathcal{H}$ on a Hilbert space $\mathcal{H}$ are the eigenvalues of the compact selfadjoint operator $(T^*T)^{\frac{1}{2}}$. The operator T is called *trace class* if $\|T\|_1 = \sum_{j=1}^\infty \lambda_j < \infty$, or equivalently, $\sum_{j=1}^\infty |(Th_j, h_j)| < \infty$ for any orthonormal set $\{h_j\}_{j=1}^\infty$ in $\mathcal{H}$. The collection $\mathcal{C}_1(\mathcal{H})$ of trace class operators on $\mathcal{H}$ is an *operator ideal* and Banach space with the norm $\|\cdot\|_1$. The references [53,127] are encyclopedias of trace class operators.

In the case of an infinite dimensional separable Hilbert space $\mathcal{H}$, the *trace* of $T \in \mathcal{C}_1(\mathcal{H})$ is defined to be

$$\text{tr}(T) = \sum_{j=1}^\infty (Th_j, h_j)$$

with respect to any orthonormal basis $\{h_j\}_{j=1}^\infty$ of $\mathcal{H}$ [127, Theorem 3.1]. *Lidskii's equality* asserts that $\text{tr}(A)$ is actually the sum of the eigenvalues of the compact operator T [127, Theorem 3.7].

Suppose that $T_k : L^2([0,1]) \to L^2([0,1])$ is a bounded linear operator with an integral kernel $k : [0,1] \times [0,1] \to \mathbb{C}$, that is, k is a measurable function such that for every $f \in L^2([0,1])$, the integral $\int_0^1 |k(x,y)f(y)| \, dy$ is finite for almost all $x \in [0,1]$ and $(T_k f)(x) = \int_0^1 k(x,y) f(y) \, dy$ for almost all $x \in [0,1]$.

Any trace class operator $T : L^2([0,1]) \to L^2([0,1])$ is an integral operator with distinguished kernel k for which (3.1) is valid due to the representation

$$T : h \longmapsto \sum_{j=1}^\infty \lambda_j \phi_j (h, \psi_j), \quad h \in L^2([0,1]), \tag{3.2}$$

with respect to the L^2-inner product $(f,g) = \int_0^1 f(x)\overline{g(x)} \, dx$, $f, g \in L^2([0,1])$, and the singular values $\{\lambda_j\}_{j=1}^\infty$ of T. The sets $\{\phi_j\}$ and $\{\psi_j\}$ of vectors are orthonormal in $L^2([0,1])$, so the representation (3.1) is valid for the distinguished integral kernel k defined by

$$k(x,y) = \sum_{j=1}^\infty \lambda_j \phi_j(x) \overline{\psi_j(y)} \tag{3.3}$$

for all $x, y \in [0, 1]$ for which the right-hand side is absolutely convergent. In particular, if $T_k : L^2([0, 1]) \to L^2([0, 1])$ is a trace class linear operator with a continuous integral kernel k on $[0, 1]^2$, then (3.1) holds [127, Theorem 3.9].

C. Brislawn observed in [18] that if $T : L^2([0, 1]) \to L^2([0, 1])$ is a trace class linear operator and k_0 is any integral kernel of T, then the kernel $k = \lim_{\epsilon \to 0+} \varphi_\epsilon * k_0$ has the property that $T = T_k$ and equation (3.1) holds. Here $\varphi_\epsilon(x) = \epsilon^{-2} \varphi(x/\epsilon)$, $x \in \mathbb{R}^2$, $\epsilon > 0$, for some nonnegative function φ on $\mathbb{R}^2$ that is zero outside $[-1, 1]^2$ and with the property that $\int_{\mathbb{R}^2} \varphi(x) \, dx = 1$ and φ has an integrable radially decreasing majorant: the characteristic function $\varphi = \chi_{[-\frac{1}{2}, \frac{1}{2}]^2}$ will do.

The convolution $u * v$ of $u, v \in L^1(\mathbb{R}^n)$ is defined for almost all $x \in \mathbb{R}^n$ by the formula

$$u * v(x) = \int_{\mathbb{R}^n} u(x - y) v(y) \, dy.$$

The convolution $u * v$ of a function $u \in L^1(\mathbb{R}^2)$ with $v \in L^1([0, 1]^2)$ is defined almost everywhere in $[0, 1]^2$ by setting v equal to zero outside $[0, 1]^2$.

Then the map $k_0 \longmapsto \lim_{\epsilon \to 0+} \varphi_\epsilon * k_0$ is a *smoothing operator* for which the value of $k = \lim_{\epsilon \to 0+} \varphi_\epsilon * k_0$ at a point (x, x) of the diagonal is defined by averages in $[0, 1]^2$ about (x, x) for almost every $x \in [0, 1]$. A related idea appears in [53, Theorem 8.4].

It is clear that this idea need not be confined to trace class operators.

Example 3.1 ([18, Example 3.2]). The Volterra operator T is defined by

$$(Tf)(x) = \int_0^x f(y) \, dy, \quad x \in [0, 1], \text{ for } f \in L^2([0, 1]).$$

Then T is defined by the integral kernel $k_0 = \chi_{\{y < x\}}$. The (lattice) positive linear map $T : L^2([0, 1]) \to L^2([0, 1])$ is a Hilbert-Schmidt operator but not trace class: it has singular values $\lambda_n = 2/(\pi(2n + 1))$, $n = 1, 2, \ldots$.

Going back to the roots of functional analysis, the singular values λ are calculated by solving $T^*Tv = \lambda^2 v$. The integral kernel of T^*T is $(x, y) \mapsto 1 - x \vee y$. Because $(Tv)' = v$, the eigenfunction v satisfies $\lambda^2 v'' = -v$ and the boundary conditions require $\cos(1/\lambda) = 0$.

For the regularised kernel $k = \lim_{\epsilon \to 0+} \varphi_\epsilon * k_0$ defined above, we have $k = k_0$ off the diagonal in $]0, 1[^2$ and $k(x, x) = 1/2$ for all $x \in]0, 1[$, so $\int_0^1 k(x, x) \, dx = \frac{1}{2}$.

Some basic facts about the Hardy-Littlewood maximal operator in the unit square are gathered in Section 3.2 and these are applied in Section 3.3 to integrable functions to produce a Banach function space $L^1(\rho)$ embedded in $L^1([0,1]^2)$. Functions belonging to a certain closed subspace of $L^1(\rho)$ have the property that the set of its Lebesgue points has full linear measure on the diagonal of $[0,1]^2$ and in this sense, they are *traceable*. For an operator T whose integral kernel is of this class, the bilinear integral $\int_\Sigma \langle T, dm \rangle$ converges. In Section 3.4, absolute integral operators $T : X \to X$ acting on a Banach function space X over a σ-finite measure space are considered. Now that convolution is unavailable, the same idea with respect to the martingale maximal function is applied. In the case that $X = L^2([0,1])$, the dyadic martingale on $[0,1]$ yields the same results as in Section 3.3. Section 3.4 ends with a short discussion of other 'generalised traces' in recent literature.

3.2 The Hardy-Littlewood maximal operator

The Lebesgue measure on $\mathbb{R}$ is denoted by λ. The Lebesgue measure of a Borel subset B of $\mathbb{R}^n$ is sometimes written as $|B|$ and it will be understood to apply to expression like 'almost everywhere' and 'almost all' with respect to subsets of $\mathbb{R}^n$. The *centred Hardy-Littlewood maximal function* of $f \in L^1([0,1]^2)$ is given by

$$M(f)(x) = \sup_{r>0} \frac{\int_{C_r} |f(x+t)| \, dt}{|C_r|}, \quad x \in [0,1]^2. \tag{3.4}$$

In the formula above, the function f is put equal to zero outside the square $[0,1]^2$ and $C_r = [-r,r] \times [-r,r]$ for $r > 0$. The maximal function $M(f)$ is equivalent to the maximal function obtained by averaging over centred disks [54, Exercise 2.1.3], but for the purposes of the present discussion it is convenient to emphasise the product structure of the unit square. According to Lebesgue's differentiation theorem [54, Corollary 2.1.16], if $f \in L^1([0,1]^2)$ we have

$$\lim_{r \to 0+} \frac{\int_{C_r} f(x+t) \, dt}{|C_r|} = f(x) \tag{3.5}$$

for almost all $x \in [0,1]^2$, so that $|f| \le M(f)$ almost everywhere and the set L_f of *Lebesgue points* $x \in [0,1]^2$ of f where

$$\lim_{r \to 0+} \frac{\int_{C_r} |f(x+t) - f(x)| \, dt}{|C_r|} = 0$$

has full measure in $[0,1]^2$.

Let $\phi : \,]-1,1[\,\to [0,\infty[$ be a continuous function with compact support and $\int_{-1}^{1} \phi(t)\,dt = 1$. For the function $\varphi : \mathbb{R}^2 \to \mathbb{R}$ defined by $\varphi(x,y) = \phi(x)\phi(y)$, for $x,y \in \,]-1,1[$ and zero outside $]-1,1[^2$, we set $\varphi_\epsilon(x) = \epsilon^{-2}\varphi(x/\epsilon)$, $x \in \mathbb{R}^2$, $\epsilon > 0$. Then a variant of Lebesgue's differentiation theorem for an integrable function f shows that $\varphi_\epsilon * f \to f$ in $L^p([0,1]^2)$ for $1 \le p < \infty$ and almost everywhere as $\epsilon \to 0+$ [54, Corollary 2.1.17].

We are interested in the class of bounded linear operators $T_k : L^2([0,1]) \to L^2([0,1])$ with a distinguished kernel $k : [0,1]^2 \to \mathbb{C}$ for which $|k|$ also defines a bounded linear operator $T_{|k|} : L^2([0,1]) \to L^2([0,1])$ (absolute integral operators) and the intersection $L_k \cap \mathrm{diag}$ of the Lebesgue set L_k of k with the diagonal $\mathrm{diag} = \{(x,x) : x \in [0,1]\}$ has full linear measure. Because constant functions belong to $L^2([0,1])$, the kernel k necessarily belongs to $L^1([0,1]^2)$, so we first look at a subspace of $L^1([0,1]^2)$ consisting of functions f for which $L_f \cap \mathrm{diag}$ has full linear measure.

3.3 The Banach function space of traceable functions

Let $(\Sigma, \mathcal{B}, \mu)$ be a σ-finite measure space. The space of all μ-equivalence classes of scalar functions measurable with respect to $\mathcal{B}$ is denoted by $L^0(\mu)$. It is equipped with the topology of convergence in μ-measure over sets of finite measure and vector operations pointwise μ-almost everywhere. The following definition has already been mentioned in Example 2.1.

Definition 3.1 (Banach function space). Any Banach space X that is a subspace of $L^0(\mu)$ with the properties that

(i) X is an order ideal of $L^0(\mu)$, that is, if $g \in X$, $f \in L^0(\mu)$ and $|f| \le |g|$ μ-a.e., then $f \in X$, and

(ii) if $f,g \in X$ and $|f| \le |g|$ μ-a.e., then $\|f\|_X \le \|g\|_X$,

is called a *Banach function space* (based on $(\Sigma, \mathcal{B}, \mu)$). The set of $f \in X$ with $f \ge 0$ μ-a.e. is written as X_+.

The map $J : [0,1] \to [0,1]^2$ defined by $J(x) = (x,x)$, $x \in [0,1]$, maps $[0,1]$ homeomorphically onto diag. For $f \in L^1([0,1]^2)$, the extended real number $\rho(f) \in [0,\infty]$ is defined by $\rho(f) = \|f\|_1 + \int_0^1 M(f) \circ J(x)\,dx$ with $M(f)$ given by equation (3.4).

Proposition 3.1. *The space $L^1(\rho) = \{f \in L^1([0,1]^2) : \rho(f) < \infty\}$ with norm ρ is a Banach function space continuously embedded in $L^1([0,1]^2)$.*

Proof. Properties (i) and (ii) follow from the observation that $M(f) \leq M(g)$ everywhere if $|f| \leq |g|$ almost everywhere on $[0,1]^2$. According to [96, Proposition 2.6.2], it is enough to prove that $L^1(\rho)$ has the Riesz-Fischer property. Suppose that $f_j \geq 0$ almost everywhere for $j = 1, 2, \ldots$ and $\sum_{j=1}^{\infty} \rho(f_j) < \infty$. Then monotone convergence ensures that $\sum_{j=1}^{\infty} f_j$ converges almost everywhere in $[0,1]^2$ and in $L^1([0,1]^2)$ to a nonnegative integrable function f and $M(f) \leq \sum_{j=1}^{\infty} M(f_j)$ everywhere on $[0,1]^2$ and so $\rho(f) \leq \sum_{j=1}^{\infty} \rho(f_j)$. Consequently, the function space $L^1(\rho)$ possesses the Riesz-Fischer property. The inequality $\|f\|_1 \leq \rho(f)$ ensures that the inclusion of $L^1(\rho)$ in $L^1([0,1]^2)$ is continuous. $\qquad\square$

Suppose that $f \in L^1(\rho)$. By [54, Corollary 2.1.12], there exists $C > 0$ independent of f such that $\sup_{\epsilon>0} |(\varphi_\epsilon * f)(x)| \leq CM(f)(x)$ for every $x \in [0,1]^2$, so if we let $\tilde{f} = \limsup_{\epsilon\to0+}(\varphi_\epsilon * f)$ on $[0,1]^2$, then $\tilde{f} = f$ almost everywhere on $[0,1]^2$ by [54, Corollary 2.1.17], $|\tilde{f} \circ J| \leq CM(f) \circ J$ and

$$\int_0^1 |\tilde{f}(x,x)|\,dx \leq C \int_0^1 M(f) \circ J(x)\,dx < \infty,$$

so in this sense, elements of $L^1(\rho)$ possess an integrable trace $\int_0^1 |\tilde{f}(x,x)|\,dx$ on diag $\subset [0,1]^2$. However, the mapping $f \longmapsto \int_0^1 \tilde{f}(x,x)\,dx$, $f \in L^1(\rho)$, is only sublinear, so next we examine a subspace for which the lim sup can be replaced by a genuine limit almost everywhere on diag.

If u and v are two real valued functions defined on $[0,1]$, the tensor product $u \otimes v : [0,1]^2 \to \mathbb{R}$ of u and v is defined by $(u \otimes v)(x,y) = u(x)v(y)$, $x \in [0,1]$. A similar notation is used for the equivalence classes of functions so that $[u \otimes v] \circ J := [u.v]$. Then $L^\infty([0,1]) \otimes L^\infty([0,1])$ denotes the linear space of all finite linear combinations of elements $u \otimes v$ with $u, v \in L^\infty([0,1])$. Each element f of the finite tensor product $L^\infty([0,1]) \otimes L^\infty([0,1])$ is essentially bounded on $[0,1]^2$ and $M(f) \leq \|f\|_\infty$, so $f \in L^1(\rho)$ and $f \circ J \in L^\infty([0,1])$.

Let $L^\infty([0,1]) \widehat{\otimes}_\rho L^\infty([0,1])$ denote the norm closure of $L^\infty([0,1]) \otimes L^\infty([0,1])$ in the Banach function space $L^1(\rho)$.

Proposition 3.2. *Let $f \in L^1([0,1]^2)$. Then f belongs to the Banach space*
$$L^\infty([0,1]) \widehat{\otimes}_\rho L^\infty([0,1])$$
if and only if $\varphi_\epsilon * f \to f$ in $L^1(\rho)$ as $\epsilon \to 0+$. If $f \in L^\infty([0,1]) \widehat{\otimes}_\rho L^\infty([0,1])$ then $(\varphi_\epsilon * f) \circ J$ converges a.e. on $[0,1]$ and in $L^1([0,1])$ as $\epsilon \to 0+$.

Proof. By an application of the Cauchy-Schwarz inequality and the L^2-bound for the Hardy-Littlewood maximal operator [54, Theorem 2.1.6], there exists $C > 0$ such that if $u, v \in L^2([0,1])$, then

$$\int_0^1 M(u \otimes v)(x,x)\,dx \leq \int_0^1 M(u)(x)M(v)(x)\,dx$$

$$\leq C\|u\|_2\|v\|_2. \tag{3.6}$$

Here $M(u)$ and $M(v)$ are the one-dimensional maximal functions of u and v defined as in formula (3.4).

Suppose first that $f = u \otimes v$ for $u, v \in L^\infty([0,1])$. Then $\varphi_\epsilon * f = (\phi_\epsilon * u) \otimes (\phi_\epsilon * v)$ because $\varphi = \phi \otimes \phi$ and so

$$\int_0^1 M(\varphi_\epsilon * f - f)(x,x)\,dx \leq C \int_0^1 M(\phi_\epsilon * u - u)(x)M(v)(x)\,dx$$

$$+ \int_0^1 M(\phi_\epsilon * v - v)(x)M(u)(x)\,dx$$

$$\leq C'(\|\phi_\epsilon * u - u\|_2\|v\|_2 + \|\phi_\epsilon * v - v\|_2\|u\|_2)$$

$$\to 0$$

as $\epsilon \to 0+$ by the Cauchy-Schwartz intequality and the L^2-bound for the Hardy-Littlewood maximal operator. Consequently, $\varphi_\epsilon * f \to f$ in $L^1(\rho)$ as $\epsilon \to 0+$ when f is a linear combination of products of functions belonging to $L^\infty([0,1])$. There exists $C > 0$ such that $\|\varphi_\epsilon * f\|_1 \leq C\|f\|_1$ for every $f \in L^1([0,1]^2)$ and

$$\int_0^1 M(\varphi_\epsilon * f)(x,x)\,dx \leq C \int_0^1 M(f)(x,x)\,dx, \quad \epsilon > 0. \tag{3.7}$$

To check the inequality (3.7), suppose that $\psi = \pi^{-1}\chi_{D_1}$ for the unit disk D_1 centred at zero in $\mathbb{R}^2$ and let $\tilde{\varphi}$ be the least decreasing radial majorant of φ. Because φ is continuous with compact support, $\tilde{\varphi}$ is integrable on $\mathbb{R}^2$. Then $\tilde{\varphi}_\epsilon * \psi_\delta$ is a radial function for which

$$2\pi \int_0^\infty r(\tilde{\varphi}_\epsilon * \psi_\delta)(re_1)\,dr = \|\tilde{\varphi}_\epsilon * \psi_\delta\|_{L^1(\mathbb{R}^2)}$$

$$= \|\tilde{\varphi}_\epsilon\|_{L^1(\mathbb{R}^2)}\|\psi_\delta\|_{L^1(\mathbb{R}^2)}$$

$$= \|\tilde{\varphi}\|_{L^1(\mathbb{R}^2)}.$$

As in the proof of [54, Theorem 2.1.10], there exists $C' > 0$ such that $\sup_{\epsilon,\delta>0} \tilde{\varphi}_\epsilon * \psi_\delta * |f| \leq C'M(f)$. Because the maximal function (3.4) is equivalent to the maximal function for centred disks, there exists $C \geq 1$

such that $M(\varphi_\epsilon * f) \leq M(\tilde{\varphi}_\epsilon * f) \leq CM(f)$ from which the inequality (3.7) follows.

Consequently, the linear map $f \longmapsto \varphi_\epsilon * f$, $f \in L^1(\rho)$, is continuous on $L^1(\rho)$ for each $\epsilon > 0$ so that if $f \in L^\infty([0,1])\hat{\otimes}_\rho L^\infty([0,1])$, then $\varphi_\epsilon * f \to f$ in $L^1(\rho)$ as $\epsilon \to 0+$. Because $\varphi_\epsilon * f \in C([0,1]^2)$ and $C([0,1]) \otimes C[0,1])$ is dense in $C([0,1]^2)$ in the uniform norm, it follows that $\varphi_\epsilon * f \in L^\infty([0,1])\hat{\otimes}_\rho L^\infty([0,1])$ for each $\epsilon > 0$, and the limit of $\varphi_\epsilon * f$ in $L^1(\rho)$ as $\epsilon \to 0+$ also belongs to $L^\infty([0,1])\hat{\otimes}_\rho L^\infty([0,1])$.

Let $T_* f = \sup_{\epsilon>0} |\varphi_\epsilon * f| \circ J$ for $f \in L^1(\rho)$. Then $T_* : L^1(\rho) \to L^1([0,1])$ is uniformly continuous. An argument similar to the proof of [54, Theorem 2.1.14] shows that $(\varphi_\epsilon * f) \circ J$ converges almost everywhere and in $L^1([0,1])$ as $\epsilon \to 0+$ for each $f \in L^1(\rho)$. $\qquad\square$

Let $f \in L^\infty([0,1])\hat{\otimes}_\rho L^\infty([0,1])$ and set $\tilde{f} = \lim_{\epsilon\to 0+} \varphi_\epsilon * f$ wherever the limit exists in $[0,1]^2$ and zero elsewhere. Writing $f^\#$ for the corresponding function with ϕ replaced by $\chi_{[-\frac{1}{2},\frac{1}{2}]}$, it follows from equation (3.5) that $f^\# = \tilde{f}$ almost everywhere on $[0,1]^2$ and $f^\# \circ J = \tilde{f} \circ J$ almost everywhere on $[0,1]$, because the last equality certainly holds when f belongs to the dense subspace $L^\infty([0,1]) \otimes L^\infty([0,1])$. In particular, $\tilde{f} \circ J \in L^1([0,1])$ and the integral $\int_B \tilde{f} \circ J(x)\, dx$, $B \in \mathcal{B}([0,1])$, does not depend on the choice of the function φ.

Example 3.2. For a continuous function f on $[0,1]^2$ equal to zero on $\mathbb{R}^2 \setminus [0,1]^2$, the continuous functions $\varphi_\epsilon * f$ converge uniformly to f on compact subsets of $]0,1[^2$ [54, Theorem 1.2.19 (2)], so that $f \in L^\infty([0,1])\hat{\otimes}_\rho L^\infty([0,1])$ and $\tilde{f} = f$. Hence, $C([0,1]^2)$ and $C([0,1])\otimes C([0,1])$ are dense in $L^\infty([0,1])\hat{\otimes}_\rho L^\infty([0,1])$.

Functions belonging to $W^{1,1}(\mathbb{R}^2)$ or the space $L^{\alpha,p}(\mathbb{R}^2)$ of Bessel potentials on $\mathbb{R}^2$ also admit a trace on $\mathrm{diag}(\mathbb{R}^2)$ if $p, \alpha p > 1$, see [1, Section 6.2].

A result of T. Carleman [22] shows there exists a continuous periodic function $\phi : \mathbb{R} \to \mathbb{C}$ with period one such that

$$\sum_{n\in\mathbb{Z}} |\hat{\phi}(n)|^p = \infty$$

for all $p < 2$. If $k(x,y) = \phi(x-y)$, then k is a continuous kernel, $M(k) \circ J \leq \|\phi\|_\infty$ and $k(x,x) = \phi(0)$ for all $x \in [0,1]$ and so $\int_0^1 k(x,x)\, dx = \phi(0)$, although the Hilbert-Schmidt operator T_k with kernel k is not a trace class operator. Because $k \leq \|\phi\|_\infty$ and a constant function is the kernel of a finite rank operator, the trace class operators do not form a lattice ideal in the Banach lattice of Hilbert-Schmidt operators despite being an *operator ideal* in $\mathcal{L}(L^2([0,1]))$.

Example 3.3. The kernel $\chi_{\{y<x\}}$ of the Volterra integral operator in Example 3.1 belongs to the Banach space $L^\infty([0,1])\widehat{\otimes}_\rho L^\infty([0,1])$ and the same holds true for the function $\chi_{\{y\le x\}}$ which differs from $\chi_{\{y<x\}}$ on diag, a set of measure zero in $[0,1]^2$.

If $T : \mathbb{R}^2 \to \mathbb{R}^2$ is a nonsingular linear transformation, then there exists $c_T > 0$ such that $\rho(f \circ T) \le c_T \rho(f)$ if both f and $f \circ T$ are supported by $[0,1]^2$ because the collection $\{TC_r : r > 0\}$ is itself a regular family of sets whose associated maximal function is equivalent to the one defined for cubes by formula (3.4). Furthermore, if $g \in L^\infty([0,1]) \otimes L^\infty([0,1])$, then $(\varphi \circ T)_\epsilon * g \to g$ in $L^1(\rho)$ as $\epsilon \to 0+$, hence $\varphi_\epsilon * (g \circ T^{-1})$ converges to $g \circ T^{-1}$ in $L^1(\rho)$ as $\epsilon \to 0+$ as well. Taking g to be the characteristic functions of squares and T to be rotation through $\pi/4$ gives $\chi_{\{y<x\}} \in L^\infty([0,1])\widehat{\otimes}_\rho L^\infty([0,1])$.

Proposition 3.3. *Every element of $L^\infty([0,1])\widehat{\otimes}_\rho L^\infty([0,1])$ has a representative function $f : [0,1]^2 \to \mathbb{R}$ for which there exist numbers $c_j \in \mathbb{R}$ and Borel subsets A_j, B_j of $[0,1]$, $j = 1, 2, \ldots$, such that*

$$\sum_{j=1}^{\infty} |c_j|(|A_j|.|B_j| + |A_j \cap B_j|) < \infty$$

and $f(x) = \sum_{j=1}^{\infty} c_j \chi_{A_j \times B_j}(x)$ for every $x \in [0,1]^2$ such that the sum $\sum_{j=1}^{\infty} |c_j|\chi_{A_j \times B_j}(x)$ is finite. In particular, $f \circ J = \sum_{j=1}^{\infty} c_j \chi_{A_j \cap B_j} = \tilde{f} \circ J$ almost everywhere.

Proof. Let $\mu = \lambda \otimes \lambda + \lambda \circ J^{-1}$ for Lebesgue measure λ on the Borel σ-algebra $\mathcal{B}$ of $[0,1]$. If $[f_0] \in L^\infty([0,1])\widehat{\otimes}_\rho L^\infty([0,1])$, then let $f = f_0$ on $[0,1]^2 \setminus \text{diag}$ and set $f \circ J = \lim_{\epsilon \to 0+}(\varphi_\epsilon * f_0) \circ J$ wherever the limit exists and zero otherwise. By [54, Corollary 2.1.12], there exists $C > 0$ such that $\sup_{\epsilon>0} |(\varphi_\epsilon * f_0)(x)| \le CM(f_0)(x)$ for every $x \in [0,1]^2$. Because $\rho([f_0]) < \infty$, f is μ-integrable. The statement now follows from the argument of [85, Proposition 2.13], as follows.

Let $\mathcal{K}$ denote the space of $(\mathcal{B} \otimes \mathcal{B})$-simple functions on $[0,1] \times [0,1]$ and for any function $f : [0,1] \times [0,1] \to \mathbb{C}$, let

$$\rho_\mathcal{K}(f) = \inf\left\{\sum_{j=1}^{\infty} \mu(|f_j|) : f = \sum_{j=1}^{\infty} f_j\right\}.$$

The infimum is taken over all sums with $f_j \in \mathcal{K}$, $j = 1, 2, \ldots$, and

$$\sum_{j=1}^{\infty} \mu(|f_j|) < \infty,$$

such that $f(x) = \sum_{j=1}^{\infty} f_j(x)$ wherever the sum converges absolutely. Similarly, let

$$\rho_{\mathcal{B} \times \mathcal{B}}(f) = \inf \left\{ \sum_{j=1}^{\infty} |c_j| \mu(A_j \times B_j) : f = \sum_{j=1}^{\infty} c_j \chi_{A_j \times B_j} \right\},$$

where $f(x) = \sum_{j=1}^{\infty} c_j \chi_{A_j \times B_j}$ for every $x \in [0,1] \times [0,1]$ such that the sum $\sum_{j=1}^{\infty} |c_j| \chi_{A_j \times B_j}(x)$ is finite. The infimum of $\rho_{\mathcal{K}}(f)$ is taken over a larger collection of functions than for $\rho_{\mathcal{B} \times \mathcal{B}}(f)$, so $\rho_{\mathcal{K}}(f) \leq \rho_{\mathcal{B} \times \mathcal{B}}(f)$. A complex valued function may be decomposed into its real and imaginary parts, so suppose that f is real valued and $\rho_{\mathcal{K}}(f) < \infty$. Monotone convergence shows that $\mu(|f|) \leq \rho_{\mathcal{K}}(f)$. Writing $f = f^+ - f^-$ in its positive $f^+ = f \vee 0$ and negative $f^- = (-f) \vee 0$ parts, there exist monotonically increasing sequences $f_j^{\pm}$, $j = 1, 2, \ldots$, of $(\mathcal{B} \otimes \mathcal{B})$-simple functions such that

$$f = \sum_{j=1}^{\infty} (f_{j+1}^+ - f_j^+) + f_1^+ - \left(\sum_{j=1}^{\infty} (f_{j+1}^- - f_j^-) + f_1^- \right)$$

pointwise on $[0,1] \times [0,1]$ and

$$\mu(|f|) = \sum_{j=1}^{\infty} \mu(f_{j+1}^+ - f_j^+) + \sum_{j=1}^{\infty} \mu(f_{j+1}^- - f_j^-) + \mu(|f_1|).$$

Consequently, $\rho_{\mathcal{K}}(f) \leq \mu(|f|)$ and the equality $\rho_{\mathcal{K}}(f) = \mu(|f|)$ follows. If f is a $(\mathcal{B} \otimes \mathcal{B})$-measurable function with $\mu(|f|) = \infty$, then clearly $\rho_{\mathcal{K}}(f) = \infty$ too. The representation is established once we show that $\rho_{\mathcal{K}}(f) = \rho_{\mathcal{B} \times \mathcal{B}}(f)$.

The vector space $\mathcal{H}$ of all simple functions $f = \sum_{j=1}^{n} c_j \chi_{A_j \times B_j}$ with $c_j \in \mathbb{C}$, $A_j \in \mathcal{B}$, $B_j \in \mathcal{B}$, $j = 1, \ldots, n$ and $n = 1, 2, \ldots$, is dense for both the seminorms $\rho_{\mathcal{K}}$ and $\rho_{\mathcal{B} \times \mathcal{B}}$, so it suffices to establish the equality for all $f \in \mathcal{H}$. We may also suppose that the rectangles $A_j \times B_j$, $j = 1, 2, \ldots, n$, are pairwise disjoint in the representation of $f \in \mathcal{H}$ as above. Then $\rho_{\mathcal{K}}(f) = \sum_{j=1}^{n} |c_j| \mu(A_j) \nu(B_j) \geq \rho_{\mathcal{B} \times \mathcal{B}}(f)$, so $\rho_{\mathcal{K}}(f) = \rho_{\mathcal{B} \times \mathcal{B}}(f)$. $\square$

The linear map $J^* f = \tilde{f} \circ J$, $f \in L^{\infty}([0,1]) \widehat{\otimes}_{\rho} L^{\infty}([0,1])$, is uniquely determined by the Banach function space $L^{\infty}([0,1]) \widehat{\otimes}_{\rho} L^{\infty}([0,1])$.

Corollary 3.1. *There exists a unique continuous linear map*

$$J^* : L^{\infty}([0,1]) \widehat{\otimes}_{\rho} L^{\infty}([0,1]) \to L^1([0,1])$$

such that $J^*(u \otimes v) = u.v$ for every $u, v \in L^{\infty}([0,1])$.

It follows from Theorem 1.8, that the *projective tensor product*

$$L^2([0,1]) \widehat{\otimes}_\pi L^2([0,1])$$

is the set of all sums

$$k = \sum_{j=1}^\infty \phi_j \otimes \psi_j \text{ a.e., with } \sum_{j=1}^\infty \|\phi_j\|_2 \|\psi_j\|_2 < \infty. \qquad (3.8)$$

The norm of $k \in L^2([0,1]) \widehat{\otimes}_\pi L^2([0,1])$ is given by

$$\|k\|_\pi = \inf \left\{ \sum_{j=1}^\infty \|\phi_j\|_2 \|\psi_j\|_2 \right\}$$

where the infimum is taken over all sums for which the representation (3.8) holds. The Banach space $L^2([0,1]) \widehat{\otimes}_\pi L^2([0,1])$ is actually the completion of the algebraic tensor product $L^2([0,1]) \otimes L^2([0,1])$ with respect to the projective tensor product norm [123, Section 6.1]. The estimate (3.6) establishes the following result.

Proposition 3.4. *The projective tensor product $L^2([0,1]) \widehat{\otimes}_\pi L^2([0,1])$ embeds onto a proper dense subspace of $L^\infty([0,1]) \widehat{\otimes}_\rho L^\infty([0,1])$.*

There is a one-to-one correspondence between the space of trace class operators acting on $L^2([0,1])$ and $L^2([0,1]) \widehat{\otimes}_\pi L^2([0,1])$, so that the trace class operator T_k has an integral kernel $k \in L^2([0,1]) \widehat{\otimes}_\pi L^2([0,1])$ given, for example, by formula (3.3). If the integral kernel k defined by equation (3.8) has the property that

$$k(x,y) = \sum_{j=1}^\infty \phi_j(x)\psi_j(y)$$

for all $x, y \in [0,1]$ such that the sum $\sum_{j=1}^\infty |\phi_j(x)\psi_j(y)|$ is finite, then k is the integral kernel of a trace class operator T_k and the equality

$$\text{tr}(T_k) = \sum_{j=1}^\infty \int_0^1 \phi_j(x)\psi_j(x)\,dx = \int_0^1 k(x,x)\,dx$$

holds. Moreover, if K is *any* integral kernel of the trace class operator T_k, then $\tilde{K}(x,x) = k(x,x)$ for almost all $x \in [0,1]$, see [18, Theorem 3.1] or the proof of Theorem 3.3 below in the case of $L^2(\mu)$ for a σ-finite measure μ.

The representation of Proposition 3.3 for elements of the Banach space $L^\infty([0,1]) \widehat{\otimes}_\rho L^\infty([0,1])$ may be viewed as a substitute for the representation (3.8) of an element of the projective tensor product $L^2([0,1]) \widehat{\otimes}_\pi L^2([0,1])$.

Example 3.4. Another way to view the trace $\text{tr}(T_k)$ of a trace class operator $T_k : L^2([0,1]) \to L^2([0,1])$ with an integral kernel k is as a type of *bilinear integral* with respect to the $L^2([0,1])$-valued vector measure $m : B \longmapsto \chi_B$, $B \in \mathcal{B}([0,1])$. For example, if $k = \sum_{j=1}^{n} \chi_{B_j} \otimes f_j$ for Borel subsets B_j of $[0,1]$ and $f_j \in L^2([0,1])$, $j = 1,\ldots,n$ and $\Phi_k : [0,1] \to L^2([0,1])$ is the $L^2([0,1])$-valued simple function defined by $\Phi_k(x) = \sum_{j=1}^{n} \chi_{B_j}(x).f_j$, $x \in [0,1]$, then

$$\int_B \langle \Phi_k, dm \rangle = \sum_{j=1}^{n} \int_{B \cap B_j} f_j(x)\, dx = \int_B k(x,x)\, dx$$

and $\int_B \Phi_k \otimes dm = (\chi_B \otimes 1).k \in L^2([0,1]) \otimes L^2([0,1])$ for $B \in \mathcal{B}([0,1])$.

By Proposition 3.4 the bilinear integrals $\int_B \langle \Phi_k, dm \rangle$ and $\int_B \Phi_k \otimes dm$ also makes sense for $k \in L^2([0,1]) \widehat{\otimes}_\pi L^2([0,1])$ where T_k is a trace class operator and

$$\text{tr}(T_k) = \int_0^1 \langle \Phi_k, dm \rangle$$

independently of the integral kernel k representing the operator T_k.

Example 3.5. On the other hand, if $k \in L^1([0,1]^2)$, then by Fubini's Theorem, the function $\Phi_k(x) = f(x,\cdot)$ has values in $L^1([0,1])$ for almost all $x \in [0,1]$ and $\int_0^1 \Phi_k \otimes dm = k$ is an element of $L^1([0,1]) \widehat{\otimes}_\pi L^1([0,1]) \equiv L^1([0,1]^2)$, see Proposition 2.5 above.

Now suppose that the bilinear integral $\int_0^1 \Phi_k \otimes dm \equiv k$ belongs to the subspace $L^\infty([0,1]) \widehat{\otimes}_\rho L^\infty([0,1])$ of $L^1([0,1]^2)$. Then $\int_B \langle \Phi_k, dm \rangle$ is defined for each Borel set B contained in $[0,1]$ by the formula

$$\int_B \langle \Phi_k, dm \rangle = \int_0^1 \left(J^* \int_B \Phi_k \otimes dm \right) d\lambda, \tag{3.9}$$

where the linear map J^* is defined immediately preceding Corollary 3.1 above. In this sense, the linear space $L^\infty([0,1]) \widehat{\otimes}_\rho L^\infty([0,1])$ consists of the *traceable* elements of $L^1([0,1]^2)$, that is, the kernel k of the operator T_k is associated with a "generalised trace" of T_k whenever k belongs to $L^\infty([0,1]) \widehat{\otimes}_\rho L^\infty([0,1])$.

Example 3.6. Now let us recast Example 3.1 in the present setting of vector integration theory. We have a bounded linear operator $T_k : L^2([0,1]) \to L^2([0,1])$ whose integral kernel k need not be the integral kernel of a trace class operator acting on $L^2([0,1])$.

In the notation of Section 2.3, suppose that $X = L^2([0,1])$, $Y = \mathbb{C}$, $E = L^1([0,1]^2)$ and $F = L^\infty([0,1])\widehat{\otimes}_\rho L^\infty([0,1])$. Then

$$L^2([0,1]) \equiv \mathcal{L}(X,Y) = X'.$$

Suppose that Φ_k takes values in $\mathcal{L}(X,\mathbb{C}) \equiv L^2([0,1]) = X$, that is, k is a Carleman kernel [60, Section 11]. Then the vector measure $m :$ $\mathcal{B}([0,1]) \to X$ has infinite variation on any Borel set with positive Lebesgue measure. The inclusion $L^2([0,1]) \otimes L^2([0,1]) \subset F$ is valid and the map $f \longmapsto \int_0^1 (J^* f)(x)\, dx$, $f \in F$, is a continuous linear extension of the duality map

$$u \otimes v \longmapsto \int_0^1 u(x)v(x)\, dx. \quad u, v \in L^2([0,1]).$$

In the situation of Example 3.1, we have $\int_0^1 \Phi_k \otimes dm \in F$ and the singular bilinear integral $\int_0^1 \langle \Phi_k, dm\rangle$ defined by formula (3.9 equals $\frac{1}{2}$.

As a matter of notation, if $T : L^2([0,1]) \to L^2([0,1])$ has an integral kernel k belonging to $L^\infty([0,1])\widehat{\otimes}_\rho L^\infty([0,1])$, then the integral $\int_0^1 \langle \Phi_k, dm\rangle$ exists and is independent of any integral kernel k representing T, so it makes sense to write $\int_0^1 \langle T, dm\rangle$ instead of the more cumbersome notation $\int_0^1 \langle \Phi_k, dm\rangle$ and we shall do so forthwith.

3.4 Traceable operators on Banach function spaces

It is clear that the ideas of the preceding section are concerned mainly with the order properties of the Banach function space $L^2([0,1])$, although the smoothing operators $k \longmapsto \varphi_\epsilon * k$, $k \in L^1([0,1]^2)$, $\epsilon > 0$, depend on the group structure of $\mathbb{R}^2$. For a σ-finite measure space $(\Sigma, \mathcal{B}, \mu)$, the same result is achieved by taking the maximal function with respect to a suitable *filtration* $\langle \mathcal{E}_n\rangle_{n\in\mathbb{N}}$ for which $\mathcal{B} = \bigvee_n \mathcal{E}_n$, that is, $\langle \mathcal{E}_n\rangle_{n\in\mathbb{N}}$ is an increasing family of σ-algebras contained in the σ-algebra $\mathcal{B}$ and the smallest σ-algebra containing $\mathcal{E}_n$ for every $n \in \mathbb{N}$ is $\mathcal{B}$.

The filtration determined by dyadic partitions $\langle \mathcal{E}_n\rangle_{n\in\mathbb{N}}$ of $\mathbb{R}$ localised to $[0,1]$ gives the results of Section 3.3 above, that is, for each $n = 1, 2, \ldots$, the algebra $\mathcal{E}_n$ is the collection of all finite unions of sets $[(k-1)/2^n, k/2^n)$, $k = 1, \ldots, 2^n$. The right-hand endpoint constitutes a negligible set.

Let X be a Banach function space based on the σ-finite measure space $(\Sigma, \mathcal{B}, \mu)$, as in Definition 3.1. A continuous linear operator $T : X \to X$ is called *positive* if $T : X_+ \to X_+$. The collection of all positive continuous linear operators on X is written as $\mathcal{L}_+(X)$. If the real and imaginary parts of a continuous linear operator $T : X \to X$ can be written as the difference of two positive operators, it is said to be *regular*. The *modulus* $|T|$ of a regular operator T is defined by

$$|T|f = \sup_{|g| \leq f} |Tg|, \quad f \in X_+.$$

The collection of all regular operators is written as $\mathcal{L}_r(X)$ and it is given the norm $T \longmapsto |||T|||$, $T \in \mathcal{L}_r(X)$ under which it becomes a Banach lattice [96, Proposition 1.3.6].

A continuous linear operator $T : X \to X$ has an *integral kernel* k if $k : \Sigma \times \Sigma \to \mathbb{C}$ is a Borel measurable function such that $T = T_k$ for the operator given by

$$(T_k f)(x) = \int_\Sigma k(x, y) f(y) \, d\mu(y), \quad \mu\text{-almost all } x \in \Sigma,$$

in the sense that, for each $f \in X$, we have $\int_\Sigma |k(x, y) f(y)| \, d\mu(y) < \infty$ for μ-almost all $x \in \Sigma$ and the map $x \longmapsto \int_\Sigma k(x, y) f(y) \, d\mu(y)$ is an element of X. If $T_k \geq 0$, then $k \geq 0$ $(\mu \otimes \mu)$-a.e. on $\Sigma \times \Sigma$ [96, Theorem 3.3.5].

A continuous linear operator T is an *absolute integral operator* if it has an integral kernel k for which $T_{|k|}$ is a bounded linear operator on $L^2(\mu)$. Then $|T_k| = T_{|k|}$ [96, Theorem 3.3.5] and the kernel k is $(\mu \otimes \mu)$-integrable on any product set $A \times B$ with finite measure. The collection of all absolute integral operators is a lattice ideal in $\mathcal{L}_r(X)$ [96, Theorem 3.3.6].

Suppose that $T \in \mathcal{L}(X)$ has an integral kernel $k = \sum_{j=1}^n f_j \otimes \chi_{A_j}$ that is an X-valued simple function with $\mu(A_j) < \infty$. Then it is natural to view

$$\int_\Sigma \langle T, dm \rangle := \sum_{j=1}^n \int_{A_j} f_j \, d\mu = \int_\Sigma k(x, x) \, d\mu(x)$$

as a bilinear integral. Our aim is to extend the integral to a wider class of absolute integral operators acting on the Banach function space X.

Suppose that for each $n = 1, 2, \ldots$, the collection $\mathcal{P}_n$ of sets belonging to the σ-algebra $\mathcal{B}$ is a countable partition of Σ into sets with finite measure such that $\mathcal{P}_{n+1}$ is a refinement of $\mathcal{P}_n$ for each $n = 1, 2, \ldots$, that is, every element of $\mathcal{P}_n$ is the union of elements of $\mathcal{P}_{n+1}$. Then the σ-algebra $\mathcal{E}_n$ generated by the partition $\mathcal{P}_n$ of Σ is the collection of all unions of elements of $\mathcal{P}_n$, so that $\mathcal{E}_n \subset \mathcal{E}_{n+1}$ for $n = 1, 2, \ldots$. Suppose that $\mathcal{B} = \bigvee_n \mathcal{E}_n$, the smallest σ-algebra containing all $\mathcal{E}_n$, $n = 1, 2, \ldots$. It follows that $\mathcal{B}$ is countably generated. The filtration $\langle \mathcal{E}_n \rangle_{n \in \mathbb{N}}$ is denoted by $\mathcal{E}$.

Suppose that $k \geq 0$ is a Borel measurable function defined on $\Sigma \times \Sigma$ that is integrable on every set $U \times V$ for $U, V \in \mathcal{P}_1$. For each $x \in \Sigma$, the set $U_n(x)$ is the unique element of the partition $\mathcal{P}_n$ containing x. For each $n = 1, 2, \ldots$, the conditional expectation $k_n = \mathbb{E}(k|\mathcal{E}_n \otimes \mathcal{E}_n)$, as in Subsection 4.1.1 below, can be represented for μ-almost all $x, y \in \Sigma$ as

$$\mathbb{E}(k|\mathcal{E}_n \otimes \mathcal{E}_n)(x, y) = \frac{1}{\mu(U_n(x))\mu(U_n(y))} \int_{U_n(x)} \int_{U_n(y)} k(s, t) \, d\mu(s)d\mu(t)$$

$$= \sum_{U, V \in \mathcal{P}_n} \frac{\int_{U \times V} k \, d(\mu \otimes \mu)}{\mu(U)\mu(V)} \chi_{U \times V}(x, y). \tag{3.10}$$

The point here is that the formula above defines a *distinguished* element of the conditional expectation $\mathbb{E}(k|\mathcal{E}_n \otimes \mathcal{E}_n)$ that possesses a *trace* on the diagonal of $\Sigma \times \Sigma$, that is, $B \longmapsto \mathbb{E}(\chi_B|\mathcal{E}_n \otimes \mathcal{E}_n)(x, y)$, $B \in \mathcal{B}$, $x, y \in \Sigma$, is a *regular conditional measure* [16, Definition 10.4.1].

Let $\mathcal{N}$ be the set of all $x \in \Sigma$ for which there exists $n = 1, 2, \ldots$ such that $\mu(U_n(x)) = 0$. Then $\mu(U_m(x)) = 0$ for all $m > n$ because $\mathcal{P}_m$ is a refinement of $\mathcal{P}_n$ if $m > n$. Moreover, $\mathcal{N}$ is μ-null because

$$\mathcal{N} \subset \bigcup_{n=1}^{\infty} \bigcup \{U \in \mathcal{P}_n : \mu(U) = 0\}.$$

If $0 \le k_1 \le k_2$ $(\mu \otimes \mu)$-a.e., then

$$\mathbb{E}(k_1|\mathcal{E}_n \otimes \mathcal{E}_n)(x, y) \le \mathbb{E}(k_2|\mathcal{E}_n \otimes \mathcal{E}_n)(x, y), \quad n = 1, 2, \ldots,$$

for all $(x, y) \in \mathcal{N}^c \times \mathcal{N}^c$. In particular,

$$\mathbb{E}(k_1|\mathcal{E}_n \otimes \mathcal{E}_n)(x, x) \le \mathbb{E}(k_2|\mathcal{E}_n \otimes \mathcal{E}_n)(x, x), \quad n = 1, 2, \ldots,$$

for all $x \in \mathcal{N}^c$ and the representation

$$\mathbb{E}(k|\mathcal{E}_n \otimes \mathcal{E}_n)(x, x) = \sum_{U \in \mathcal{P}_n} \frac{\int_{U \times U} k \, d(\mu \otimes \mu)}{\mu(U)^2} \chi_U(x)$$

on the diagonal is valid μ-almost everywhere. Although

$$\text{diag}(\Sigma \times \Sigma) = \{(x, x) : x \in \Sigma\}$$

may be a set of $(\mu \otimes \mu)$-measure zero, the application of the conditional expectation operators $k \longmapsto \mathbb{E}(k|\mathcal{E}_n \otimes \mathcal{E}_n)$, $n = 1, 2, \ldots$, has the effect of *regularising* k. By an appeal to the Martingale Convergence Theorem [16, Theorem 10.2.3], the function k_n converges $(\mu \otimes \mu)$-a.e. to k as $n \to \infty$.

For any Borel measurable function $f : \Sigma \times \Sigma \to \mathbb{C}$ that is integrable on every set $U \times V$ for $U, V \in \mathcal{P}_1$, let

$$M_{\mathcal{E}^2}(f)(x, y) = \sup_{n \in \mathbb{N}} \mathbb{E}(|f||\mathcal{E}_n \otimes \mathcal{E}_n)(x, y), \quad x, y \in \Sigma, \tag{3.11}$$

be the maximal function associated with the martingale $\langle \mathbb{E}(|f||\mathcal{E}_n \otimes \mathcal{E}_n) \rangle_{n \in \mathbb{N}}$. The maximal function associated with dyadic partitions $\mathcal{P}_n$ of $[0,1)$ into intervals $[(k-1)/2^n, k/2^n)$, $k = 1, 2, \ldots, 2^n$ of length 2^{-n} is equivalent to the maximal function considered in Section 3.3 [54, Exercise 2.1.12].

Let $\mathfrak{C}_1(\mathcal{E}, X)$ denote the collection of absolute integral operators $T_k : X \to X$ whose integral kernels k have the property that $\mathbb{E}(|f||\mathcal{E}_1 \otimes \mathcal{E}_1)$ takes finite values and

$$\int_\Sigma M_{\mathcal{E}^2}(k)(x,x)\, d\mu(x) < \infty.$$

Where convenient, if k is the integral kernel of T, the maximal function $M_{\mathcal{E}^2}(k)$ is also written as $M_{\mathcal{E}^2}(T)$. The map $J : \Sigma \to \Sigma \times \Sigma$ defined by $J(x) = (x,x)$ for $x \in \Sigma$ maps Σ bijectively onto $\operatorname{diag}(\Sigma \times \Sigma)$.

Theorem 3.1. *The space $\mathfrak{C}_1(\mathcal{E}, X)$ is a lattice ideal in $\mathcal{L}_r(X)$, that is, if $S, T \in \mathcal{L}_r(X)$, $|S| \leq |T|$ and $T \in \mathfrak{C}_1(\mathcal{E}, X)$, then $S \in \mathfrak{C}_1(\mathcal{E}, X)$. Moreover, $\mathfrak{C}_1(\mathcal{E}, X)$ is a Dedekind complete Banach lattice with the norm $\| \cdot \|_{\mathfrak{C}_1(\mathcal{E}, X)}$ defined by*

$$\|T\|_{\mathfrak{C}_1(\mathcal{E}, X)} = \||T|\| + \int_\Sigma M_{\mathcal{E}^2}(T) \circ J\, d\mu, \qquad T \in \mathfrak{C}_1(\mathcal{E}, X). \qquad (3.12)$$

Proof. If $S, T \in \mathcal{L}_r(X)$, $|S| \leq |T|$ and $T \in \mathfrak{C}_1(\mathcal{E}, X)$, then S is an absolute integral operator by [96, Theorem 3.3.6]. If k_1 is the integral kernel of S and k_2 is the integral kernel of T, then by [96, Theorem 3.3.5], the inequality $|k_1| \leq |k_2|$ holds $(\mu \otimes \mu)$-a.e. . Then $|M_{\mathcal{E}^2}(k_1)(x,x)| \leq |M_{\mathcal{E}^2}(k_2)(x,x)|$ for all $x \in \Sigma$, so that

$$\int_\Sigma M_{\mathcal{E}^2}(k_1) \circ J\, d\mu \leq \int_\Sigma M_{\mathcal{E}^2}(k_2) \circ J\, d\mu < \infty.$$

Hence $S \in \mathfrak{C}_1(\mathcal{E}, X)$ and $\|S\|_{\mathfrak{C}_1(\mathcal{E}, X)} \leq \|T\|_{\mathfrak{C}_1(\mathcal{E}, X)}$.

To show that $\mathfrak{C}_1(\mathcal{E}, X)$ is complete in its norm, suppose that

$$\sum_{j=1}^\infty \left(\||T_j|\| + \int_\Sigma M_{\mathcal{E}^2}(T_j) \circ J\, d\mu \right) < \infty$$

for $T_j \in \mathfrak{C}_1(\mathcal{E}, X)$. Then $T = \sum_{j=1}^\infty T_j$ in the space of regular operators on X. The inequality $|T| \leq \sum_{j=1}^\infty |T_j|$ ensures that T is an absolute integral operator with kernel k by [96, Theorem 3.3.6] and $|k| \leq \sum_{j=1}^\infty |k_j|$ $(\mu \otimes \mu)$-a.e.

Suppose first that X is a real Banach function space. Each positive part T_j^+ of T_j, $j = 1, 2, \ldots$ has an integral kernel k_j^+. By monotone convergence, there exists a set of full μ-measure on which $\mathbb{E}(k^+|\mathcal{E}_n \otimes \mathcal{E}_n)(x,x) \leq$

$\sum_{j=1}^{\infty} \mathbb{E}(k_j^+ | \mathcal{E}_n \otimes \mathcal{E}_n)(x, x)$ for each $n = 1, 2, \ldots$. Taking the supremum and applying the monotone convergence theorem pointwise and under the sum shows that $M_{\mathcal{E}^2}(k^+)(x, x) \leq \sum_{j=1}^{\infty} M_{\mathcal{E}^2}(k_j^+)(x, x)$ for μ-almost all $x \in \Sigma$ and $\int_{\Sigma} M_{\mathcal{E}^2}(k^+) \circ J \, d\mu < \infty$. Applying the same argument to T^- and then the real and imaginary parts of T ensures that $T \in \mathfrak{C}_1(\mathcal{E}, X)$ and

$$\||T\|| + \int_{\Sigma} M_{\mathcal{E}^2}(T) \circ J \, d\mu \leq \sum_{j=1}^{\infty} \left(\||T_j\|| + \int_{\Sigma} M_{\mathcal{E}^2}(T_j) \circ J \, d\mu \right).$$

Dedekind completeness is inherited from $\mathcal{L}_r(X)$ [96, Theorem 1.3.2] and $L^1(\mu)$ [96, Example v) p. 9]. $\qquad\square$

As in Section 3.3, we may define $\tilde{k} = \limsup_{n \to \infty} \mathbb{E}(k | \mathcal{E}_n \otimes \mathcal{E}_n)$ for the integral kernel k of an operator $T \in \mathfrak{C}_1(\mathcal{E}, X)$ so that $\int_{\Sigma} \tilde{k} \circ J \, d\mu \leq \|T\|_{\mathfrak{C}_1(\mathcal{E}, X)}$. The same function $\tilde{k} : \Sigma \times \Sigma \to \overline{\mathbb{R}}$ is obtained for any integral kernel k associated with the operator T. The integral $\int_{\Sigma} \tilde{k} \circ J \, d\mu$ is denoted as $\int_{\Sigma} \langle T, dm \rangle$, which is the notation used in [66] and is consistent with the notation following Example 3.6, provided that dyadic partitions of $[0, 1]$ are used.

We next see when the limsup can be replaced by a genuine limit, as is the case for trace class operators on $L^2(\mu)$.

In order that there are sufficiently many finite rank operators $T : X \to X$ with an integral kernel, we assume that both X and the Köthe dual

$$X^\times = \{ g \in L^0(\mu) : \int_{\Sigma} |fg| \, d\mu < \infty \text{ for all } f \in X \}$$

of X be order dense in $L^0(\mu)$, see [96, Theorem 3.3.7]. Then $X \cap X^\times$ is order dense in $L^0(\mu)$.

We suppose that the filtration $\mathcal{E} = \{\mathcal{E}_n\}_{n=1}^{\infty}$ is based on partitions $\mathcal{P}_n$, $n = 1, 2, \ldots$, consisting of sets of finite positive measure whose characteristic functions belong to $X \cap X^\times$. Hence, if $n = 1, 2, \ldots$ and $A \in \mathcal{P}_n$, then

i) $0 < \mu(A) < \infty$,
ii) $\chi_A \in X$ and
iii) $\int_A |f| \, d\mu < \infty$ for all $f \in X$.

Suppose also that the conditional expectation $\mathbb{E}_n : f \longmapsto \mathbb{E}(f | \mathcal{E}_n)$, $f \in X$, is a bounded linear operator on the Banach function space X for each $n = 1, 2, \ldots$ and $\mathbb{E}_n \to Id$ in the strong operator topology as $n \to \infty$, as is the case for $X = L^p(\mu)$, $1 \leq p < \infty$.

It is a simple matter to check that

$$\int_\Sigma f.\mathbb{E}(g|\mathcal{E}_n) = \int_\Sigma \mathbb{E}(f|\mathcal{E}_n).g, \quad f \in X, \ g \in X^\times,$$

so $\mathbb{E}(g|\mathcal{E}_n) \to g$ in the weak topology $\sigma(X^\times, X)$ for each $g \in X^\times$. Suppose also that $\int_\Sigma M_{\mathcal{E}^2}(f \otimes g) \circ J \, d\mu < \infty$ for all $f \in X$ and $g \in X^\times$. Then by dominated convergence, $\mathbb{E}_n(f).\mathbb{E}_n(g) \to f.g$ μ-a.e. and in $L^1(\mu)$ for every $f \in X$ and $g \in X^\times$.

The closure in $\mathfrak{C}_1(\mathcal{E}, X)$ of the collection of all finite rank operators $T : X \to X$ with integral kernels of the form

$$k = \sum_{j=1}^n f_j \otimes g_j, \tag{3.13}$$

for $f_j \in X$, $g_j \in X^\times$ for $j = 1, \ldots, n$ and $n = 1, 2, \ldots$, is denoted by $X \widehat{\otimes}_\mathcal{E} X^\times$.

The following statement is the martingale analogue of Proposition 3.2, proved along the same lines. We assume below that the conditional expectation operators $\mathbb{E}_n$, $n = 1, 2, \ldots$, enjoy the properties mentioned above.

Proposition 3.5. *Suppose that k is the integral kernel of the operator $T \in \mathfrak{C}_1(\mathcal{E}, X)$ and T_n has the integral kernel $\mathbb{E}(k|\mathcal{E}_n \otimes \mathcal{E}_n)$ for $n = 1, 2, \ldots$. If $T \in X \widehat{\otimes}_\mathcal{E} X^\times$ then $T_n \to T$ in the strong operator topology as $n \to \infty$ and $\mathbb{E}(k|\mathcal{E}_n \otimes \mathcal{E}_n) \circ J$ converges a.e. on Σ and in $L^1(\mu)$ as $n \to \infty$.*

Proof. Suppose first that $T : X \to X$ is a finite rank linear operator with a kernel k given by formula (3.13). Then $k_n = \mathbb{E}(k|\mathcal{E}_n \otimes \mathcal{E}_n)$ converges to k $(\mu \otimes \mu)$-a.e. as $n \to \infty$, by the Martingale Convergence Theorem [16, Theorem 10.2.3]. First we need to check that $T_n f \to T f$ in X as $n \to \infty$ for each $f \in X$ by noting that

$$\mathbb{E}(u \otimes v|\mathcal{E}_n \otimes \mathcal{E}_n) = \mathbb{E}(u|\mathcal{E}_n) \otimes \mathbb{E}(v|\mathcal{E}_n), \quad u \in X, \ v \in X^\times,$$

for each $n = 1, 2, \ldots$, and appealing to the convergence properties of the conditional expectation operator $\mathbb{E}_n$ as $n \to \infty$. Clearly $\mathbb{E}(k|\mathcal{E}_n \otimes \mathcal{E}_n) \circ J$ converges a.e. on Σ and in $L^1(\mu)$ as $n \to \infty$ because k is a sum of product functions and $\int_\Sigma M_{\mathcal{E}^2}(f \otimes g) \circ J \, d\mu < \infty$ for all $f \in X$ and $g \in X^\times$.

For a general operator $T \in \mathfrak{C}_1(\mathcal{E}, X)$ with kernel k, the equality $\mathbb{E}_n T \mathbb{E}_n = T_n$ holds, so $T_n \to T$ in the strong operator topology on $\mathcal{L}(X)$. The inequality $|\mathbb{E}_n T \mathbb{E}_n| \le \mathbb{E}_n |T| \mathbb{E}_n$ in the Banach lattice of regular operators and the uniform boundedness principle together with the inequality

$$M_{\mathcal{E}^2}(\mathbb{E}_n \otimes \mathbb{E}_n(k)) \le M_{\mathcal{E}^2}(k), \quad n = 1, 2, \ldots,$$

ensures that the map $T \longmapsto \mathbb{E}_n T \mathbb{E}_n$, $T \in \mathfrak{C}_1(\mathcal{E}, X)$, is uniformly bounded on the Banach lattice $\mathfrak{C}_1(\mathcal{E}, X)$ for $n = 1, 2, \ldots$. Consequently, the convergence a.e. of $\mathbb{E}(k | \mathcal{E}_n \otimes \mathcal{E}_n) \circ J$ on Σ and in $L^1(\mu)$ as $n \to \infty$ follows by approximation of k by elements of $X \otimes X^\times$. $\qquad\square$

For $T \in X \widehat{\otimes}_\varepsilon X^\times$, we have $\int_\Sigma \langle T, dm \rangle = \lim_{n \to \infty} \int_\Sigma \mathbb{E}(k | \mathcal{E}_n \otimes \mathcal{E}_n) \circ J \, d\mu$. The space $\mathfrak{C}_1(\mathcal{E}, X)$ is itself a Banach lattice and a lattice ideal in the Banach lattice of regular operators on X.

The closure of $X \otimes X^\times$ in $\mathfrak{C}_1(\mathcal{E}, X)$ with respect to the locally convex Hausdorff topology associated with the seminorms

$$T \longmapsto \||T|f\| + \int_\Sigma M_{\mathcal{E}^2}(T) \circ J \, d\mu, \qquad T \in \mathfrak{C}_1(\mathcal{E}, X),$$

for $f \in X$ could also be used, but it is not a metrisable topology if X is infinite dimensional.

We state the analogue of Proposition 3.3 in the present setting.

Proposition 3.6. *Every element $T \in X \widehat{\otimes}_\varepsilon X^\times$ has a representative kernel $k : \Sigma \times \Sigma \to \mathbb{R}$ for which there exist numbers $c_j \in \mathbb{C}$ and sets A_j, B_j from $\mathcal{B}$, $j = 1, 2, \ldots$, such that*

$$\sum_{j=1}^\infty |c_j| (\mu(A_j) . \mu(B_j) + \mu(A_j \cap B_j)) < \infty$$

and $k(x) = \sum_{j=1}^\infty c_j \chi_{A_j \times B_j}(x)$ for every $x \in \Sigma \times \Sigma$ such that the sum $\sum_{j=1}^\infty |c_j| \chi_{A_j \times B_j}(x)$ is finite. In particular, $f \circ J = \sum_{j=1}^\infty c_j \chi_{A_j \cap B_j} = \tilde{f} \circ J$ almost everywhere.

Remark 3.1. The limit $\tilde{k} \circ J$ is independent μ-a.e. of the filtration $\mathcal{E}$ if the kernel k is sufficiently 'regular', as in the following cases.

a) $k \in L^2(\mu) \widehat{\otimes}_\pi L^2(\mu)$—see the proof of Theorem 3.3 below.

b) μ is a finite regular Borel measure on a compact Hausdorff space K and $k : K \times K \to \mathbb{C}$ is continuous. Then $\tilde{k}(x, x) = k(x, x)$ for μ-almost all $x \in K$, because $C(K) \otimes C(K)$ is dense in $C(K \times K)$ by the Stone-Weierstrass Theorem.

c) Functions belonging to the closure of $L^\infty(\mu) \otimes L^\infty(\mu)$ in $L^\infty(\mu \otimes \mu)$.

d) $k = \chi_{\{x < y\}}$ on $[0, 1]^2$ as in Example 3.1, because for any measurable subset U of $[0, 1]$, by symmetry $|(U \times U) \cap \{x < y\}| = |(U \times U) \cap \{y < x\}|$.

e) If μ is a σ-finite Borel measure on a Souslin space Σ (see [125, Chapter II] and Section 3.4.1 below), suppose that k is the *regular image* $f \circ T$ of a function f belonging $L^\infty(\mu) \otimes L^\infty(\mu)$.

Then k belongs to $L^1(\mu) \widehat{\otimes}_{\mathcal{E}^2} L^\infty(\mu)$ and the limit $\tilde{k} \circ J$ is independent μ-a.e. of a Lusin μ-filtration $\mathcal{E}$ provided that $T : \Sigma \times \Sigma \to \Sigma \times \Sigma$ is injective, Lusin $(\mu \otimes \mu)$-measurable and compatible with $\mathcal{E}^2$, see [125] and Section 3.4.1 below. Nevertheless, the subset $\{T(x,x) : x \in \Sigma\}$ of the Cartesian product $\Sigma \times \Sigma$ may be somewhat irregular.

The integral kernels of operators are often continuous functions, so we make a formal statement of the observation b) above. The following result addresses the fact that unless a compact Hausdorff space K is metrisable, the Borel σ-algebra $\mathcal{B}(K)$ of K need not be countably generated.

A *Radon measure* μ on a Hausdorff space is an inner regular and locally finite Borel measure [16, Definition 7.1.1].

Proposition 3.7. *Let μ be a σ-finite Radon measure on a Hausdorff space Σ. Let $k : \Sigma \times \Sigma \to \mathbb{C}$ be a continuous function. Then there exists a filtration $\mathcal{E}(k) = \langle \mathcal{E}_n \rangle_{n \in \mathbb{N}}$ based on partitions, such that k is $(\mu \otimes \mu)$-measurable with respect to the σ-algebra $\sigma(\mathcal{E}(k)) = \bigvee_{n=1}^\infty (\mathcal{E}_n \otimes \mathcal{E}_n)$ and $\mathbb{E}(k|\mathcal{E}_n \otimes \mathcal{E}_n) \circ J$ converges μ-a.e. to $k \circ J$.*

Proof. If $\Sigma = K$ is a compact Hausdorff space and $k = u \otimes v$ for continuous function $u, v \in C(K)$, then $\mathbb{E}(k|\mathcal{E}_n \otimes \mathcal{E}_n) \circ J = \mathbb{E}(u|\mathcal{E}_n).\mathbb{E}(v|\mathcal{E}_n)$ for each $n = 1, 2, \ldots$, according to formula (3.10). Because

$$\sup_{x,n} \mathbb{E}(|k - k'||\mathcal{E}_n \otimes \mathcal{E}_n)(x,x) \le \|k - k'\|_\infty$$

for each $k' \in C(K) \otimes C(K)$ and the Martingale Convergence Theorem ensures that $\lim_{n \to \infty} \mathbb{E}(k'|\mathcal{E}_n \otimes \mathcal{E}_n)(x,x) = \mathbb{E}(k'|\sigma(\mathcal{E}))(x,x)$ for μ-almost all $x \in K$, an application of the Stone-Weierstrass Theorem shows for any $m = 1, 2, \ldots$, there exists $k' \in C(K) \otimes C(K)$ such that $\|k - k'\|_\infty < 1/(2m)$ and

$$|k(x,x) - \mathbb{E}(k|\mathcal{E}_n \otimes \mathcal{E}_n)(x,x)| \le |k(x,x) - \mathbb{E}(k'|\sigma(\mathcal{E} \otimes \mathcal{E}))(x,x)| +$$
$$|\mathbb{E}(k'|\sigma(\mathcal{E}))(x,x) - \mathbb{E}(k'|\mathcal{E}_n \otimes \mathcal{E}_n)(x,x)| + \mathbb{E}(|k - k'||\mathcal{E}_n \otimes \mathcal{E}_n)(x,x)|.$$

We may choose a filtration $\mathcal{E}(k)$ such that k' is $\sigma(\mathcal{E}(k) \otimes \mathcal{E}(k))$-measurable for each such k' and then $\limsup_{n \to \infty} |k(x,x) - \mathbb{E}(k|\mathcal{E}_n \otimes \mathcal{E}_n)(x,x)| < 1/m$ for all $x \in K$ outside a μ-null set $N_m \subset K$, so $\lim_{n \to \infty} \mathbb{E}(k|\mathcal{E}_n \otimes \mathcal{E}_n)(x,x) = k(x,x)$ for all $x \in \bigcap_{m=1}^\infty (K \setminus N_m)$.

For the case of a σ-finite Radon measure on a Hausdorff space, the filtration $\mathcal{E}(k)$ is constructed on each set with finite measure and pieced together, so the intersection of the set of points $x \in \Sigma$ where $\lim_{n \to \infty} \mathbb{E}(k|\mathcal{E}_n \otimes \mathcal{E}_n)(x,x) \ne k(x,x)$ on each compact set is μ-null. $\square$

For a general Radon measure μ, the complete and locally determined measure space $(\Sigma, \mathcal{S}, \mu)$ [46, 411 H (b)] associated with the Radon measure μ is decomposable [46, 415A], so Proposition 3.7 is valid in this case if we allow uncountable partitions.

3.4.1 *Lusin filtrations*

Let $\mathcal{B}$ be a σ-algebra of subsets of a nonempty set Σ. We have already borrowed the following notion from probability theory: a *filtration* $\langle \mathcal{E}_t \rangle_{t \in I}$ is a family of sub-σ-algebras of $\mathcal{B}$ increasing with respect to the directed index set $I = \mathbb{N}$ or $I = \mathbb{R}_+$. The σ-algebra generated by a family $\mathcal{F}$ of sets belonging to the σ-algebra $\mathcal{B}$ is denoted by $\sigma(\mathcal{F})$. The relative σ-algebra $\{ B \cap E : B \in \mathcal{B} \}$ on a subset E of Σ is denoted by $\mathcal{B} \cap E$. The collection of all partitions of Σ is directed by refinement, that is, a partition $\mathcal{P}$ of Σ is *finer* than a partition $\mathcal{Q}$ if each element of $\mathcal{P}$ is a subset of some element of $\mathcal{Q}$.

The filtration $\mathcal{E} = \langle \mathcal{E}_n \rangle_{n \in \mathbb{N}}$ is assumed above to be constructed from an increasing sequence $\langle \mathcal{P}_n \rangle_{n \in \mathbb{N}}$ of countable partitions of Σ into measurable sets. The essential property of the filtration $\mathcal{E}$ is that there exists a natural regular conditional measure [16, Definition 10.4.1] $B \longmapsto \mathbb{E}(\chi_B | \mathcal{E}_n)$, $B \in \mathcal{B}$, for each $n = 1, 2, \dots$.

It is helpful to give a name to our earlier construction to see where it sits in descriptive set theory.

Definition 3.2. A *Lusin filtration* $\mathcal{E} = \langle \mathcal{E}_n \rangle_{n=1}^{\infty}$ of the measurable space $(\Sigma, \mathcal{B})$ is a filtration for which there exists an increasing sequence $\mathcal{P}_n^{\mathcal{E}}$, $n = 1, 2, \dots$, of countable partitions of Σ by sets in $\mathcal{B}$ with $\mathcal{E}_n = \sigma(\mathcal{P}_n^{\mathcal{E}})$ for each $n = 1, 2, \dots$ such that $\mathcal{B} = \bigvee_n \mathcal{E}_n$. A Lusin filtration $\mathcal{E}$ is called a *strict Lusin filtration* if, in addition,

  i) for each $x, y \in \Sigma$, $x \neq y$, there exist $n = 1, 2, \dots$ and $U, V \in \mathcal{P}_n^{\mathcal{E}}$ with $U \neq V$ such that $x \in U$ and $y \in V$ and
  ii) each decreasing sequence from $\bigcup_{n \in \mathbb{N}} \mathcal{P}_n^{\mathcal{E}}$ has nonempty intersection.

Given a Lusin filtration $\mathcal{E}$ of Σ, for every $n = 1, 2, \dots$, each element x of Σ belongs to a unique set $U_n(x)$ of the partition $\mathcal{P}_n^{\mathcal{E}}$ of Σ. The collection of sets $\mathcal{U}(x) = \{ U_n(x) : n \in \mathbb{N} \}$ is a neighbourhood base at $x \in \Sigma$ for the topology $\tau_{\mathcal{E}}$ on Σ.

As is usual in analysis, the possibility of discarding sets of measure zero needs to be accommodated. We achieved this earlier by discarding all points x for which $U_n(x)$ has measure zero for some $n = 1, 2, \ldots$.

Definition 3.3. Let $(\Sigma, \mathcal{B}, \mu)$ be a σ-finite measure space and $\mathcal{E} = \langle \mathcal{E}_n \rangle_{n=1}^{\infty}$ a filtration. Then $\mathcal{E}$ is called a *Lusin μ-filtration* if there exists a set Σ_0 with full μ-measure and an increasing sequence $\mathcal{P}_n^{\mathcal{E}}$, $n = 1, 2, \ldots$, of countable partitions of Σ_0 with $\mathcal{E}_n = \sigma(\mathcal{P}_n^{\mathcal{E}})$ for each $n = 1, 2, \ldots$, such that $\mathcal{B} \cap \Sigma_0 = \bigvee_n \mathcal{E}_n$ and $0 < \mu(U_n(x)) < \infty$ for each $x \in \Sigma_0$ and $n = 1, 2, \ldots$. A Lusin μ-filtration $\mathcal{E}$ is called a *strict Lusin μ-filtration* if, in addition,

   i) for each $x, y \in \Sigma_0$, $x \neq y$, there exist $n = 1, 2, \ldots$ and $U, V \in \mathcal{P}_n^{\mathcal{E}}$ with $U \neq V$ such that $x \in U$ and $y \in V$ and

   ii) each decreasing sequence from $\bigcup_{n \in \mathbb{N}} \mathcal{P}_n^{\mathcal{E}}$ has nonempty intersection.

Lemma 3.1. *A σ-finite measure space $(\Sigma, \mathcal{B}, \mu)$ has a Lusin μ-filtration if $\mathcal{B}$ is countably generated.*

Proof. Let $\mathcal{U} = \{U_1, U_2, \ldots\}$ be a countable subfamily of $\mathcal{B}$ such that $\mathcal{B} = \sigma(\mathcal{U})$. An increasing family of countable partitions $\mathcal{P}_n$, $n = 1, 2, \ldots$, is defined recursively by setting $\mathcal{P}_1$ equal to a partition of Σ into sets of finite μ-measure and

$$\mathcal{P}_{j+1} = \{P \cap U_j, P \setminus U_j : P \in \mathcal{P}_j\}$$

for $j = 1, 2, \ldots$. As before, set $\Sigma_0 = \Sigma \setminus \mathcal{N}$. $\qquad\qquad\square$

The terminology is associated with certain ideas from descriptive set theory. According to [125, Chapter II], a separable Hausdorff topological space is called *Polish* if its topology is metrisable and complete in the associated metric. The continuous *injective* image of a Polish space is said to be a *Lusin space*. A Hausdorff topological space is called a *Souslin space* if it is the continuous image of a Polish space.

A Lusin space Σ is characterised by the existence of a strict Lusin filtration, whose partitions form a *strict subdivision* in the terminology of L. Schwartz [125, Definition 5, p. 97]. By [125, Lemma 9, p. 98], associated with any strict Lusin filtration $\mathcal{E}$, there is a Polish space $(P, d_{\mathcal{E}})$, a Hausdorff topology $\tau_{\mathcal{E}}$ on Σ and a continuous injective map $j : P \to \Sigma$, so that $(\Sigma, \tau_{\mathcal{E}})$ is a Lusin space and $\mathcal{B}$ is the Borel σ-algebra of $(\Sigma, \tau_{\mathcal{E}})$. Conversely, an infinite Polish space is Borel isomorphic to either $\mathbb{R}$ or $\mathbb{N}$, so any Lusin space (Σ, τ) (or standard Borel space [16, Definition 6.2.10]) possesses a strict Lusin filtration obtained from one in $\mathbb{R}$ or $\mathbb{N}$. Moreover, by [125, Lemma 9, p. 98], the strict Lusin filtration $\mathcal{E}$ may be chosen so that the associated Hausdorff topology $\tau_{\mathcal{E}}$ is stronger than the Lusin topology τ of Σ.

The following observation follows from the results of [125, Chapter II].

Theorem 3.2. *Every σ-finite Borel measure μ on a Souslin space possesses a strict Lusin μ-filtration.*

Proof. A finite Borel measure on a Souslin space is a Radon measure [125, Theorem 10, p. 122] supported by the countable union of compact metrisable sets [125, Theorem 3, p. 96, Corollary 2, p. 106], which is itself a Lusin space [125, Corollary 2, p. 106]. By [125, Lemma 9, p. 98], each piece of a σ-finite Borel measure has a strict Lusin μ-filtration. $\square$

A Lusin μ-filtration $\mathcal{E}$ is associated with a natural topology $\tau_{\mathcal{E}}$ on Σ_0 for which $\mathcal{U}_{\mathcal{E}}(x) = \{U_n(x)\}_n$ is a neighbourhood base for $\tau_{\mathcal{E}}$ for $x \in \Sigma_0$ and Σ_0 supports μ. Let $\mathcal{U}_{\mathcal{E}'}(x)$ be a neighbourhood base at $x \in \Sigma_0$ for another Lusin μ-filtration $\mathcal{E}'$, enlarging $\mathcal{N}$ if necessary.

Suppose that there exist constants $c_1, c_2 > 0$ such that for every $x \in \Sigma_0$, the following two conditions hold:

(i) for every $U \in \mathcal{U}_{\mathcal{E}}(x)$, there exists $V \in \mathcal{U}_{\mathcal{E}'}(x)$ such that $U \subset V$ and $\mu(V) \leq c_1 \mu(U)$,

(ii) for every $V \in \mathcal{U}_{\mathcal{E}'}(x)$, there exists $U \in \mathcal{U}_{\mathcal{E}}(x)$ such that $V \subset U$ and $\mu(U) \leq c_2 \mu(V)$.

Then for each $f \in L^1(\mu \otimes \mu)$, the inequalities

$$c_2^2 M_{\mathcal{E}'^2}(f) \leq M_{\mathcal{E}^2}(f) \leq c_1^2 M_{\mathcal{E}'^2}(f)$$

hold on $\Sigma_0 \times \Sigma_0$ for the maximal functions defined by equation (3.11) with respect to either filtration $\mathcal{E}$ and $\mathcal{E}'$. Consequently, for any other filtration $\mathcal{E}'$ such that $\mathcal{E}$ and $\mathcal{E}'$ satisfy (i) and (ii), the maximal functions $M_{\mathcal{E}^2}(f)$ and $M_{\mathcal{E}'}(f)$ are equivalent and so $\mathfrak{C}_1(\mathcal{E}, X) = \mathfrak{C}_1(\mathcal{E}', X)$.

If we have a given strict Lusin μ-filtration $\mathcal{E}$, then the Hausdorff topological space $(\Sigma_0, \tau_{\mathcal{E}})$ is a Lusin space for which $\mathcal{B}$ is the associated Borel σ-algebra on Σ_0 [125, Theorem 5, p. 101], so there exists a metric $d_{\mathcal{E}}$ on Σ_0 whose topology is stronger than $\tau_{\mathcal{E}}$ and $(\Sigma_0, d_{\mathcal{E}})$ is complete and separable. Then $\mathcal{B}$ is also the Borel σ-algebra for the metric $d_{\mathcal{E}}$ [125, Corollary 2, p. 101]. In this case there also exists a metric maximal function $M_{d_{\mathcal{E}}}(f)$ defined by

$$M_{d_{\mathcal{E}}}(f)(x,y) = \sup_{r>0} \frac{\int_{B_r(x) \times B_r(y)} f \, d(\mu \otimes \mu)}{\mu(B_r(x))\mu(B_r(y))}, \quad x, y \in \Sigma_0,$$

comparable to the maximal function $M_{\mathcal{E}^2}$ if (i) and (ii) are satisfied with respect to the neighbourhood base of balls in the metric $d_{\mathcal{E}}$.

3.4.2   *Connection with other generalised traces*

An axiomatic treatement of traces on operator ideals is given in [108, 109] with recent updates in [110, 111]. The starting point is the Calkin theorem [110, Theorem 2.2] which asserts that the collection of all operator ideals on a separable Hilbert space $\mathcal{H}$ is in one-to-one correspondence with symmetric sequence ideals. The correspondence is obtained from the singular values of operators in the ideal.

A *trace* on an operator ideal $\mathfrak{U}(\mathcal{H})$ then corresponds to a unitarily invariant linear functional on $\mathfrak{U}(\mathcal{H})$ or, equivalently, a symmetric linear functional on the corresponding sequence ideal [110, Theorem 6.2]. A particular example that has assumed importance recently because of noncommutative geometry is the *Dixmier trace* defined on the Marcinkiewicz operator ideal. The Dixmier trace is an example of a singular trace because it vanishes on all finite rank operators, see [21] for example.

By contrast, for the purposes of this chapter, the emphasis with the Hardy-Littlewood maximal function approach to traces is on the *Banach lattice* of all absolute integral operators T on a Banach function space X, so that $T \geq 0$ implies $\int_\Sigma \langle T, dm \rangle \geq 0$—just what is needed in the proof of the Cwikel-Lieb-Rozenbljum inequality for dominated semigroups in Chapter 7. A result of D. Lewis [92] shows that on an infinite dimensional Hilbert space, the collection of all Hilbert-Schmidt operators is the only Banach operator ideal isomorphic to a Banach lattice, despite the observation that a symmetric sequence ideal is itself a Riesz space. For a choice of Banach limit $\omega \in (\ell^\infty)'$, the map

$$T \longmapsto \int_\Sigma \omega(\{\mathbb{E}(k|\mathcal{E}_n \otimes \mathcal{E}_n) \circ J\}_{n=1}^\infty) \, d\mu, \quad T \in \mathfrak{C}_1(\mathcal{E}, X),$$

is continuous and linear on $\mathfrak{C}_1(\mathcal{E}, X)$, so there may be many possible choices of a continuous trace on the whole Banach lattice $\mathfrak{C}_1(\mathcal{E}, X)$ depending on ω. As Proposition 3.5 indicates, all such choices of a continuous trace agree on the closed subspace $X \widehat{\otimes}_\mathcal{E} X^\times$ of the lattice ideal $\mathfrak{C}_1(\mathcal{E}, X)$ in $\mathcal{L}_r(X)$.

The *Selberg trace formula* also relates regularised traces (geometric information) to asymptotic estimates for eigenvalues (spectral information) of a Laplacian, see [9] for a survey of this deep subject.

3.5   Hermitian positive operators

Suppose that $(\Sigma, \mathcal{B}, \mu)$ is a σ-finite measure space. On a complex Hilbert space $\mathcal{H}$, we call a bounded linear operator $T : \mathcal{H} \to \mathcal{H}$ *hermitian positive* if $(Tu, u) \geq 0$ for all $u \in \mathcal{H}$. A trace condition ensures that an hermitian positive bounded linear operator on $L^2(\mu)$ is trace class.

Theorem 3.3. *Let $T_k : L^2(\mu) \to L^2(\mu)$ be an hermitian positive integral operator with kernel k and let $\mathcal{E} = \langle \mathcal{E}_n \rangle_{n \in \mathbb{N}}$ be a Lusin μ-filtration for which $\mathbb{E}(|k| | \mathcal{E}_n \otimes \mathcal{E}_n)$ has finite values for each $n = 1, 2, \ldots$. The operator T_k is trace class if and only if $\int_\Sigma M_{\mathcal{E}^2}(k)(x, x) \, d\mu(x) < \infty$. If T_k is trace class, then the formula*

$$\operatorname{tr}(T_k) = \int_\Sigma \tilde{k}(x, x) \, d\mu(x) \qquad (3.14)$$

holds with respect to the integral kernel $\tilde{k}$ of the operator T_k defined by

$$\tilde{k} = \lim_{n \to \infty} \mathbb{E}(k | \mathcal{E}_n \otimes \mathcal{E}_n),$$

wherever the limit exists. Moreover, $\int_\Sigma M_{\mathcal{E}^2}(k)(x, x) \, d\mu(x) \leq 4\operatorname{tr}(T_k)$.

Versions of Theorem 3.3 are well-known from the work of J. Weidmann [135], Gohberg-Krein [51], C. Brislawn [18,19] and [127, Theorem 2.12] for the continuous case. In these works, the term *integral operator* is replaced by *Hilbert-Schmidt operator*—where, at least locally, the integral kernel k belongs to $L^2(\mu \otimes \mu)$. Nuclear operators between L^p-spaces and Banach spaces have been studied recently along similar lines in [23, 24, 31–33, 44].

The assumption in Theorem 3.3 that T_k is just an integral operator with kernel k is a significant relaxation of the condition that $k \in L^2(\mu \otimes \mu)$, because we do not even assume that T_k is locally a *compact* linear operator—this conclusion is actually a consequence of Theorem 3.3 where the compactness of the bounded linear operator T_k follows from the trace condition $\int_\Sigma M_{\mathcal{E}^2}(k)(x, x) \, d\mu(x) < \infty$ and positivity on $L^2(\mu)$. The simple example of the Volterra integral operator described in Example 3.1 shows that the assumption that T_k is an *hermitian positive* operator cannot be omitted.

The essential ingredients of the proof of Theorem 3.3 below are the notion of a Lusin μ-filtration mentioned above, the Martingale Convergence Theorem [16, Theorem 10.3.13] and the non-commutative Fatou lemma [127, Theorem 2.7 (d)] asserting that the closed unit ball of $\mathcal{C}_1(\mathcal{H})$ is sequentially closed in the weak operator topology on $\mathcal{L}(\mathcal{H})$, given here for later reference.

Proposition 3.8. *Let $\mathcal{H}$ be a separable Hilbert space. If $A_n \in \mathcal{C}_1(\mathcal{H})$ converges weakly to an operator $A \in \mathcal{L}(\mathcal{H})$ as $n \to \infty$ and $\sup_n \|A_n\|_1 < \infty$, then $A \in \mathcal{C}_1(\mathcal{H})$ and $\|A\|_1 \leq \sup_n \|A_n\|_1$.*

Given an integral operator $T_k : L^2(\mu) \to L^2(\mu)$ with kernel k as in Theorem 3.3 and a Lusin μ-filtration $\mathcal{E}$, it is a simple matter [60, Lemma 7.4] to divide up the sets in the partition $\mathcal{P}_1^{\mathcal{E}}$ so that k is integrable on $U \times V$ for all sets U, V belonging to $\mathcal{P}_1^{\mathcal{E}}$—we shall assume this has already been done so that $k_n = \mathbb{E}(k|\mathcal{E}_n \otimes \mathcal{E}_n)$ is a finite function for each $n = 1, 2, \ldots$, as has been assumed in Theorem 3.3.

Proof of Theorem 3.3. The conditional expectation operator $E_n : f \longmapsto \mathbb{E}(f|\mathcal{E}_n)$ is a contraction on $L^2(\mu)$ given by

$$(E_n f)(x) = \mathbb{E}(f|\mathcal{E}_n)(x) = \sum_{U \in \mathcal{P}_n^{\mathcal{E}}} \chi_U(x) \frac{\int_U f \, d\mu}{\mu(U)},$$

so that E_n is an absolute integral operator with kernel

$$e_n(x, y) = \sum_{U \in \mathcal{P}_n^{\mathcal{E}}} \frac{\chi_U(x)\chi_U(y)}{\mu(U)}.$$

Hence $e_n(x, y) = e_n(y, x)$ for all $x, y \in \Sigma_0$, a set of full measure in Σ. The background of conditional expectation operators in probability theory is outlined in Section 4.1.1 below.

Applying the Fubini-Tonelli Theorem, for each $n = 1, 2, \ldots$, a glance at formula (3.10) shows that

$$(E_n T_k E_n f)(x) = \sum_{U, V \in \mathcal{P}_n^{\mathcal{E}}} \chi_V(x) \frac{\int_V \int_U k(x_2, x_1) \, d\mu(x_1) d\mu(x_2)}{\mu(U)\mu(V)} \int_U f \, d\mu$$

$$= (T_{k_n} f)(x)$$

for all $f \in L^2(\mu)$ and $x \in \Sigma_0$. The interchange of integrals is valid provided that k is integrable on sets $U \times V$ with $U, V \in \mathcal{P}_n^{\mathcal{E}}$, guaranteed by the assumption above about the Lusin μ-filtration $\mathcal{E}$. Consequently, $T_{k_n} = E_n T_k E_n$ is a bounded linear operator on $L^2(\mu)$ with norm bounded by the norm of T_k (compare [24, Theorem 3.9] in the Hilbert-Schmidt setting).

Let $\mathcal{P}_n^{\mathcal{E}} = \{U_{n,1}, U_{n,2}, \ldots\}$ be an enumeration of the partition $\mathcal{P}_n^{\mathcal{E}}$. If $\mathcal{P}_{n,m}^{\mathcal{E}}$ is its truncation to length m and $\mathcal{E}_{n,m} = \sigma(\mathcal{P}_{n,m}^{\mathcal{E}})$, then the conditional expectation operators $E_{n,m} : f \longmapsto \mathbb{E}(f|\mathcal{E}_{n,m})$ converge to E_n in the strong operator topology of $\mathcal{L}(L^2(\mu))$ as $m \to \infty$. Each operator $E_{n,m} T_k E_{n,m}$ has finite rank and

$$(E_{n,m} T_k E_{n,m} u, u) = (T_k E_{n,m} u, E_{n,m} u) \geq 0$$

for $k = 1, 2, \ldots$ and $u \in L^2(\mu)$. Because finite rank operators are trace class, the equality

$$\|E_{n,m} T_k E_{n,m}\|_1 = \operatorname{tr}(E_{n,m} T_k E_{n,m}) = \int_\Sigma k_{n,m}(x, x) \, d\mu(x)$$

holds for $k_{n,m} = \mathbb{E}(k|\mathcal{E}_{n,m}^2)$ by the elementary version of formula (3.14).

The first equality is valid because the finite rank operator $E_{n,m}T_k E_{n,m}$ is hermitian positive, so the trace norm of $E_{n,m}T_k E_{n,m}$ is just the sum of its eigenvalues.

It is worthwhile looking at the simple proof from linear algebra of formula (3.14) in the basic case of finite rank operators. The formula

$$k_{n,m} = \sum_{U,V \in \mathcal{P}_{n,m}^{\mathcal{E}}} \frac{\int_{U \times V} k \, d(\mu \otimes \mu)}{\mu(U)\mu(V)} \chi_{U \times V}$$

expresses the matrix of the finite rank operator $E_{n,m}T_k E_{n,m}$ as

$$\left\{ \frac{\int_{U \times V} k \, d(\mu \otimes \mu)}{\mu(U)^{\frac{1}{2}}\mu(V)^{\frac{1}{2}}} \right\}_{U,V \in \mathcal{P}_{n,m}^{\mathcal{E}}}$$

with respect to the finite orthonormal subset $\mathcal{O} = \{\mu(U)^{-\frac{1}{2}}\chi_U : U \in \mathcal{P}_{n,m}^{\mathcal{E}}\}$ of $L^2(\mu)$. The conditional expectation operator $E_{n,m}$ is the orthogonal projection onto the subspace spanned by $\mathcal{O}$. Then

$$\operatorname{tr}(E_{n,m}T_k E_{n,m}) = \sum_{U \in \mathcal{P}_{n,m}^{\mathcal{E}}} \frac{\int_{U \times U} k \, d(\mu \otimes \mu)}{\mu(U)}$$

$$= \int_{\Sigma} k_{n,m}(x,x) \, d\mu(x),$$

which is just formula (3.14) in this elementary setting. The representation

$$k_{n,m}(x,x) = \sum_{U \in \mathcal{P}_{n,m}^{\mathcal{E}}} \chi_U(x) \frac{\int_U \int_U k(x_2,x_1) \, d\mu(x_1) d\mu(x_2)}{\mu(U)^2}$$

$$= \sum_{U \in \mathcal{P}_{n,m}^{\mathcal{E}}} \chi_U(x) \mu(U)^{-2}(T_k \chi_U, \chi_U)$$

shows that $0 \le k_{n,m}(x,x) \le k_n(x,x) \le M_{\mathcal{E}^2}(k)(x,x)$ for all $x \in \Sigma_0$ and the inequality

$$\operatorname{tr}(E_{n,m}T_k E_{n,m}) \le \int_{\Sigma} M_{\mathcal{E}^2}(k)(x,x) \, d\mu(x)$$

holds for all $n,m = 1,2,\ldots$. Taking $m \to \infty$ and applying the non-commutative Fatou lemma, Proposition 3.8, each operator T_{k_n} is trace class and

$$\operatorname{tr}(T_{k_n}) \le \int_{\Sigma} M_{\mathcal{E}^2}(k)(x,x) \, d\mu(x).$$

Again $\operatorname{tr}(T_{k_n})$ is equal to the trace norm $\|T_{k_n}\|_1$ of T_{k_n} because each operator T_{k_n} is positive. By the Martingale Convergence Theorem, $E_n \to I$ in the strong operator topology of $\mathcal{L}(L^2(\mu))$, so $T_{k_n} \to T_k$ in the strong operator topology as $n \to \infty$. Another application of the non-commutative Fatou lemma shows that T_k is trace class and

$$\operatorname{tr}(T_k) \leq \int_\Sigma M_{\mathcal{E}^2}(k)(x,x)\, d\mu(x).$$

The operator T_k has the representation (3.2) and

$$M_{\mathcal{E}^2}(k) \circ J \leq \sum_{j=1}^\infty \lambda_j M_{\mathcal{E}}(\phi_j) M_{\mathcal{E}}(\psi_j)$$

and $\int_\Sigma M_{\mathcal{E}}(\phi_j) M_{\mathcal{E}}(\psi_j)\, d\mu \leq 4$ for $j = 1, 2, \ldots$ by the Cauchy-Schwartz inequality and the L^2-norm equivalence of $M_{\mathcal{E}}(f)$ and f the L^2-norm equivalence of $M_{\mathcal{E}}(f)$ and f [16, Corollary 10.3.11]. Then $\int_\Sigma M_{\mathcal{E}^2}(k)(x,x)\, d\mu(x) \leq 4\|T_k\|_1 = 4\operatorname{tr}(T_k)$ by the isometry between $\mathcal{C}_1(\mathcal{H})$ and $L^2(\mu)\widehat{\otimes}_\pi L^2(\mu)$ [107, Theorem 8.3.3].

Furthermore, by the Martingale Convergence Theorem and dominated convergence

$$\int_\Sigma \tilde{k} \circ J\, d\mu = \sum_{j=1}^\infty \lambda_j \int_\Sigma \lim_{n \to \infty} \mathbb{E}(\phi_j | \mathcal{E}_n).\mathbb{E}(\overline{\psi_j} | \mathcal{E}_n)\, d\mu$$

$$= \sum_{j=1}^\infty \lambda_j (\phi_j, \psi_j) = \operatorname{tr}(T_k).$$

$\square$

Remark 3.2. i) If the σ-algebra $\mathcal{B}$ is not itself countably generated, there may exist a smaller countably generated σ-algebra $\mathcal{B}_k$ for which the integral kernel k is $(\mathcal{B}_k \otimes \mathcal{B}_k)$-measurable. If not, then T_k is certainly not a trace class operator because a function with a representation (3.3) is measurable with respect to the σ-algebra generated by a countable family of measurable product functions. A Lusin μ-filtration may then be constructed for $(\Sigma, \mathcal{B}_k)$.

ii) In the case of a Radon measure μ on a Hausdorff space Σ, a *continuous* kernel k is the kernel of a Hilbert-Schmidt operator on each product $K \times K$ of a compact set K with itself (provided that K has positive measure), because k is uniformly bounded on $K \times K$. Then $\int_K \tilde{k}(x,x)\, d\mu = \int_K k(x,x)\, d\mu$ is equal to the trace of the positive operator $Q(K)T_k Q(K)$ by Proposition 3.7. Now we may appeal to the inner regularity of μ [127, Theorem 2.12].

Example 3.7. Suppose that $T : L^2([0,1]) \to L^2([0,1])$ is the operator of convolution with respect to a bounded periodic function $\varphi : \mathbb{R} \to \mathbb{C}$ with period one. Then

$$\int_0^1 M_{\mathcal{E}^2}(\varphi \circ u)(x,x)\, dx \le \|\varphi\|_\infty$$

for the mapping $u : (x,y) \longmapsto x - y$.

The Fourier transform $\hat{\ }: L^2([0,1]) \to \ell^2$ diagonalises the normal opera- tor T. By the Riemann-Lebesgue lemma the eigenvalues $\hat{\varphi}(n)$ of T satisfy $\hat{\varphi}(n) \to 0$ as $n \to \infty$, so T is necessarily compact. In fact, T is a Hilbert-Schmidt operator so $\hat{\varphi} \in \ell^2$. However, it is possible to choose a continuous φ for which $\hat{\varphi} \notin \ell^p$ for any $p < 2$, see [127, p. 24] and [22].

Chapter 4

Stochastic integration

In this Chapter, we describe how the approach to the stochastic integration of Banach space valued processes developed by J.A.M.van Neerven, M. Veraar and L. Weis [133] is related to bilinear integration as described in Definition 2.2.

So far, our treatment of bilinear integration has involved mainly functional analysis and measure theory. We start with a brief discussion of probability theory and stochastic processes. Although we do not need all the material found in [75], it provides an account of stochastic processes and stochastic integration in greater depth than provided here.

4.1 Background on probability and discrete processes

Probability theory uses the language of measure theory. Although probability and stochastic analysis is now part of the toolbag of most analysts, we give a brief résumé in this section.

A *probability measure space* $(\Omega, \mathcal{F}, P)$ is a measure space for which $P : \mathcal{F} \to [0,1]$ is a probability measure. Thus, $\mathcal{F}$ is a σ-algebra of subsets of Ω and P is a nonnegative σ-additive set function for which $P(\Omega) = 1$.

The set Ω is called the *sample space* and $\mathcal{F}$ is the collection of *events*, so that $P(A)$ is the probability that the event $A \in \mathcal{F}$ occurs. Some basic examples illustrate the ideas.

Example 4.1. i) Suppose that a fair coin is tossed n consecutive times for some $n = 1, 2, \ldots$. The possible outcomes is the set of all ordered n-tuples such as $HTHH \ldots TH$ for which H denotes an outcome of heads in a particular toss and T for tails. Representing H by 1 and T by 0, we can take the sample space Ω to be the Cartesian product $\{0,1\}^n = \{0,1\} \times \cdots \times \{0,1\}$ of the two-point set $\{0,1\}$ with itself n times.

The probability $P\{\omega\}$ of any particular outcome $\omega \in \Omega$ is $\frac{1}{2^n}$, so $\mathcal{F}$ is the collection of all subsets of Ω and

$$P(A) = \sum_{\omega \in A} \frac{1}{2^n} = \frac{\#A}{2^n}$$

for any subset A of Ω. The number of elements of the finite set A is denoted by $\#A$. The set function P is clearly σ-additive. Because Ω has 2^n elements, it follows that $P(\Omega) = 1$.

ii) Suppose that a fair coin is tossed infinitely many times. The possible outcomes is the set of all infinite sequences such as $HTHH\ldots$, so that we can take the sample space Ω to be the Cartesian product $\prod_{n=1}^{\infty}\{0,1\}$. Now the probability $P\{\omega\}$ of any particular outcome $\omega \in \Omega$ is zero. Here we take $\mathcal{F}$ to be the Borel σ-algebra of Ω (in the product topology, where $\{0,1\}$ has the discrete topology) and P is the product $\otimes_{n=1}^{\infty}\mu$ of the measures μ with $\mu(A) = \#A/2$ for $A \subset \{0,1\}$.

Let $(\Omega, \mathcal{F}, P)$ be a probability measure space. A *random variable* ξ : $\Omega \to \mathbb{R}$ is a function which is measurable with respect to the σ-algebras $\mathcal{F}$ on Ω and the Borel σ-algebra $\mathcal{B}(\mathbb{R})$ on $\mathbb{R}$. More accurately, it is the equivalence class $[\xi] \in L^0(P)$ of all measurable functions equal to ξ P-almost everywhere, as in Section 1.2, but we rarely need to make this distinction.

Notation: The set $\{\xi \in B\} = \{\omega : \xi(\omega) \in B\} = \xi^{-1}(B)$ is an element of the σ-algebra $\mathcal{F}$ for each $B \in \mathcal{B}(\mathbb{R})$.

The σ-algebra $\sigma(\xi) = \{\xi^{-1}(B) : B \in \mathcal{B}(\mathbb{R})\}$ is the smallest σ-algebra for which the random variable ξ is measurable. Similarly, if $\{\xi_i : i \in I\}$ is a collection of random variables, then $\sigma(\xi_i : i \in I)$ is the σ-algebra *generated* by $\{\xi_i : i \in I\}$, that is, the smallest σ-algebra containing all events $\{\xi_i \in B\}$ for $i \in I$ and $B \in \mathcal{B}(\mathbb{R})$.

The *distribution* P_ξ of a random variable $\xi : \Omega \to \mathbb{R}$ is the Borel measure defined by $P_\xi(B) = P\{\xi \in B\}$ for all $B \in \mathcal{B}(\mathbb{R})$. The *distribution function* F_ξ of ξ is defined by

$$F_\xi(x) = P\{\xi \leq x\}, \quad x \in \mathbb{R}.$$

Note: If $f : \mathbb{R} \to \mathbb{R}$ is a function, then

$$\int_{\mathbb{R}} f \, dF_\xi = \int_{\mathbb{R}} f \, dP_\xi = \int_{\Omega} f \circ \xi \, dP$$

in the sense that if one of the integrals exists, then they all do and are equal to each other.

A random variable ξ has an *absolutely continuous* distribution if there exists $f_\xi : \mathbb{R} \to [0, \infty)$ such that

$$P\{\xi \in B\} = \int_B f_\xi(x)\, dx, \quad B \in \mathcal{B}(\mathbb{R}).$$

Then f_ξ is called the *density* of ξ. By the *Radon-Nikodym Theorem* [60, Theorem 19.23], this is the same as saying that the distribution P_ξ of ξ is zero on Borel sets of Lebesgue measure zero in $\mathbb{R}$.

The *expectation* of an integrable random variable ξ is

$$\mathbb{E}\xi = \int_\Omega \xi\, dP.$$

If ξ is square integrable, then its *variance* is defined by $\operatorname{var}\xi = \int_\Omega (\xi - \mathbb{E}\xi)^2\, dP$.

Example 4.2. A random variable $X : \Omega \to \mathbb{R}$ is said to be *normally distributed* if there exist $\mu \in \mathbb{R}$ and $\sigma > 0$ such that

$$P_X(B) = (2\pi\sigma)^{-1/2} \int_B e^{-(x-\mu)^2/(2\sigma)}\, dx$$

for all $B \in \mathcal{B}(\mathbb{R})$. Then

$$
\begin{aligned}
\mathbb{E}X &= \int_\Omega X\, dP \\
&= \int_\mathbb{R} x\, dP_X \quad \text{by the change of variables formula,} \\
&= (2\pi\sigma)^{-1/2} \int_\mathbb{R} x e^{-(x-\mu)^2/(2\sigma)}\, dx \\
&= \mu(2\pi\sigma)^{-1/2} \int_\mathbb{R} e^{-(x-\mu)^2/(2\sigma)}\, dx + (2\pi\sigma)^{-1/2} \int_\mathbb{R} t e^{-t^2/(2\sigma)}\, dt \\
&= \mu.
\end{aligned}
$$

Similarly, $\operatorname{var}X = \sigma$.

We shall also have occasion to consider random vectors $\xi : \Omega \to \mathbb{R}^d$, for which we can write $\xi = (\xi_1, \dots, \xi_d)$ for real valued random variables $\xi_1, \dots, \xi_d$. Then

$$P_{\xi_1, \dots, \xi_d}(B) = P\{(\xi_1, \dots, \xi_d) \in B\}, \quad B \in \mathcal{B}(\mathbb{R}^d),$$

is the *joint distribution* of the random variables $\xi_1, \dots, \xi_d$. It is a Borel probability measure on $\mathbb{R}^d$.

Sometimes we consider random variables ξ with a *discrete distribution* so that we can write $P_\xi = \sum_{j=1}^{\infty} c_j \delta_{x_j}$, where $c_j \geq 0$, $\sum_{j=1}^{\infty} c_j = 1$ and $c_j = P\{\xi = x_j\}$ is the *mass* at x_j. The Dirac measure at x_j has been denoted by δ_{x_j}. Then

$$P\{\xi \in B\} = \sum_{x_j \in B} P\{\xi = x_j\}, \quad B \in \mathcal{B}(\mathbb{R}^d).$$

4.1.1 *Conditional probability and expectation*

Let $A, B \in \mathcal{F}$ be two events with $P(B) \neq 0$. The *conditional probability* $P(A|B)$ of A given B is defined by the formula

$$P(A|B) = \frac{P(A \cap B)}{P(B)}.$$

This idea recurs frequently. The event B gives us extra information with which to judge the probability of the occurrence of the event A.

Hence it makes sense to call two sets $A, B \in \mathcal{F}$ *independent* if

$$P(A \cap B) = P(A)P(B).$$

A finite collection of events $A_1, \ldots, A_n \in \mathcal{F}$ is called *independent* if

$$P(A_{j_1} \cap \cdots \cap A_{j_k}) = P(A_{j_1}) \cdots P(A_{j_k})$$

for any subset $\{j_1, \ldots, j_k\}$ of indices $\{1, \ldots, n\}$.

Two random variables ξ, η are *independent* if the sets $\{\xi \in A\}$ and $\{\eta \in B\}$ are independent for each $A, B \in \mathcal{B}(\mathbb{R})$. Similarly, the random variables $\xi_1, \ldots, \xi_n$ are *independent* if for all $B_1, \ldots, B_n \in \mathcal{B}(\mathbb{R})$, the events $\{\xi \in B_1\}, \ldots, \{\xi \in B_n\}$ are independent. A collection of random variables is independent if any finite subfamily is.

Example 4.3. Suppose that a fair coin is tossed infinitely many times with 1 representing heads and 0 representing tails at each toss. Let $\xi_n : \Omega \to \{0, 1\}$ be the random variable representing the outcome of the nth toss for $n = 1, 2, \ldots$. Then

$$P(\{\xi_{j_1} = n_1\} \cap \cdots \cap \{\xi_{j_k} = n_k\}) = \frac{1}{2^k} = P\{\xi_{j_1} = n_1\} \cdots P\{\xi_{j_k} = n_k\}$$

$$\tag{4.1}$$

for all distinct indices $j_1, \ldots, j_k = 1, 2, \ldots$, all $n_1, \ldots, n_k \in \{0, 1\}$ and all $k = 1, 2, \ldots$. The infinite collection $\{\xi_n : n = 1, 2, \ldots\}$ of random variables is therefore independent.

On a technical note, as mentioned above, P is the infinite product $\otimes_{n=1}^{\infty} \mu$ of probability measures. This means that P is the additive set function such that $P(E)$ is defined by Equation (4.1) for any event

$$E = \{\xi_{j_1} = n_1\} \cap \cdots \cap \{\xi_{j_k} = n_k\}.$$

Some work needs to be done to prove that P is σ-additive on the algebra generated by the sets E [60, Theorem 22.8, p. 433].

Two integrable random variables ξ, η are *uncorrelated* if $\mathbb{E}(\xi\eta) = \mathbb{E}(\xi)\mathbb{E}(\eta)$. Independent integrable random variables are necessarily uncorrelated.

Two σ-algebras $\mathcal{G}$, $\mathcal{H}$ of sets belonging to $\mathcal{F}$ (sub-σ-algebras of $\mathcal{F}$) are *independent* if for every $A \in \mathcal{G}$ and $B \in \mathcal{H}$, the events A and B are independent. A random variable ξ is *independent* of a σ-algebra $\mathcal{G}$ if $\sigma(\xi)$ and $\mathcal{G}$ are independent σ-algebras. A similar definition applies for n random variables $\xi_1, \ldots, \xi_n$ being independent of m σ-algebras $\mathcal{G}_1, \ldots, \mathcal{G}_m$.

The existence of the conditional expectation of an integrable random variable with respect to a sub-σ-algebra is an immediate consequence of the Radon-Nikodym Theorem [60, Theorem 19.23]. First we look at the significance of the concept in some simple examples.

Let $(\Omega, \mathcal{F}, P)$ be a probability measure space and let $B \in \mathcal{F}$ be an event that occurs with nonzero probability. The *conditional expectation* $\mathbb{E}(\xi|B)$ of a random variable $\xi \in L^1(P)$ given B is the number

$$\mathbb{E}(\xi|B) = \frac{\int_B \xi \, dP}{P(B)}.$$

The conditional expectation $\mathbb{E}(\xi|B)$ is just the expected value of $\xi : B \to \mathbb{R}$ with respect to the probability $P' = P/P(B)$ defined on subsets of B.

Suppose that $\xi \in L^1(P)$ and η is a discrete random variable with values belonging to the set $\{y_1, y_2, \ldots\}$. The *conditional expectation* $\mathbb{E}(\xi|\eta)$ of ξ given the random variable η is the random variable $\mathbb{E}(\xi|\eta) : \Omega \to \mathbb{R}$ defined by

$$\mathbb{E}(\xi|\eta)(\omega) = \mathbb{E}(\xi|\{\eta = y_j\}), \quad \text{if } \eta(\omega) = y_j,$$

for $\omega \in \Omega$ and $j = 1, 2, \ldots$.

If instead we think of $\mathbb{E}(\xi|\eta)$ as $\mathbb{E}(\xi|\sigma(\eta))$, the conditional expectation with respect to the σ-algebra generated by η, then we arrive at the following generalisation. Let $(\Omega, \mathcal{F}, P)$ be a probability measure space and let $\mathcal{G}$ be a σ-algebra contained in $\mathcal{F}$ and $\xi \in L^1(P)$. The *conditional expectation* $\mathbb{E}(\xi|\mathcal{G})$ of ξ given $\mathcal{G}$ is a random variable such that

1) $\mathbb{E}(\xi|\mathcal{G})$ is $\mathcal{G}$-measurable and

2) $\int_A \mathbb{E}(\xi|\mathcal{G})\, dP = \int_A \xi\, dP$ for all $A \in \mathcal{G}$.

The measure $\mu : \mathcal{G} \to \mathbb{R}$ defined by $\mu(A) = \int_A \xi\, dP$ for all $A \in \mathcal{G}$ is absolutely continuous with respect to P on $\mathcal{G}$, that is, $\mu(A) = 0$ for all $A \in \mathcal{G}$ such that $P(A) = 0$, so by the Radon Nikodym theorem, such a random variable $\mathbb{E}(\xi|\mathcal{G})$ exists and any two versions are equal P-a.e.

If η is a discrete random variable, then $\mathbb{E}(\xi|\eta)$ satisfies conditions 1) and 2) above with respect to the σ-algebra $\mathcal{G} = \sigma(\eta)$. Hence, $\mathbb{E}(\xi|\sigma(\eta)) = \mathbb{E}(\xi|\eta)$.

Theorem 4.1. (Properties of conditional expectation)

Let $(\Omega, \mathcal{F}, P)$ be a probability measure space and let $\mathcal{G}$ be a σ-algebra contained in $\mathcal{F}$. Up to a set of P-measure zero, we have

Linearity
1) *for each $\xi, \eta \in L^1(P)$ and $a, b \in \mathbb{R}$,*

$$\mathbb{E}(a\xi + b\zeta|\mathcal{G}) = a\mathbb{E}(\xi|\mathcal{G}) + b\mathbb{E}(\zeta|\mathcal{G});$$

Preservation of expectation
2) $\mathbb{E}(\mathbb{E}(\xi|\mathcal{G})) = \mathbb{E}(\xi)$, *for each $\xi \in L^1(P)$;*
The random variable ξ is "known"
3) $\mathbb{E}(\xi\zeta|\mathcal{G}) = \xi\mathbb{E}(\zeta|\mathcal{G})$, *for each $\mathcal{G}$-measurable ξ with $\xi, \zeta \in L^1(P)$;*
Independence
4) $\mathbb{E}(\xi|\mathcal{G}) = \mathbb{E}(\xi)$, *if $\xi \in L^1(P)$ is independent of $\mathcal{G}$;*
Tower property
5) $\mathbb{E}(\mathbb{E}(\xi|\mathcal{G})|\mathcal{H}) = \mathbb{E}(\xi|\mathcal{H})$, *for each $\xi \in L^1(P)$ and σ-algebra $\mathcal{H} \subset \mathcal{G}$;*
Positivity
6) $\mathbb{E}(\xi|\mathcal{G}) \geq 0$ *for each $\xi \in L^1(P)$ with $\xi \geq 0$ a.e.*

The mapping $\xi \mapsto \mathbb{E}(\xi|\mathcal{G})$ is a selfadjoint projection on $L^2(P)$. We have already appealed to this general property in Section 3.5 for the elementary case of the σ-algebra associated with a countable partition of sets, where it is simple to verify directly.

4.1.2 Discrete Martingales

Let $(\Omega, \mathcal{F}, P)$ be a probability space. As is usual in probability theory, we shall often fail to mention the underlying probability space. A *filtration* on Ω is a sequence $\mathcal{F}_1 \subset \mathcal{F}_2 \subset \cdots \subset \mathcal{F}$ of σ-algebras. A sequence $\xi_1, \xi_2, \ldots$ of random variables is *adapted* to a filtration $\mathcal{F}_1, \mathcal{F}_2, \ldots$, if ξ_n is $\mathcal{F}_n$-measurable for each $n = 1, 2, \ldots$.

Example 4.4. Let $\xi_1, \xi_2, \ldots$ be random variables and $\mathcal{F}_n = \sigma(\xi_1, \ldots, \xi_n)$, $n = 1, 2, \ldots$. Then $\mathcal{F}_1, \mathcal{F}_2, \ldots$ is a filtration to which the random variables $\xi_1, \xi_2, \ldots$ are adapted.

Let $\mathcal{F}_1, \mathcal{F}_2, \ldots$ be a filtration. A sequence $\xi_1, \xi_2, \ldots$ of random variables is called a *martingale* with respect to the filtration $\mathcal{F}_1, \mathcal{F}_2, \ldots$ if

1) ξ_n is integrable for each $n = 1, 2, \ldots$;
2) the sequence $\xi_1, \xi_2, \ldots$ of random variables is *adapted* to $\mathcal{F}_1, \mathcal{F}_2, \ldots$;
3) $\mathbb{E}(\xi_{n+1}|\mathcal{F}_n) = \xi_n$ almost surely for each $n = 1, 2, \ldots$.

The sequence $\xi_1, \xi_2, \ldots$ of random variables is called a *supermartingale* if instead of 3) we have

$$\mathbb{E}(\xi_{n+1}|\mathcal{F}_n) \leq \xi_n, \quad \text{a.s.}$$

for each $n = 1, 2, \ldots$ and a *submartingale* if

$$\mathbb{E}(\xi_{n+1}|\mathcal{F}_n) \geq \xi_n, \quad \text{a.s.}$$

for each $n = 1, 2, \ldots$.

Notice that from 3) and properties of conditional expectation, we have

$$\mathbb{E}(\xi_{n+1}) = \mathbb{E}(\mathbb{E}(\xi_{n+1}|\mathcal{F}_n)) = \mathbb{E}(\xi_n), \quad n = 1, 2, \ldots,$$

so that for a martingale $\xi_1, \xi_2, \ldots$, the expectation $\mathbb{E}(\xi_n) = \mathbb{E}(\xi_1)$ of ξ_n, $n = 1, 2, \ldots$, is necessarily constant.

Example 4.5. i) Let $\eta_1, \eta_2, \ldots$ be independent integrable random variables with zero mean and set $\xi_n = \eta_1 + \cdots + \eta_n$ and $\mathcal{F}_n = \sigma(\eta_1, \ldots, \eta_n)$ for each $n = 1, 2, \ldots$. Then $\xi_1, \xi_2, \ldots$ is adapted to $\mathcal{F}_1, \mathcal{F}_2, \ldots$ and

$$\mathbb{E}(|\xi_n|) \leq \mathbb{E}(|\eta_1|) + \cdots + \mathbb{E}(|\eta_n|) < \infty$$

for each $n = 1, 2, \ldots$. To see that $\xi_1, \xi_2, \ldots$ is a martingale, we note that

$$\mathbb{E}(\xi_{n+1}|\mathcal{F}_n) = \mathbb{E}(\eta_{n+1}|\mathcal{F}_n) + \mathbb{E}(\xi_n|\mathcal{F}_n)$$

$$= \mathbb{E}(\eta_{n+1}|\mathcal{F}_n) + \xi_n \quad (\text{because } \xi_n \text{ is } \mathcal{F}_n\text{-measurable})$$

$$= \mathbb{E}(\eta_{n+1}) + \xi_n$$

$$\qquad (\text{because } \eta_{n+1} \text{ is independent of } \xi_j \text{ for } 1 \leq j \leq n)$$

$$= \xi_n \quad (\text{because } \eta_{n+1} \text{ has zero mean}).$$

ii) Let $\xi \in L^1(P)$ and let $\mathcal{F}_1, \mathcal{F}_2, \ldots$ be any filtration. If we set $\xi_n = \mathbb{E}(\xi | \mathcal{F}_n)$ for $n = 1, 2, \ldots$, then $\xi_1, \xi_2, \ldots$ is a martingale.

iii) Let P be Lebesgue measure on $[0, 1)$ and let $\xi \in L^1(P)$. Let Φ_n be the partition $[(k - 1)/2^n, k/2^n)$, $k = 1, \ldots, 2^n$ of $[0, 1)$ into disjoint intervals and let $\mathcal{F}_n$ be the algebra generated by Φ_n for each $n = 1, 2, \ldots$. Then

$$\xi_n = 2^{-n} \sum_{A \in \Phi_n} \chi_A \int_A \xi \, dP, \quad n = 1, 2, \ldots,$$

is a martingale with respect to $\mathcal{F}_1, \mathcal{F}_2, \ldots$. Moreover, $\xi_n \to \xi$ in $L^1(P)$ as $n \to \infty$. Martingales of this type were used in Section 3.5 above.

Gambling. The most compelling way to view a discrete parameter martingale is as a mathematical model of gambling over an infinite number of rounds of a game.

Let $\eta_1, \eta_2, \ldots$ be integrable random variables where η_n represents the *winning or losses per unit stake* in the nth round. Then $\xi_n = \eta_1 + \cdots + \eta_n$ will be the total winnings or losses per unit stake after n rounds, $n = 1, 2, \ldots$.

As we saw in Example i) above, if $\eta_1, \eta_2, \ldots$ are independent with zero mean, then $\xi_1, \xi_2, \ldots$ is a martingale. We would like to allow the possibility that the winnings or losses per unit stake η_n depend on the preceding outcomes $\eta_1, \eta_2, \ldots, \eta_{n-1}$, possibly by using some gambling strategy that depends on what happened before the nth bet.

Let $\mathcal{F}_1, \mathcal{F}_2, \ldots$ be the filtration defined by $\mathcal{F}_n = \sigma(\eta_1, \ldots, \eta_n)$ for $n = 1, 2, \ldots$. Set $\mathcal{F}_0 = \{\emptyset, \Omega\}$ and $\xi_0 = 0$. For each $n = 1, 2, \ldots$, the σ-algebra $\mathcal{F}_{n-1}$ represents the *accumulated knowledge about the game* after $n - 1$ rounds. Then

Martingale:	$\mathbb{E}(\xi_n | \mathcal{F}_{n-1}) = \xi_{n-1}$ a.s.	*fair game;*
Submartingale:	$\mathbb{E}(\xi_n | \mathcal{F}_{n-1}) \geq \xi_{n-1}$ a.s.	*favourable to you;*
Supermartingale:	$\mathbb{E}(\xi_n | \mathcal{F}_{n-1}) \leq \xi_{n-1}$ a.s.	*unfavourable to you.*

So, for a submartingale say, for almost all *trials* $\omega \in \Omega$, the expected total outcome $\mathbb{E}(\xi_n | \mathcal{F}_{n-1})(\omega)$ after the nth round, given accumulated knowledge $\mathcal{F}_{n-1}$ about the preceding rounds, exceeds the total outcome $\xi_{n-1}(\omega)$ after the $(n-1)$th round. In this sense, the game is favourable to the player because using knowledge of the preceding outcomes increases the players winnings.

It is natural to call $\alpha_1, \alpha_2, \ldots$ a *gambling strategy*, if for each $n = 1, 2, \ldots$, the random variable α_n is $\mathcal{F}_{n-1}$-measurable, that is, for a given trial $\omega \in \Omega$, a bet of size $\alpha_n(\omega)$ is made in the nth round, depending on the accumulated knowledge $\mathcal{F}_{n-1}$ gained in the preceding rounds.

So, let $\alpha_1, \alpha_2, \ldots$ be a gambling strategy and let

$$\zeta_n = \alpha_1 \eta_1 + \cdots + \alpha_n \eta_n$$
$$= \alpha_1 (\xi_1 - \xi_0) + \cdots + \alpha_n (\xi_n - \xi_{n-1})$$

be the *total winnings* at the nth round of the game. This finite sum is, in fact, a *stochastic integral* of the type we shall consider in greater generality later on.

Now suppose that we only have a finite sum K of money to bet within any round: $|\alpha_n| \leq K$ for $n = 1, 2, \ldots$. Then appealing to the properties of conditional expectation, we have

$$\mathbb{E}(\zeta_n | \mathcal{F}_{n-1}) = \mathbb{E}(\zeta_{n-1} + \alpha_n(\xi_n - \xi_{n-1}) | \mathcal{F}_{n-1})$$
$$= \zeta_{n-1} + \alpha_n \mathbb{E}((\xi_n - \xi_{n-1}) | \mathcal{F}_{n-1})$$
$$= \zeta_{n-1} + \alpha_n \left(\mathbb{E}(\xi_n | \mathcal{F}_{n-1}) - \xi_{n-1} \right)$$

because ζ_{n-1}, α_n and ξ_{n-1} are all $\mathcal{F}_{n-1}$-measurable. Consequently, if $\xi_1, \xi_2, \ldots$ is a martingale (respectively, a submartingale or supermartingale), then so is $\zeta_1, \zeta_2, \ldots$. This interpretation underlies our consideration of stochastic integration later.

Although mathematical finance has spurred on much development in stochastic modelling, recent history has demonstrated that the concealment of past information in, say, complex financial derivatives can lead to catastrophic failures in financial markets. At this time of life, the author prefers rational central planning as a means of organising complex social systems, despite having fallen out of favour in current political thinking.

4.1.3 *Discrete stopping times*

Let $\mathcal{F}_1, \mathcal{F}_2, \ldots$ be the filtration. A **stopping time** $\tau : \Omega \to \{1, 2, \ldots\} \cup \{\infty\}$ (with repect to the filtration $\mathcal{F}_1, \mathcal{F}_2, \ldots$) is a random variable such that $\{\tau = n\} \in \mathcal{F}_n$ for each $n = 1, 2, \ldots$.

Example 4.6. Suppose that we toss a coin with stakes \$1, starting with \$5 and stopping when we win \$10 or lose the lot. Then the number τ of tosses at which we stop is a stopping time.

To see this, let ξ_n be the total winnings at the nth toss of the coin. Then $\tau = \min\{n : \xi_n = 0 \text{ or } 10\}$ is the **first hitting time** of the set $\{0, 10\}$ and we have

$$\{\tau = n\} = \{0 < \xi_1 < 10\} \cap \cdots \cap \{0 < \xi_{n-1} < 10\} \cap \{\xi_n = 0 \text{ or } 10\} \in \mathcal{F}_n.$$

The random variable ξ_τ defined by $\xi_\tau(\omega) = \xi_{\tau(\omega)}(\omega)$ for $\omega \in \Omega$ has values 0 or 10 with equal probability.

If τ is a stopping time and $\xi_1, \xi_2, \ldots$ is a sequence of random variables adapted to the filtration $\mathcal{F}_1, \mathcal{F}_2, \ldots$, then the **sequence** $\xi_{\tau \wedge n}$, $n = 1, 2, \ldots$, **stopped at** τ is given by

$$\xi_{\tau \wedge n}(\omega) = \xi_{\tau(\omega) \wedge n}(\omega), \qquad \text{for all } \omega \in \Omega \text{ and } n = 1, 2, \ldots.$$

Then $\xi_{\tau \wedge n}$, $n = 1, 2, \ldots$ is also adapted to the filtration $\mathcal{F}_1, \mathcal{F}_2, \ldots$. Moreover, if $\xi_1, \xi_2, \ldots$ is a martingale (resp. submartingale, supermartingale), then so is the stopped sequence $\xi_{\tau \wedge n}$, $n = 1, 2, \ldots$, as may be seen by defining the gambling strategy

$$\alpha_n = \begin{cases} 1 \text{ if } \tau \geq n \\ 0 \text{ if } \tau < n, \end{cases}$$

and observing that $\xi_{\tau \wedge n} = \alpha_1(\xi_1 - \xi_0) + \cdots + \alpha_n(\xi_n - \xi_{n-1})$ for $n = 1, 2, \ldots$. Because $\xi_{\tau \wedge n}$, $n = 1, 2, \ldots$, is a martingale, we necessarily have

$$\mathbb{E}(\xi_{\tau \wedge n}) = \mathbb{E}(\xi_{\tau \wedge 1}) = \mathbb{E}(\xi_1), \quad \text{for all } n = 1, 2, \ldots.$$

Suppose that $P\{\tau < \infty\} = 1$ so that $\tau \wedge n \nearrow \tau$ a.s. as $n \to \infty$. Taking $n \to \infty$, we would hope that the expected value $\mathbb{E}(\xi_\tau)$ of the random variable ξ_τ defined by $\xi_\tau(\omega) = \xi_{\tau(\omega)}(\omega)$ for all $\omega \in \Omega$ is equal to the expected value $\mathbb{E}(\xi_1)$ of the initial winnings. The following example illustrates that some care needs to be exercised.

Example 4.7 (St. Petersburg lottery). Suppose that a coin is tossed and initially \$1 is bet on heads. After that, the strategy is to double the stakes if tails occurs or quit with a win.

Let $\eta_j = 1$ for a heads and $\eta_j = -1$ for a tails on the jth toss. The gambling strategy is then

$$\alpha_n = \begin{cases} 2^{n-1} \text{ if } \eta_1 = \cdots = \eta_{n-1} = -1, \\ 0 \qquad \text{otherwise} \end{cases}$$

and $\tau = \min\{n : \eta_n = 1\}$ is the associated stopping time.

If we set $\zeta_n = \eta_1 + 2\eta_2 + \cdots + 2^{n-1}\eta_n$ for $n = 1, 2, \ldots$, then $\zeta_{\tau \wedge n}$ is the winnings after the nth round and $\zeta_{\tau \wedge n}$, $n = 1, 2, \ldots$, is a martingale. We know that $P(\tau < \infty) = 1$, because the probability of all tails is zero. But $\zeta_\tau = 1$ because $-1 - 2 - \cdots - 2^{n-1} + 2^n = 1$ and $\mathbb{E}(\zeta_{\tau \wedge n}) = \mathbb{E}(\zeta_1) = 0$ for all $n = 1, 2, \ldots$, so that

$$1 = \mathbb{E}(\zeta_\tau) \neq \lim_{n \to \infty} \mathbb{E}(\zeta_{\tau \wedge n}) = \mathbb{E}(\zeta_1) = 0.$$

The problem is that although the martingale $\zeta_{\tau \wedge n}$, $n = 1, 2, \ldots$, converges almost surely to ζ_τ as $n \to \infty$ (because $P(\tau < \infty) = 1$), it does not converge in $L^1(P)$ because the stakes are increasing exponentially fast.

Theorem 4.2 (Optional Stopping Theorem). *Let ξ_n, $n = 1, 2, \ldots$ be a martingale and τ a stopping time with respect to a filtration $\mathcal{F}_1, \mathcal{F}_2, \ldots$, such that*

i) $\tau < \infty$ *a.s.,*
ii) ξ_τ *is integrable and*
iii) $\mathbb{E}(\xi_n 1_{\{\tau > n\}}) \to 0$ *as $n \to \infty$.*

Then $\mathbb{E}(\xi_\tau) = \mathbb{E}(\xi_1)$.

Proof. Writing $\xi_\tau = \xi_{\tau \wedge n} + (\xi_\tau - \xi_n) 1_{\{\tau > n\}}$, we see that

$$\mathbb{E}(\xi_\tau) = \mathbb{E}(\xi_{\tau \wedge n}) + \mathbb{E}\left(\xi_\tau 1_{\{\tau > n\}}\right) - \mathbb{E}\left(\xi_n 1_{\{\tau > n\}}\right).$$

The last term approaches zero as $n \to \infty$ by condition iii). The equality $\mathbb{E}(\xi_{\tau \wedge n}) = \mathbb{E}(\xi_1)$ holds for each $n = 1, 2, \ldots$, because $\xi_{\tau \wedge n}$, $n = 1, 2, \ldots$, is a martingale.

By assumption i), we have $1_{\{\tau > n\}} \searrow 0$ a.s., so by monotone convergence and assumption ii), it follows that $\mathbb{E}\left(\xi_\tau 1_{\{\tau > n\}}\right) \to 0$ as $n \to \infty$. Hence $\mathbb{E}(\xi_\tau) = \mathbb{E}(\xi_1)$.

4.2 Stochastic processes

From now on, we suppose that $(\Omega, \mathcal{F}, P)$ is a *complete probability space*. This means that: Ω is a set, $\mathcal{F}$ a σ-algebra of sets of Ω, P a probability measure on Ω such that each subset of a P-null set also belongs to $\mathcal{F}$. Recall that if $(\Omega, \mathcal{F}, P)$ is not complete, we may extend $\mathcal{F}$ to $\overline{\mathcal{F}}$ by including all subsets of null sets, and define P on $\overline{\mathcal{F}}$ in the obvious way, so that $(\Omega, \overline{\mathcal{F}}, P)$ is complete.

Suppose Γ is an index set, $\{X_\gamma : \gamma \in \Gamma\}$ a family of random variables. The σ-algebra $\sigma\{X_\gamma : \gamma \in \Gamma\}$ is the smallest σ-algebra of subsets of Ω such that X_γ is measurable for all $\gamma \in \Gamma$. This is the σ-algebra *generated* by the random variables X_γ, $\gamma \in \Gamma$.

A *filtration* is a family $\{\mathcal{F}_t : t \in \mathbb{R}_+\}$ of sub-σ-algebras of $\mathcal{F}$ such that $\mathcal{F}_s \subseteq \mathcal{F}_t$, for all $s < t$.

A filtration $\{\mathcal{F}_t : t \in \mathbb{R}_+\}$ is called a *standard filtration* if

(1) $\mathcal{F}_t = \mathcal{F}_{t+} := \cap_{s>t}\mathcal{F}_s$ for all $t > 0$ (*right continuity*);
(2) $\mathcal{F}_0$ contains all the P-null sets   (*completeness*).

Example 4.8. Let $\mathcal{F}_n$ be generated by $\left\{\left[\dfrac{j}{2^n}, \dfrac{j+1}{2^n}\right] : j \in \mathbb{N}\right\}$ and null sets. Set $\mathcal{F}_t = \mathcal{F}_{[t]}$. This is complete but not right continuous.

A (d-dimensional) *stochastic process* is a function $X : I \times \Omega \longrightarrow \mathbb{R}^d$ where I is an interval in $\mathbb{R}^+$ and for every $t \in I$, the function $X(t, \cdot)$ is measurable.

We say X is *measurable* if X is $\mathcal{B}(I) \otimes \mathcal{F}$-measurable. If $0 \in I$, then we say $x \in \mathbb{R}^d$ that is the *initial value* of X if $X(0, \cdot) = x$ a.s. We denote it by $\{X_t\}$ where $X_t = X(t, \cdot)$.

Given an increasing family $\{\mathcal{F}_t : t \in I\}$ of σ-algebras the process is *adapted* to $\mathcal{F}_t$ if X_t is $\mathcal{F}_t$-measurable for all $t \in I$.

A stochastic process X is said to be *left continuous* (respectively, *right continuous*) if the function $t \longmapsto X(t, \omega)$ is left continuous (respectively, right continuous) on I for every $\omega \in \Omega$.

Two stochastic processes $X = \{X_t\}$ and $Y = \{Y_t\}$ are *versions* of each other if

$$P(X_t = Y_t) = 1 \quad \text{for all } t \in I.$$

They are *indistinguishable* if

$$P(X_t = Y_t \quad \text{for all } t \in I) = 1.$$

If X_t and Y_t are indistinguishable, they must be versions of each other, but versions are not necessarily indistinguishable.

Remark. If X_t and Y_t are continuous versions of each other, they are indistinguishable. Almost always we work with $I = \mathbb{R}_+$ and $d = 1$.

4.3 Brownian motion

The strange spontaneous movements of small particles suspended in liquid were first studied by Robert Brown, an English botanist, in 1828, although they had apparently been noticed much earlier by other scientists. L. Bachelier gave the first mathematical description of the phenomenon in 1900, going so far as to note the Markov property of the process.

In 1905 A. Einstein and, independently and around the same time, M. V. Smoluchowski developed physical theories of Brownian motion which could be used, for example, to determine molecular diameters. The mathematical theory of Brownian motion was invented in 1923 by N. Wiener, and accordingly the Brownian motion that we will be working with is frequently called the Wiener process.

A process $B = \{B_t : t \in \mathbb{R}_+\}$ is called a *Brownian motion* in $\mathbb{R}$ if

(i) For $0 \leq s < t < \infty$, $B_t - B_s$ is a normally distributed random variable with mean zero and variance $t - s$.

Recall that X is *normally distributed* with variance σ and mean zero if
$$P(X \in A) = \frac{1}{\sqrt{2\pi\sigma}} \int_A e^{-\frac{|x|^2}{2\sigma}} dx.$$

(ii) For all $0 \leq t_0 < t_1 \ldots < t_\ell < \infty$ and $\ell = 1, 2, \ldots$, the set
$$\{B_{t_0}; B_{t_k} - B_{t_{k-1}}; \quad k = 1, \ldots, \ell\}$$
is a finite collection of independent random variables.

Note that by (i) $B_t - B_s$ depends only on $t - s$ (temporal homogeneity). A *Brownian motion* in $\mathbb{R}^d$ is a d-tuple
$$B = \{B_t\} = \{B_t^1, B_t^2 \ldots B_t^d : t \in \mathbb{R}_+\}$$
where each $\{B_t^i\}$ is a Brownian motion on $\mathbb{R}$ and the B^i's are independent.

We say B *starts at* x if $B_0 = x$ a.s.

If $B_0 = x$ a.s., then
$$P_t(x, A) = P(B_t \in A) = \frac{1}{(2\pi t)^{d/2}} \int_A e^{-\frac{|x-y|^2}{2t}} dy$$
for $t > 0, x \in \mathbb{R}^d$ and $A \in \mathcal{B}(\mathbb{R}^d)$. It follows that
$$P_t(x_0 + x, x_0 + A) = P_t(x, A), \quad \text{for all } x_0 \in \mathbb{R}^d \quad \text{(spatial homogeneity)}.$$

Property (ii) is called *independence of increments* of B. A more exact idea of the modulus of continuity of a typical Brownian path is given by the following theorem of A. Khintchine and P. Lévy

Theorem 4.3 (Local law of the iterated logarithm). *For each $t \geq 0$,*
$$P\left\{\omega : \limsup_{h \to 0+} \frac{B_{t+h}(\omega) - B_t(\omega)}{\sqrt{2h \log \log\left(\frac{1}{h}\right)}} = 1\right\} = 1.$$

Corollary 4.1. *Let $0 < \epsilon < \frac{1}{2}$. For P-almost all $\omega \in \Omega$, the following condition holds: there exists $C_\epsilon > 0$ such that for each $t > 0$, there exists $\delta_{t,\epsilon} > 0$ such that*
$$|B_u(\omega) - B_t(\omega)| \leq C_\epsilon |u - t|^\epsilon, \qquad \text{for all } |u - t| < \delta_{t,\epsilon}.$$

Canonical construction of Brownian motion (Wiener's construction)

Let $\Omega = C([0,1])$ and $\mathcal{B}(C[0,1])$ be the σ-algebra generated by the cylinders sets $\{\omega \in \Omega : \omega(t_1) \in B_1, \ldots, \omega(t_n) \in B_n\}$ for given $B_1, \ldots, B_n \in \mathcal{B}(\mathbb{R})$, $0 \le t_1 < t_2 < \ldots < t_n \le 1$ and $n = 1, 2, \ldots$.

Consider the stochastic process $(W_t)_{t \in [0,1]}$ on Ω such that

(1) $W_t(\omega) = \omega(t)$
(2) $W_0 = 0$ a.e.
(3) W_t is $\mathcal{F}_t$-measurable and $W_t - W_s$ is independent of $\mathcal{F}_s$ for every $0 < s < t$.
(4) $P\{W_t - W_s \in A\} = [2\pi(t - s)]^{-1/2} \int_A \exp(\frac{-x^2}{2(t-s)})\, dx$.

This is called the *canonical 1-dimensional Brownian motion* and the measure P is called *Wiener measure*. The work here goes into proving that P is actually σ-additive on $\mathcal{B}(C([0,1]))$.

We may take $\mathcal{F}_t$ to be σ-algebra $\{W_s : s \le t\}$ duly completed. Wiener proved that the Wiener measure exists and may be constructed as a limit of induced measures of random walks.

4.3.1 *Some properties of Brownian paths*

The ensemble of Brownian travellers moves according to subtle and remarkable principles. Some feeling for the nature of the process may be obtained from consideration of the following facts.

4.3.1.1 *Unbounded variation*

Let B be standard Brownian motion. For almost all ω, the sample paths $t \to B_t(\omega)$ are of unbounded variation on any interval, see Theorem 4.3.

4.3.1.2 *Non-differentiability*

An immediate consequence of the local law of the iterated logarithm, Theorem 4.3, is that for a fixed t, and almost all ω, $B_t(\omega)$ is not differentiable at t. It can be proved that in fact $B_t(\omega)$ is nowhere differentiable for almost all ω. (And thus, in the sense of Wiener measure P on $C([0, \infty))$, 'almost all' continuous functions are nowhere differentiable.)

4.3.1.3 *Law of the iterated logarithm*

$$P\left\{\omega : \limsup_{t\to\infty} \frac{B_t(\omega)}{\sqrt{2t\log\log t}} = 1\right\} = 1$$

(A. Khintchine).

This shows that the Brownian traveller wanders off to infinity, and gives an idea of the speed with which she does so. However, she still returns to any neighbourhood of the origin infinitely many times.

4.4 Stochastic integration of vector valued processes

As mentioned above, the total winnings of a gambling game is a type of stochastic integral. For continuous time processes, a stochastic integral is a random process of the form

$$\int Y_t dX_t. \tag{4.2}$$

Unfortunately, the most important examples of processes like a Brownian motion process do not have the property that $t \longmapsto X_t(\omega)$ has finite variation for almost all sample points ω, so the interpretation of the 'integral' above is problematic from the point of view of measure theory.

Rather than repeat the accepted theory of stochastic integration of random processes that may be found in the monograph [75], we shall interpret the stochastic integral (4.2) as a type of singular bilinear integral of a random process Y with respect to a random measure dX_t. Our approach has the advantage of being able to treat processes Y with values in a Banach space E that has the necessary geometric property (UMD spaces).

Let $(\Sigma, \mathcal{S}, \mu)$, $(\Omega, \mathcal{E}, \mathbb{P})$ be probability measure spaces. Let $L^0(\Omega, \mathcal{E}, \mathbb{P})$ be the space of real valued random variables with respect to $\mathbb{P}$ equipped with convergence in $\mathbb{P}$-measure. A mapping

$$W : \mathcal{S} \to L^0(\Omega, \mathcal{E}, \mathbb{P})$$

is called a *Gaussian random measure* on $(\Sigma, \mathcal{S}, \mu)$ if

(a) for all disjoint sets $A_1, A_2, \ldots$ from $\mathcal{S}$, we have

$$W\Big(\bigcup_{n=1}^{\infty} A_n\Big) = \sum_{n=1}^{\infty} W(A_n),$$

where the sum converges with probability one;

(b) the random variables $W(A_1), \ldots, W(A_n)$ are independent for all disjoint sets $A_1, \ldots, A_n$ belonging to $\mathcal{S}$ and all $n = 1, 2, \ldots$;

(c) for every $A \in \mathcal{E}$, the random variable $W(A)$ has a normal distribution with mean zero and variance $\mu(A)$.

Example 4.9. Let $\Sigma = \mathbb{R}_+$ with μ Lebesgue measure on $\mathbb{R}_+$. If W_t is Wiener process, then we write $W((s, t]) = W_t - W_s$ for $0 \le s < t$. For each $t > 0$, let $\mathcal{F}_t$ be the σ-algebra generated by the random variables $\{W_s\}_{0 \le s \le t}$.

According to property (i) of Brownian motion, the random variable $W((s, t])$ is Gaussian with mean zero and variance $t - s$. Property (ii) of Brownian motion establishes that the random variables $W(A_1), \ldots, W(A_n)$ are independent for all disjoint half-open intervals $A_1, \ldots, A_n$ and all $n = 1, 2, \ldots$.

For any finite union A of half-open intervals, $W(A)$ is defined by additivity and it can be checked directly that $W(A)$ is a Gaussian random variable with mean zero and variance $\mu(A)$. The extension of W to the δ-ring of all Borel sets with finite Lebesgue measure is facilitated by the following lemma concerning Gaussian vectors.

Lemma 4.1 ([117], Lemma 2.1). *Let X_n be symmetric Gaussian random vectors such that $X_n \to X$ in probability $\mathbb{P}$. Then X is a symmetric Gaussian random vector and, for every $0 < p < \infty$, $\mathbb{E}\|X_n - X\|^p \to 0$ as $n \to \infty$.*

Thus, a Gaussian random measure W is a vector measure in any of the spaces $L^p(\Omega, \mathcal{E}, \mathbb{P})$ for $0 \le p < \infty$.

Let E be a Banach space and $1 \le p < \infty$. The Banach space of all equivalence classes of strongly $\mathbb{P}$-measurable functions $f : \Omega \to E$ for which $\int_\Omega \|f\|^p \, d\mathbb{P} < \infty$ is denoted by $L^p(\Omega, \mathcal{E}, \mathbb{P}; E)$ or just $L^p(\mathbb{P}; E)$. The norm is given by

$$f \longmapsto \left(\int_\Omega \|f\|^p \, d\mathbb{P} \right)^{1/p}, \quad f \in L^p(\mathbb{P}; E).$$

The relative topology of $L^p(\mathbb{P}; E)$ on $L^p(\Omega, \mathcal{E}, \mathbb{P}) \otimes E$ is *completely separated*, so we may consider the integral of an E-valued function $f : \Sigma \to E$ with respect to W in $L^p(\mathbb{P}; E)$, in the sense of Definition 2.2. Then for each $\xi \in E'$, the scalar function $\langle f, \xi \rangle : \sigma \mapsto \langle f(\sigma), \xi \rangle$, $\sigma \in \Sigma$, is W-integrable and by the Itô isometry [75], the random variable $\int_\Sigma \langle f, \xi \rangle \, dW$ is Gaussian with variance $\int_\Sigma |\langle f, \xi \rangle|^2 \, d\mu$.

By [117, Corollary 4.2], an E-valued function $f : \Sigma \to E$, integrable with respect to W in $L^p(\mathbb{P}; E)$ as in Definition 2.2 is necessarily integrable with respect to the Gaussian random measure W in the sense of [117, Definition 2.1]. The converse statement follows from the proof of [117, Theorem 4.1] and [117, Corollary 2.1].

It follows from the lemma above, that if $f : \Sigma \to E$ is integrable with respect to W in $L^p(\mathbb{P}; E)$ for some $1 \leq p < \infty$, then it is integrable with respect to W in $L^p(\mathbb{P}; E)$ for every $1 \leq p < \infty$. A function $f : \Sigma \to E$ is integrable with respect to W in $L^p(\mathbb{P}; E)$ for some $1 \leq p < \infty$, if and only if it is strongly measurable and stochastically integrable with respect to W as in [132, Definition 2.1]. It follows from [117, Proposition 6.1] that W has finite E-semivariation in $L^p(\mathbb{P}; E)$ if and only if E is a Banach space of type 2. The domain of the function f just considered does not depend on sample points $\omega \in \Omega$, that is, f is not a *random process*, so the integral

$$\int_\Sigma f dW = \int_\Sigma f \otimes dW \in E \hat{\otimes}_p L^p(\mathbb{P}) \equiv L^p(\mathbb{P}; E)$$

is a decoupled bilinear integral of the type considered in Chapter 2.

We shall now see that when f is a *progressively measurable* random E-valued process, it is possible to *decouple* the stochastic integral $\int_\Sigma f dW$ in the manner described in Chapter 2 by using special features of the Gaussian random measure W. At the same time, we can deal with vector valued random processes f for free. Our purpose is to emphasise connections with other areas of analysis featuring singular bilinear integrals, so we do not strive for the utmost generality in dealing with stochastic integration. The treatment in [75] suffices for practical applications such as financial mathematics and stochastic modelling.

Let $(\Omega', \mathcal{E}', \mathbb{P}')$ be another probability measure space. Then strongly measurable functions $\varphi : \Sigma \to L^p(\mathbb{P}'; E)$ may also be integrated with respect to W in the space $L^p(\mathbb{P}' \otimes \mathbb{P}; E)$, in the sense of Definition 2.2.

If $\varphi : \Sigma \to L^p(\mathbb{P}'; E)$ is strongly measurable, then an appeal to [117, Corollary 4.2] shows that φ is W-integrable in $L^p(\mathbb{P}' \otimes \mathbb{P}, E)$, if and only if the E-valued function $t \longmapsto \langle \varphi(t), g \rangle$, $t \in \Sigma$, is W-integrable in $L^p(\mathbb{P}; E)$ for every $g \in L^q(\mathbb{P}')$ and for every set $A \in \mathcal{S}$, there exists $(\varphi \otimes W)(A) \in L^p(\mathbb{P}' \otimes \mathbb{P}, E)$ such that $\langle (\varphi \otimes W)(A), g \rangle = \int_A \langle \varphi(t), g \rangle \, dW(t)$ $\mathbb{P}$-a.e. in $L^p(\mathbb{P}; E)$ for each $g \in L^q(\mathbb{P}')$. Here $1/p + 1/q = 1$ and $\langle \cdot, g \rangle$ denotes integration against g in the $\mathbb{P}'$-variable.

Roughly speaking, if we apply the pointwise multiplication map to the element $(\varphi \otimes W)(A)$ of $L^p(\mathbb{P} \otimes \mathbb{P}, E)$, then we obtain the stochastic integral of a strongly measurable function $\varphi : \Sigma \to L^p(\mathbb{P}; E)$ with respect to the Gaussian random measure W.

For example, if $L^2(\mathbb{P}, E)\hat{\otimes}_\pi L^2(\mathbb{P})$ denotes the projective tensor product of $L^2(\mathbb{P}, E)$ with $L^2(\mathbb{P})$ [88, 41.2], then an application of the Cauchy-Schwarz inequality shows that there exists a unique continuous linear map

$$J : L^2(\mathbb{P}, E)\hat{\otimes}_\pi L^2(\mathbb{P}) \to L^1(\mathbb{P}, E)$$

such that for every $f, g \in L^2(\mathbb{P})$ and $x \in E$, the equality $J((x.g) \otimes f)(\omega) = x.g(\omega)f(\omega)$ holds for almost every $\omega \in \Omega$, that is, J is the pointwise multiplication map that restricts a product function to the diagonal of the Cartesian product $\Omega \times \Omega$ of the sample space Ω with itself—a situation reminiscent of taking the trace of a linear operator as considered in Chapter 3.

Requiring that the bilinear integral $(\varphi \otimes W)(A)$ belongs to the projective tensor product $L^2(\mathbb{P}, E)\hat{\otimes}_\pi L^2(\mathbb{P})$ is a restrictive assumption. As we saw in Chapter 3, 'generalised traces' of bounded linear operators still make sense when we move away from projective tensor products into certain Banach function spaces.

It is remarkable that under mild assumption on the Banach space E, the multiplication map J is continuous for the relative topology of $L^2(\mathbb{P} \otimes \mathbb{P}, E)$ into $L^2(\mathbb{P}, E)$, provided it is restricted to a certain class of definite integrals $(\varphi \otimes W)(A)$ with respect to a Gaussian random measure W. The remainder of this chapter is devoted to the proof of this simple but powerful observation.

Let W be a Gaussian random measure on $\mathbb{R}_+$. For each $t > 0$, let $\mathcal{F}_t$ be the σ-algebra generated by the random variables $\{W((0, s])\}_{0 \le s \le t}$.

A function $\phi : \mathbb{R}_+ \to L^2(\mathbb{P}) \otimes E$ is said to be an *elementary progressively measurable function* if there exist times $0 < t_1 < \cdots < t_N$, vectors $x_{mn} \in E$ and sets $A_{mn} \in \mathcal{F}_{t_{n-1}}$, $n = 1, \ldots, N$, $m = 1, \ldots, M$ such that

$$\phi(t) = \sum_{n=1}^{N} \sum_{m=1}^{M} x_{mn} \chi_{A_{mn}} \cdot \chi_{(t_{n-1}, t_n]}(t), \quad t \in \mathbb{R}_+.$$

Then ϕ has values in every space $L^p(\mathbb{P}) \otimes E$ for $1 \le p \le \infty$, ϕ is W-integrable in $L^p(\mathbb{P}) \otimes E \otimes L^p(\mathbb{P})$ for every $1 \le p < \infty$ and we have

$$\int_{\mathbb{R}_+} \phi \otimes dW = \sum_{n=1}^{N} \sum_{m=1}^{M} (x_{mn}\chi_{A_{mn}}) \otimes (W((t_{n-1}, t_n])). \qquad (4.3)$$

Let X denote the linear subspace of $L^\infty(\mathbb{P}) \otimes E \otimes L^p(\mathbb{P})$ consisting of all vectors $\int_{\mathbb{R}_+} \phi \otimes dW$ with $\phi : \mathbb{R}_+ \to L^\infty(\mathbb{P}) \otimes E$ an elementary progressively measurable function. For each $1 \leq p < \infty$, let $J : L^\infty(\mathbb{P}) \otimes E \otimes L^p(\mathbb{P}) \to L^p(\mathbb{P}, E)$ be the linear map defined by $J(g \otimes x \otimes f)(\omega) = xg(\omega).f(\omega)$ for almost all $\omega \in \Omega$.

Definition 4.1. A Banach space E is called a *UMD space* (or, E has the *unconditional martingale difference property*) if for any $1 < p < \infty$, there exists $C_p > 0$ such that for any E-valued martingale difference $\{\xi_j\}_{j=1}^n$ and $n = 1, 2, \ldots$, the inequality

$$\mathbb{E} \left\| \sum_{j=1}^n \epsilon_j \xi_j \right\|_E^p \leq C_p \mathbb{E} \left\| \sum_{j=1}^n \xi_j \right\|_E^p$$

holds for every $\epsilon_j \in \{\pm 1\}$, $j = 1, \ldots, n$.

The following result is from [50, Theorems 2 and 2']. The simplified proof presented below comes from [133, Lemma 3.4]. It is given here to spell out the connection with bilinear integration in tensor products.

Theorem 4.4. *Let E be a UMD space and $1 < p < \infty$. The multiplication map J is continuous from X into $L^p(\mathbb{P}, E)$ for the relative topology of the Banach space $L^p(\mathbb{P} \otimes \mathbb{P}, E)$ on X.*

Proof. Suppose that $\int_{\mathbb{R}_+} \phi \otimes dW$ is given by formula (4.3). For each $n = 1, \ldots, N$, let $\xi_n = \sum_{m=1}^M (x_{mn} \chi_{A_{mn}})$ and define

$$d_n = (\xi_n.W((t_{n-1}, t_n])) \otimes 1$$
$$e_n = \xi_n \otimes W((t_{n-1}, t_n]).$$

Then $\sum_{n=1}^N e_n = \int_{\mathbb{R}_+} \phi \otimes dW$ and $\sum_{n=1}^N d_n = J\left(\int_{\mathbb{R}_+} \phi \otimes dW\right) \otimes 1$.

For $n = 1, \ldots, N$, let $r_{2n-1} := \frac{1}{2}(d_n + e_n)$ and $r_{2n} := \frac{1}{2}(d_n - e_n)$. We now show that $\{r_j\}_{j=1}^{2N}$ is a martingale difference sequence with respect to the filtration $\{\mathcal{G}_j\}_{j=1}^{2N}$, where

$$\mathcal{G}_{2n} = \sigma(\mathcal{F}_{t_n} \otimes \mathcal{F}_{t_n})$$

$$\mathcal{G}_{2n-1} = \sigma\Big(\mathcal{F}_{t_{n-1}} \otimes \mathcal{F}_{t_{n-1}}, (W((t_{n-1}, t_n])) \otimes 1 +$$

$$1 \otimes (W((t_{n-1}, t_n])) \Big).$$

Clearly r_j is $\mathcal{G}_{2k}$-measurable for $j = 1, \ldots, 2k$ and $k = 1, 2, \ldots, N$. Denote the expectation with respect to $\mathbb{P} \otimes \mathbb{P}$ by $\mathbb{E}_{\mathbb{P} \otimes \mathbb{P}}$. Since ξ_n is $\mathcal{F}_{t_{n-1}}$-measurable, we have

$$\mathbb{E}_{\mathbb{P} \otimes \mathbb{P}}(d_n + e_n | \mathcal{G}_{2n-2}) = \mathbb{E}_{\mathbb{P} \otimes \mathbb{P}}(d_n | \mathcal{F}_{t_{n-1}} \otimes \mathcal{F}_{t_{n-1}}) + \mathbb{E}_{\mathbb{P} \otimes \mathbb{P}}(e_n | \mathcal{F}_{t_{n-1}} \otimes \mathcal{F}_{t_{n-1}}) = 0,$$

so that $\mathbb{E}_{\mathbb{P}\otimes\mathbb{P}}(r_{2n-1}|\mathcal{G}_{2n-2}) = 0$. On the other hand,

$$d_n + e_n = (\xi_n \otimes 1).\big((W(t_n) - W(t_{n-1})) \otimes 1 + 1 \otimes (W(t_n) - W(t_{n-1}))\big),$$

so r_{2n-1} is $\mathcal{G}_{2n-1}$-measurable and we have

$$\begin{aligned}
2\mathbb{E}_{\mathbb{P}\otimes\mathbb{P}}(r_{2n}|\mathcal{G}_{2n-1}) &= \mathbb{E}_{\mathbb{P}\otimes\mathbb{P}}(d_n - e_n|\mathcal{G}_{2n-1}) \\
&= (\xi_n \otimes 1).\mathbb{E}_{\mathbb{P}\otimes\mathbb{P}}\big(W((t_{n-1}, t_n])\big) \otimes 1 \\
&\qquad -1 \otimes \big(W((t_{n-1}, t_n])\big)|\mathcal{G}_{2n-1}) \\
&= 0.
\end{aligned}$$

The last equality follows from the observation that

$$f = \big(W((t_{n-1}, t_n])\big) \otimes 1$$

and

$$g = 1 \otimes \big(W((t_{n-1}, t_n])\big)$$

are i.i.d. random variables independent of $\mathcal{F}_{t_{n-1}} \otimes \mathcal{F}_{t_{n-1}}$ and $\mathcal{G}_{2n-1} = \sigma\big(\mathcal{F}_{t_{n-1}} \otimes \mathcal{F}_{t_{n-1}}, f + g\big)$. Consequently,

$$\begin{aligned}
\mathbb{E}_{\mathbb{P}\otimes\mathbb{P}}(f - g|\mathcal{G}_{2n-1}) &= \mathbb{E}_{\mathbb{P}\otimes\mathbb{P}}(f|\mathcal{G}_{2n-1}) - \mathbb{E}_{\mathbb{P}\otimes\mathbb{P}}(g|\mathcal{G}_{2n-1}) \\
&= \frac{1}{2}(f + g) - \frac{1}{2}(f + g) = 0,
\end{aligned}$$

and it follows that $\{r_j\}_{j=1}^{2N}$ is a martingale difference sequence.

Because $\sum_{n=1}^{N} d_n = \sum_{j=1}^{2N} r_j$ and $\sum_{n=1}^{N} e_n = \sum_{j=1}^{2N} (-1)^{j+1} r_j$, the inequalities

$$C_p^{-1} \mathbb{E}_{\mathbb{P}\otimes\mathbb{P}} \left\| \sum_{n=1}^{N} e_n \right\|_E^p \leq \mathbb{E}_{\mathbb{P}\otimes\mathbb{P}} \left\| \sum_{n=1}^{N} d_n \right\|_E^p \leq C_p \mathbb{E}_{\mathbb{P}\otimes\mathbb{P}} \left\| \sum_{n=1}^{N} e_n \right\|_E^p$$

follow by appealing to the UMD property, so that

$$\begin{aligned}
C_p^{-1} \mathbb{E}_{\mathbb{P}\otimes\mathbb{P}} \left\| \int_{\mathbb{R}_+} \phi \otimes dW \right\|_E^p &\leq \mathbb{E} \left\| J \left(\int_{\mathbb{R}_+} \phi \otimes dW \right) \right\|_E^p \\
&\leq C_p \mathbb{E}_{\mathbb{P}\otimes\mathbb{P}} \left\| \int_{\mathbb{R}_+} \phi \otimes dW \right\|_E^p.
\end{aligned}$$

Hence, $J : X \to L^p(\mathbb{P}, E)$ is continuous for the relative topology of $L^p(\mathbb{P} \otimes \mathbb{P}, E)$ on the linear space

$$X = \left\{ \int_{\mathbb{R}_+} \phi \otimes dW : \phi \text{ elementary progressively measurable} \right\}.$$

$\square$

Consequently, if $\phi : \mathbb{R}_+ \to L^p(\mathbb{P}, E)$ is the limit a.e. of elementary progressively measurable functions ϕ_n, $n = 1, 2, \ldots$, such that

$$\left\{ \int_A \phi_n \otimes dW \right\}_{n=1}^{\infty}$$

converges in $L^p(\mathbb{P} \otimes \mathbb{P}, E)$ for each $A \in \mathcal{B}(\mathbb{R}_+)$, then ϕ is W-integrable in $L^p(\mathbb{P} \otimes \mathbb{P}, E)$ in the sense of Definition 2.2. Moreover, if E is a UMD Banach space, then for $t > 0$, the stochastic integral of ϕ with respect to W on the interval $(0, t]$ may be represented by $J\left(\int_{(0,t]} \phi \otimes dW \right)$ and it belongs to the space $L^p(\mathbb{P}, E)$.

We leave our discussion of stochastic integration at this point. The treatment in [133] is aimed at applications to stochastic partial differential equations and L^p-regularity. In [75], the scalar case $E = \mathbb{C}$ is treated with the Gaussian random measure W replaced by a *continuous semimartingale* X. Here the L^2 theory is applied utilising a judicious choice of stopping times and the Itô calculus is obtained.

Chapter 5

Scattering theory

Scattering theory can be considered as a part of the more general perturbation theory in physics [76]. The main idea is that detailed information about an unperturbed selfadjoint operator H_0 in a Hilbert space $\mathcal{H}$ enables us to draw conclusions about another selfadjoint operator H provided that H_0 and H differ little from one to another in an appropriate sense. One of the earliest successes of abstract scattering theory dealt was with the case where $H - H_0$ is a trace class operator acting on the Hilbert space $\mathcal{H}$. We return to this topic in Section 8.4 below where we consider Krein's spectral shift function.

In the mathematical formulation of quantum mechanics, H_0 is the free Hamiltonian and $H = H_0 + V$ is the total Hamiltonian associated with the interaction potential V obtained from the model in classical mechanics that we seek to quantise. The operator sum '$H_0 + V$' is usually interpreted as a form sum so that H is a selfadjoint operator acting in $\mathcal{H}$. The spectrum $\sigma(H) \subset \mathbb{R}$ of H must be bounded below, otherwise particle interactions may need to be taken into account.

We briefly discuss in the following sections the two main approaches to the mathematical formulation of quantum mechanical scattering theory—the time-dependent and the time-independent or stationary scattering theory.

5.1 Time-dependent scattering theory

In the time-dependent scattering theory, we consider the time evolution of an incident particle (wave packet) under the influence of the interaction with a scattering centre or with another particle by the evolution equation

$$i\frac{\partial u}{\partial t} = Hu, \quad u(0) = f.$$

The behaviour of the state u for large times is studied in terms of the free equation $i\partial u_0/\partial t = H_0 u_0$. With the appropriate assumptions on the potential V, for every vector f orthogonal to the eigenvectors of H, there exist vectors $f_0^{(\pm)}$ orthogonal to the eigenvectors of the free Hamiltonian H_0 such that

$$\lim_{t\to\pm\infty} \|u(t) - u_0(t)\| = 0,$$

if $u_0(0) = f_0^{(\pm)}$. The initial data f and $f_0^{(\pm)}$ are related by the equality

$$f = \lim_{t\to\pm\infty} e^{itH} e^{-itH_0} f_0^{(\pm)}$$

because $u(t) = e^{-itH} f$ and $u_0(t) = e^{-itH_0} f_0^{(\pm)}$ for $\pm t \geq 0$. The *wave operators*

$$W_\pm \doteq W_\pm(H, H_0) = \lim_{t\to\pm\infty} e^{itH} e^{-itH_0} P_0^{(a)} \qquad (5.1)$$

encode this property provided that the limits in the strong operator topology exist. The operator $P_0^{(a)}$ is the orthogonal projection onto the absolutely continuous subspace $\mathcal{H}_0^{(a)}$ of the operator H_0. The *scattering operator* $S = W_+^* W_-$ connects the asymptotic behaviour of a quantum system as $t \to -\infty$ and $t \to \infty$ in terms of the free problem, that is, $S : f_0^{(-)} \mapsto f_0^{(+)}$. In the context of quantum computation where $\mathcal{H}$ is a finite dimensional Hilbert space, the scattering operator represents the quantum computation associated with a potential V.

In the time-independent or stationary scattering theory, one studies solutions of the time-independent Schrödinger equation with a parameter that belongs to the continuous part of the spectrum of the total Hamiltonian operator H. These solutions lie outside the Hilbert space and are characterized by certain asymptotic properties partly motivated by physical considerations. The observables, in particular the S-operator, are obtained from the asymptotic properties of such solutions [7].

It has been well known for a long time that these two methods are mathematically very different. The connections between them has been a problem studied since the seventies. Of fundamental importance is the task to establish conditions for which the final objects of the calculations (the S operator) are identical in both cases. This question is not easy to answer because of the nature of the calculations in the stationary scattering formalism. This theory uses mathematical manipulations that must first be interpreted in some sense before they can be made rigorous so that it is possible to compare with the time-dependent method, which is a very well-developed mathematical theory.

The recent developments in scattering theory can be found, for example, in [137] and references therein. Specially important is the work developed by M. Sh. Birman and D. R. Yafaev in stationary scattering theory and for the time-dependent theory, the work developed by Werner O. Amrein, Vladimir Georgescu, J.M. Jauch and K. B. Shina (see for example [5,6,8]). We can cite also the book from Berthier [12] and the work of J. Dereziński and C. Gérard (see for example [35] and references therein).

5.2 Stationary state scattering theory

In this chapter we focus our attention on the passage from time-dependent to the stationary formalism. Our starting point is the paper from W.O. Amrein, V. Georgescu, J.M. Jauch [7]. The principal problem to solve in this passage is the following: the basic quantities in the time-dependent theory (e.g. the wave operator) will be expressed in terms of a Bochner integral of certain operators over the time available. These formulas have been known for a long time. Operators in the Bochner integral can be expressed as a spectral integrals via the Spectral Theorem. Then the passage is achieved if we are able to interchange the two integrals and evaluate the time integral. The main problem of a mathematical nature is under which conditions we can interchange the two integrals and verify that the conditions are in fact satisfied for the integrals that we encounter in scattering theory.

In order to develop this alternative approach, we have to change the definition of wave operators $W_\pm$ replacing the unitary groups by the corresponding resolvents $R_0(z) = (H_0 - z)^{-1}$ and $R(z) = (H - z)^{-1}$ ([137]). In this stationary approach, in place of the limits in time, one has to study the boundary values in a suitable topology of the resolvents as the spectral parameter z tends to the real axis.

An important advantage of the stationary approach is that it gives convenient formulas for the wave operators and the scattering matrix. The temporal asymptotics of the time-dependent Schrödinger equation is closely related to the asymptotics at large distances of solutions of the stationary

Schrödinger equation:

$$-\Delta\Psi + V(x)\Psi = \lambda\Psi. \tag{5.2}$$

In other words, from the physical point of view, in the time-dependent formalism we consider that the particle being scattered have to behave as a free particle in the far past and in the far future ($t \to \pm\infty$). In the time-independent formalism we consider that the particle being scattered has to behave as a free particle far away from the scattering centre, where the influence of the potential is negligible and the total Hamitonian is practically the free Hamitonian. In terms of boundary values of the resolvent, the scattering solution, or eigenfunction of the continuous spectrum, can be constructed using the Lippmann-Schwinger equations (see for example [137]).

More precisely, as in [7], suppose that the potential V is a real valued measurable function and $H = H_0 + V$ and H_0 are both self adjoint on the domain $\mathcal{D}(H_0)$ of H_0. The strong operator limits in the formula (5.1) for the wave operators $W_\pm$ are assumed to exist with $P_0^{(a)} = I$, that is, H_0 has absolutely continuous spectrum. Furthermore, we suppose that range$(W_+) = $ range(W_-) is the orthogonal complement of the closed linear subspace spanned by the eigenvectors of H.

Suppose that F_0 is the spectral measure of H_0 and F is the spectral measure of H, so that by the Spectral Theorem for selfadjoint operators [121, Theorem 13.30]

$$H_0 = \int_{\mathbb{R}} \lambda \, dF_0(\lambda), \quad H = \int_{\mathbb{R}} \lambda \, dF(\lambda).$$

Typical formulas that relate quantities in the stationary and time-dependent approaches are given in [7, (30), (31)]:

$$W_- = \lim_{\epsilon\downarrow 0} i\epsilon \int_{\mathbb{R}} R(\lambda - i\epsilon) \, dF_0(\lambda) = \lim_{\epsilon\downarrow 0} -i\epsilon \int_{\mathbb{R}} dF(\lambda) R_0(\lambda - i\epsilon), \tag{5.3}$$

$$W_+ = \lim_{\epsilon\downarrow 0} -i\epsilon \int_{\mathbb{R}} R(\lambda + i\epsilon) \, dF_0(\lambda) = \lim_{\epsilon\downarrow 0} i\epsilon \int_{\mathbb{R}} dF(\lambda) R_0(\lambda - i\epsilon), \tag{5.4}$$

and [7, (34), (35)]:

$$(W_- - I) \restriction_{\mathcal{D}(H_0)} = \lim_{\epsilon\downarrow 0} - \int_{\mathbb{R}} R(\lambda - i\epsilon) V \, dF_0(\lambda)$$

$$= \lim_{\epsilon\downarrow 0} \int_{\mathbb{R}} dF(\lambda) V R_0(\lambda - i\epsilon), \tag{5.5}$$

$$(W_+ - I) \upharpoonright_{\mathcal{D}(H_0)} = \lim_{\epsilon \downarrow 0} - \int_{\mathbb{R}} R(\lambda + i\epsilon) V \, dF_0(\lambda)$$

$$= \lim_{\epsilon \downarrow 0} \int_{\mathbb{R}} dF(\lambda) V R_0(\lambda - i\epsilon). \tag{5.6}$$

The limits in formulas (5.5) and (5.6) are taken in the strong operator topology of $\mathcal{L}(\mathcal{D}(H_0), \mathcal{H})$ and in terms of the spectral measure F_0 and resolvent R_0 of the free Hamiltonian, we obtain [7, (38), (39)]:

$$(W_- - I) \upharpoonright_{\mathcal{D}(H_0)} = \lim_{\epsilon \downarrow 0} - \int_{\mathbb{R}} R_0(\lambda - i\epsilon) V W_- \, dF_0(\lambda)$$

$$= \lim_{\epsilon \downarrow 0} \int_{\mathbb{R}} dF_0(\lambda) V W_- R_0(\lambda - i\epsilon), \tag{5.7}$$

$$(W_+ - I) \upharpoonright_{\mathcal{D}(H_0)} = \lim_{\epsilon \downarrow 0} - \int_{\mathbb{R}} R_0(\lambda + i\epsilon) V W_+ \, dF_0(\lambda)$$

$$= \lim_{\epsilon \downarrow 0} \int_{\mathbb{R}} dF_0(\lambda) V W_+ R_0(\lambda - i\epsilon). \tag{5.8}$$

The formulas (5.5) and (5.7) correspond to the 'Lippmann-Schwinger equations'

$$\psi(\omega, \lambda) = \psi_0(\omega, \lambda) - R(\lambda - i0) V \psi_0(\omega, \lambda) \tag{5.9}$$

$$\psi(\omega, \lambda) = \psi_0(\omega, \lambda) - R_0(\lambda - i0) V \psi(\omega, \lambda) \tag{5.10}$$

for the scattering solutions $\psi(\omega, \lambda)$ of the stationary Schrödinger equation (5.2), see [7, p. 427].

The mathematical difficulty with the 'Lippmann-Schwinger equations', as written in equations (5.9) and (5.10) above, is how to interpret the limits $R(\lambda - i0)$ and $R_0(\lambda - i0)$ when the function $\psi(\omega, \lambda)$ does not belong to the underlying Hilbert space $\mathcal{H}$ ($\lambda > 0$ belongs to the absolutely continuous spectrum of $H = -\Delta + V$ and $\psi(\omega, \lambda)$ is not an eigenfunction of the operator H). In stationary state scattering theory, the problem of interpreting the limit is circumvented by appealing to the specific structure of the Schrödinger operator $H = -\Delta + V$ and specifying the asymptotic behaviour of the scattering solutions $\psi(\omega, \lambda)$ as $|x| \to \infty$, see [137] and Section 5 below. In contrast, formulas (5.5) and (5.7) are applicable in a much wider context.

Each of the formulas (5.3)-(5.8) involve the integration of an operator valued function Φ with respect to a spectral measure P in which the values of Φ act on the values of P or the values of P act on the values of Φ by operator composition—a type of *bilinear integration*. In Section 5.4 below, we discuss in greater detail why regular bilinear integration cannot deal with the situation described above. Our exposition concentrates on formula (5.5). A similar approach suffices to deal with the other formulas.

To see how formula (5.5) reduces to the interchange of integrals, first note that for each $\psi \in \mathcal{D}(H_0)$, the left-hand side of (5.5) can be written as

$$(W_- - I)\psi = \lim_{\epsilon \downarrow 0} i \int_0^\infty e^{-\epsilon t} e^{itH} V e^{-itH_0} \psi \, dt, \qquad (5.11)$$

see [7, (16)]. The integral is a Bochner integral in the Hilbert space $\mathcal{H}$. For each $\epsilon > 0$, we have

$$\int_0^\infty e^{-\epsilon t} e^{itH} V e^{-itH_0} \psi \, dt = \int_0^\infty e^{itH} V \left(\int_\mathbb{R} e^{-i(\lambda - i\epsilon)t} \, d(F_0\psi)(\lambda) \right) dt. \tag{5.12}$$

On the right-hand side of (5.5), we have

$$\int_\mathbb{R} R(\lambda - i\epsilon) V \, dF_0(\lambda) = i \int_\mathbb{R} \left(\int_0^\infty e^{-i(\lambda - i\epsilon)t} e^{itH} \, dt \right) V \, d(F_0\psi)(\lambda),$$

so we need to show that

$$\int_0^\infty e^{itH} V \left(\int_\mathbb{R} e^{-i(\lambda - i\epsilon)t} \, d(F_0\psi)(\lambda) \right) dt$$

$$= \int_\mathbb{R} \left(\int_0^\infty e^{-i(\lambda - i\epsilon)t} e^{itH} \, dt \right) V \, d(F_0\psi)(\lambda)$$

for each $\epsilon > 0$. The values of the $\mathcal{L}(\mathcal{D}(H_0), \mathcal{H})$-valued function

$$\lambda \longmapsto \left(\int_0^\infty e^{-i(\lambda - i\epsilon)t} e^{itH} \, dt \right) V, \quad \lambda \in \mathbb{R},$$

act on the $\mathcal{D}(H_0)$-valued measure $F_0\psi$, giving rise to the bilinear integration process that has been determined in earlier chapters. The main feature here is that the integrals are 'decoupled', so it is easy to see that a type of Fubini Theorem is valid for vector valued integrals.

5.3   Time-dependent scattering theory for bounded Hamiltonians and potentials

The second basic example from [65, Section 4] is relevant to the connection between stationary-state and time-dependent scattering theory [7] where H_0 represents the free Hamiltonian operator and V represents an interaction potential. Attention is first retricted to bounded operators.

It is well known that quantum computation relies on calculations within a finite dimensional complex inner product space $\mathcal{H}$ whose dimension represents the number of qubits available [79].

Consequently the dynamics of quantum computations is determined by matrix Hamiltonians acting on $\mathcal{H}$, so that the case of bounded Hamiltonian operators and bounded interaction potentials treats the dynamics of quantum computations and their asymptotic limit. A quantum computer is particularly adapted to Monte-Carlo type approximations of Feynman path integrals, so a quantum computer may be capable of analysing, say, the Riemann hypothesis or colliding supermassive black holes at the level of quantum gravity.

At the time of writing, quantum computers do not exist and neither Feynman path integrals nor quantum gravity nor the Riemann hypothesis is well understood, so we must leave such speculation to physically inclined Philosophers of Science and Mathematics and noncommutative geometers [29].

Example 5.1. Let $(\mathcal{H}, (\cdot \mid \cdot))$ be a separable Hilbert space. Suppose H_0 and V are *bounded* selfadjoint operators. Then $H = H_0 + V$ is also a bounded selfadjoint operator and the function $f_\epsilon : \mathbb{R}_+ \times \mathbb{R} \to \mathcal{L}(\mathcal{H})$ defined for $\epsilon > 0$ by

$$f_\epsilon(t, \sigma) = e^{itH} V e^{-i(\sigma - i\epsilon)t}$$

for $t \geq 0$ and $\sigma \in \mathbb{R}$ is uniformly bounded in $\mathcal{L}(\mathcal{H})$. Let P be the spectral measure associated with the selfadjoint operator H_0 and $h \in \mathcal{H}$. Lebesgue measure on $\mathbb{R}_+$ is denoted by λ. We would like to verify the identities

$$\int_{\mathbb{R}_+ \times \mathbb{R}} f_\epsilon(t, \sigma) \, d(\lambda \otimes (Ph))(t, \sigma) = \int_0^\infty e^{-\epsilon t} e^{itH} V e^{-itH_0} h \, dt$$

$$= \int_{\mathbb{R}} \left(\int_0^\infty e^{-\epsilon t} e^{itH} V e^{-it\sigma} \, dt \right) d(Ph)(\sigma)$$

that help to establish the connection between stationary-state and time-dependent scattering theory in the case of unbounded selfadjoint operators H_0 and V [7]. The $\mathcal{H}$-valued measure $\lambda \otimes (Ph) : \mathcal{S} \to \mathcal{H}$ is given by

$$\lambda \otimes (Ph)(A \times B) = \lambda(A)(Ph)(B), \quad A \in \mathcal{B}_f(\mathbb{R}_+), \ B \in \mathcal{B}(\mathbb{R}).$$

Here $\mathcal{B}_f(\mathbb{R}_+)$ is the collection of Borel subsets of $\mathbb{R}_+$ with finite Lebesgue measure and $\mathcal{S}$ is the δ-ring generated by the collection $\{A \times B : A \in \mathcal{B}_f(\mathbb{R}_+), B \in \mathcal{B}(\mathbb{R})\}$ of product sets in $\mathbb{R}_+ \times \mathbb{R}$. We check that the projective tensor product

$$E = \mathcal{L}(\mathcal{H}) \widehat{\otimes}_\pi \mathcal{H}$$

of $\mathcal{H}$ and $\mathcal{L}(\mathcal{H})$ with the uniform operator norm is bilinear admissible for $\mathcal{H}$ in the sense of Section 2.3 above and f_ϵ is $(\lambda \otimes (Ph))$-integrable in E, with the appropriate modification for integration with respect to a vector measure defined on a δ-ring. Because a Hilbert space $\mathcal{H}$ necessarily has the approximation property [125, III.9], $\mathcal{H} \otimes \mathcal{H} \otimes \mathcal{H}$ separates points of E [88, 43.2(12)].

Let $B(t) = \{\sigma : (t,\sigma) \in B\}$ be the section at $t \geq 0$ of the Borel subset B of $\mathbb{R}_+ \times \mathbb{R}$. We check that the function

$$\Phi_\epsilon^B : t \longmapsto e^{-\epsilon t} \left(e^{itH} V\right) \otimes \left(e^{-itH_0} P(B(t))h\right), \quad t \geq 0,$$

is Bochner integrable in E. The function $t \longmapsto P(B(t))h$, $t \geq 0$, is strongly measurable so by Lusin's Theorem on each interval $[0,T]$, there exists a Borel set Σ of arbitrarily large measure on which it is continuous in $\mathcal{H}$. Then Φ_ϵ^B is continuous in E on Σ because H and H_0 are assumed to be *bounded* selfadjoint operators so that the unitary groups e^{itH} and e^{-itH_0}, $t \in \mathbb{R}$, are continuous for the *uniform operator topology*, as is easily seen from the exponential power series expansion. Consequently, Φ_ϵ^B is strongly measurable in E. Because

$$\int_0^\infty \|\Phi_\epsilon^B\|_E \, dt \leq \int_0^\infty e^{-\epsilon t} \left\|e^{itH} V\right\| \cdot \left\|e^{-itH_0} P(B(t))h\right| \, dt \leq \frac{\|V\|\|h\|}{\epsilon},$$

it follows that Φ_ϵ is Bochner integrable in E. For every $u, v, w \in \mathcal{H}$, we have

$$\left\langle \int_0^\infty \Phi_\epsilon^B \, dt, u \otimes v \otimes w \right\rangle = \int_0^\infty e^{-\epsilon t} \left(e^{itH} V u \big| v\right) \left(e^{-itH_0} P(B(t))h \big| w\right) \, dt$$

$$= \int_B \langle f_\epsilon(t,\sigma), u \otimes v \rangle \, d\left(\lambda \otimes ((Ph)\big|w)\right)(t,\sigma).$$

According to Definition 2.2, the function f_ϵ is $(\lambda \otimes (Ph))$-integrable in E and for every Borel subset B of $\mathbb{R}_+ \times \mathbb{R}$, we have

$$\int_B f_\epsilon \, d(\lambda \otimes (Ph)) = \int_0^\infty \Phi_\epsilon^B \, dt.$$

Furthermore, for each $t \geq 0$, the $\mathcal{L}(\mathcal{H})$-valued function $f_\epsilon(t, \cdot)$ is (Ph)-integrable in E the vector measure Ph and for each $B \in \mathcal{B}(\mathbb{R})$, we have

$$\int_B f_\epsilon(t,\sigma) \otimes d(Ph) = e^{-\epsilon t} e^{itH} V \otimes e^{-itH_0} P(B)h \in E,$$

so that

$$\int_\mathbb{R} f_\epsilon(t,\sigma) \, d(Ph)(\sigma) = J \int_\mathbb{R} f_\epsilon(t,\sigma) \otimes d(Ph) = e^{-\epsilon t} e^{itH} V e^{-itH_0} h.$$

For each $\sigma \in \mathbb{R}$, the function $f_\epsilon(\cdot, \sigma)$ is Bochner integrable for the uniform norm of $\mathcal{L}(\mathcal{H})$ and the $\mathcal{L}(\mathcal{H})$-valued function

$$\sigma \longmapsto \int_0^\infty f_\epsilon(t,\sigma) \, dt, \quad \sigma \in \mathbb{R},$$

is (Ph)-integrable in E. For every Borel subset C of $\mathbb{R}$, we have

$$\int_C \left(\int_0^\infty f_\epsilon(t, \sigma)\, dt \right) d(Ph) = J \int_0^\infty \Phi_\epsilon^{\mathbb{R}_+ \times C}\, dt.$$

The scalar version of Fubini's Theorem and the assumption that $\mathcal{H} \otimes \mathcal{H} \otimes \mathcal{H}$ separates points of the bilinear admissible space E ensure the identity

$$\int_\mathbb{R} \left(\int_0^\infty f_\epsilon(t, \sigma)\, dt \right) d(Ph) = \int_0^\infty \left(\int_\mathbb{R} f_\epsilon(t, \sigma)\, d(Ph) \right) dt$$

relevant to scattering theory.

We can still make sense of Example 5.1 if H_0 and V are unbounded operators, but it is clear that the auxiliary space $E = \mathcal{L}(\mathcal{H}) \widehat{\otimes}_\pi \mathcal{H}$ will no longer suffice, because the unitary groups e^{itH} and e^{-itH_0}, $t \in \mathbb{R}$, are only continuous for the *strong operator topology*—it is too much to expect that the function Φ_ϵ^B defined in the example above will be Bochner integrable in the Banach space $\mathcal{L}(\mathcal{H}) \widehat{\otimes}_\pi \mathcal{H}$. We first need to consider the approximation of $\mathcal{L}(\mathcal{H})$-valued functions in the strong operator topology.

5.4 Bilinear integrals in scattering theory

In this section we return to the situation considered in Example 5.1, but with the more physically realistic case of unbounded selfadjoint operators.

Let $(\mathcal{H}, (\cdot \mid \cdot))$ be a separable Hilbert space. Suppose $H_0 : \mathcal{D}(H_0) \to \mathcal{H}$ and $V : \mathcal{D}(V) \to \mathcal{H}$ are selfadjoint operators with respective dense domains $\mathcal{D}(H_0) \subset \mathcal{D}(V)$. We suppose that $H = H_0 + V$ is also a selfadjoint operator on $\mathcal{D}(H_0)$. For example, this is the situation for the free Hamiltonian $H_0 = -\frac{\hbar^2}{2m}\Delta$ of a particle moving in $\mathbb{R}^3$ subject to a Coulomb potential $V(x) = c/|x|$, $x \in \mathbb{R}^3 \setminus \{0\}$, $c \neq 0$.

The function $f_\epsilon : \mathbb{R}_+ \times \mathbb{R} \to \mathcal{L}(\mathcal{D}(H_0), \mathcal{H})$ given by

$$f_\epsilon(t, \sigma) = e^{itH} V e^{-i(\sigma - i\epsilon)t}$$

for $t \geq 0$ and $\sigma \in \mathbb{R}$ is defined on $\mathcal{D}(H_0)$. Let P_0 be the spectral measure associated with the selfadjoint operator H_0 and $h \in \mathcal{D}(H_0)$. As above, Lebesgue measure on $\mathbb{R}_+$ is denoted by λ. The spectral measure P_0 commutes with the selfadjoint operator H_0 in the sense that for each Borel subset B of $\mathbb{R}$, the inclusion $P_0(B)\mathcal{D}(H_0) \subset \mathcal{D}(H_0)$ is valid and $H_0 P_0(B)h = P_0(B)H_0 h$ for every $h \in \mathcal{D}(H_0)$. We give $\mathcal{D}(H_0)$ the graph norm $h \longmapsto (\|h\|_\mathcal{H}^2 + \|H_0 h\|_\mathcal{H}^2)^{\frac{1}{2}}$, $h \in \mathcal{D}(H_0)$, under which it is itself a Hilbert space.

Now we seek a suitable space E, bilinear admissible for $\mathcal{D}(\mathcal{H}_0)$, $\mathcal{H}$, for which f_ϵ is $\big(\lambda \otimes (Ph)\big)$-integrable in E with respect to the $\mathcal{D}(\mathcal{H}_0)$-valued measure $\lambda \otimes (Ph)$.

Let $\mathcal{X}, \mathcal{Y}$ be Banach spaces. For $y^* \in \mathcal{Y}^*$, we have

$$\left| \left\langle \sum_{j=1}^n T_j x_j, y^* \right\rangle \right| = \left| \left\langle \sum_{j=1}^n x_j, T_j^* y^* \right\rangle \right|$$

$$\leq \sum_{j=1}^n \|x_j\|_{\mathcal{X}} \cdot \|T_j^* y^*\|_{\mathcal{X}^*},$$

for all $T_j \in \mathcal{L}(\mathcal{X}, \mathcal{Y})$ and $x_j \in \mathcal{X}$, $j = 1, \ldots, n$ and all $n = 1, 2, \ldots$. Hence, if we let

$$\|u\|_\tau = \sup_{\|y^*\| \leq 1} \inf \left\{ \sum_{j=1}^n \|x_j\|_{\mathcal{X}} \cdot \|T_j^* y^*\|_{\mathcal{X}^*} : u = \sum_{j=1}^n T_j \otimes x_j \right\} \quad (5.13)$$

over all representations $u = \sum_{j=1}^n T_j \otimes x_j$, $n = 1, 2 \ldots$, of $u \in \mathcal{L}(\mathcal{X}, \mathcal{Y}) \otimes \mathcal{X}$, then the inequality $\|Ju\|_{\mathcal{Y}} \leq \|u\|_\tau$ holds for the product map $Ju = \sum_{j=1}^n T_j x_j$ by the Hahn-Banach Theorem. The completion of the linear space $\mathcal{L}(\mathcal{X}, \mathcal{Y}) \widehat{\otimes}_\tau \mathcal{X}$ with respect to the norm $\|\cdot\|_\tau$ is written as $\mathcal{L}(\mathcal{X}, \mathcal{Y}) \widehat{\otimes}_\tau \mathcal{X}$.

Lemma 5.1. *For each* $u = \sum_{j=1}^n T_j \otimes x_j \in \mathcal{L}(\mathcal{X}, \mathcal{Y}) \otimes \mathcal{X}$, *let* $\mathcal{T}_u = \sum_{j=1}^n x_j \otimes T_j^* \in \mathcal{L}(\mathcal{Y}^*, \mathcal{X} \widehat{\otimes}_\pi \mathcal{X}^*)$ *denote the linear map* $y^* \longmapsto \sum_{j=1}^n x_j \otimes (T_j^* y^*)$, $y^* \in \mathcal{Y}^*$.

Then the linear mapping $k : u \longmapsto \mathcal{T}_u$, $u \in \mathcal{L}(\mathcal{X}, \mathcal{Y}) \otimes \mathcal{X}$, *is the restriction to* $\mathcal{L}(\mathcal{X}, \mathcal{Y}) \otimes \mathcal{X}$ *of an isometry of* $\mathcal{L}(\mathcal{X}, \mathcal{Y}) \widehat{\otimes}_\tau \mathcal{X}$ *onto a closed linear subspace of* $\mathcal{L}(\mathcal{Y}^*, \mathcal{X} \widehat{\otimes}_\pi \mathcal{X}^*)$ *in the uniform norm.*

Proof. For $u \in \mathcal{L}(\mathcal{X}, \mathcal{Y}) \otimes \mathcal{X}$, we have $\|u\|_\tau = \|\mathcal{T}_u\|_{\mathcal{L}(\mathcal{Y}^*, \mathcal{X} \widehat{\otimes}_\pi \mathcal{X}^*)}$. $\qquad \square$

Lemma 5.2. *Let* $\mathcal{X}, \mathcal{Y}$ *be Banach spaces with* $\mathcal{Y}^*$ *norm separable. If* B *is a bounded, absolutely convex subset of* $\mathcal{L}(\mathcal{X}, \mathcal{Y}) \otimes \mathcal{X}$ *with* τ-*closure* $\overline{B}$, *then* $k\overline{B}$ *is a closed subset of* $\mathcal{L}_s(\mathcal{Y}^*, \mathcal{X} \widehat{\otimes}_\pi \mathcal{X}^*)$.

Proof. Every element v of the closure of $k\overline{B}$ in $\mathcal{L}_s(\mathcal{Y}^*, \mathcal{X} \widehat{\otimes}_\pi \mathcal{X}^*)$ can be represented as $v = \sum_{j=1}^\infty \lambda_j x_j \otimes T_j^*$ in $\mathcal{L}_s(\mathcal{Y}^*, \mathcal{X} \widehat{\otimes}_\pi \mathcal{X}^*)$, where $\sum_{j=1}^\infty |\lambda_j| < \infty$, the finite sum $\sum_{j=1}^n \lambda_j x_j \otimes T_j^*$ belongs to B for each $n = 1, 2, \ldots$ and $T_j^* \to 0$ in $\mathcal{L}_s(\mathcal{Y}^*, \mathcal{X}^*)$, $x_j \to 0$ in $\mathcal{X}$ as $j \to \infty$.

To see this, we note that $\mathcal{Y}^*$ is separable, so the relative topology of $\mathcal{L}_s(\mathcal{Y}^*, \mathcal{X} \widehat{\otimes}_\pi \mathcal{X}^*)$ on $k\overline{B}$ coincides with the relative topology of $\mathcal{X} \widehat{\otimes}_\pi \mathcal{F}$, where $\mathcal{F}$ is the metrisable locally convex space $\mathcal{L}(\mathcal{Y}^*, \mathcal{X}^*)$ endowed with the topology of pointwise convergence on a countable dense subset of the unit ball of $\mathcal{Y}^*$ [88, (1) p. 138]. Then the given representation for v follows from Theorem 1.8 applied to $\mathcal{X}\widehat{\otimes}_\pi\mathcal{F}$.

Because

$$\left\|\sum_{j=n}^\infty \lambda_j T_j \otimes x_j\right\|_\tau \leq \sup_{\|y^*\|\leq 1} \sum_{j=n}^\infty |\lambda_j| \|x_j\|_\mathcal{X} \|T_j^* y^*\|_{\mathcal{X}^*}$$

$$\leq \sup_j\{\|x_j\|_\mathcal{X}\|T_j\|_{\mathcal{L}(\mathcal{X},\mathcal{Y})}\} \sum_{j=n}^\infty |\lambda_j| \to 0$$

as $n \to \infty$, it follows that $v \in k\overline{B}$, hence $k\overline{B}$ is closed in $\mathcal{L}_s(\mathcal{Y}^*, \mathcal{X}\widehat{\otimes}_\pi\mathcal{X}^*)$. □

A Banach space $\mathcal{X}$ has the *approximation property* if $\mathcal{X}^* \otimes \mathcal{X}$ is dense in $\mathcal{L}_\kappa(\mathcal{X})$, with κ the topology of precompact convergence [123, Section III.9]. Each element $\sum_{j=1}^n x_j^* \otimes x_j$ of $\mathcal{X}^* \otimes \mathcal{X}$ defines the finite rank operator

$$x \longmapsto \sum_{j=1}^n \langle x, x_j^*\rangle x_j, \quad x \in \mathcal{X}.$$

Lemma 5.3. *The Banach space $E = \mathcal{L}(\mathcal{X}, \mathcal{Y})\widehat{\otimes}_\tau\mathcal{X}$ is bilinear admissible for the Banach spaces $\mathcal{X}$ and $\mathcal{Y}$ provided that $\mathcal{X}$ has the approximation property.*

Proof. Properties a)-c) of bilinear admissibility clearly hold, so it remains to establish property d): the family of all linear maps $x\otimes y^*\otimes x^* : T\otimes x \longmapsto \langle Tx, y^*\rangle\langle x, x^*\rangle$ for $x \in \mathcal{X}$, $x^* \in \mathcal{X}^*$ and $y^* \in \mathcal{Y}^*$ separates points of E.

Let $\mathfrak{B}_e((\mathcal{X}\otimes\mathcal{Y}^*)_s, \mathcal{X}_s^*)$ denote the linear space of all separately continuous bilinear forms on $(\mathcal{X}\otimes\mathcal{Y}^*) \times \mathcal{X}^*$ for the topologies of simple convergence $\sigma(\mathcal{X}\otimes\mathcal{Y}^*, \mathcal{L}(\mathcal{X}, \mathcal{Y}))$ and $\sigma(\mathcal{X}^*, \mathcal{X})$. It is equipped with the topology e of bi-equicontinuous convergence [88, p. 167]. Because $\mathcal{X}$ has the approximation property, the canonical linear map

$$\psi : \mathcal{L}_s(\mathcal{Y}^*, \mathcal{X}^*)\widehat{\otimes}_\pi\mathcal{X} \to \mathfrak{B}_e((\mathcal{X}\otimes\mathcal{Y}^*)_s, \mathcal{X}_s^*)$$

is one-to-one [88, 43.2 (12)].

The topology τ has a fundamental system of neighbourhoods of zero closed for the weaker topology of $\mathcal{L}_s(\mathcal{Y}^*, \mathcal{X}^*)\widehat{\otimes}_\pi\mathcal{X}$, so E embeds in $\mathfrak{B}_e((\mathcal{X}\otimes\mathcal{Y}^*)_s, \mathcal{X}_s^*)$ too [88, 18.4 (4)]. Because $(\mathcal{X} \otimes \mathcal{Y}^*) \times \mathcal{X}^*$ separates the space $\mathfrak{B}_e((\mathcal{X} \otimes \mathcal{Y}^*)_s, \mathcal{X}_s^*)$ of bilinear forms, it follows that $\mathcal{X} \otimes \mathcal{Y}^* \otimes \mathcal{X}^*$ separates points of E. □

Remark 5.1. The approximation property for $\mathcal{X}$ is needed to make sense of the integral $\int_\Omega f \, dm$ of an $\mathcal{L}(\mathcal{X}, \mathcal{Y})$-valued function f with respect to an $\mathcal{X}$-valued measure m. Similarly, the approximation property is needed to define the trace of a nuclear operator on $\mathcal{X}$ in the case $\mathcal{Y} = \mathbb{C}$. All Banach spaces of practical interest, including Hilbert spaces, possess the approximation property, see [94].

The following result improves Theorem 4.4 of [49] in the sense that we obtain a stronger form of integrability in the conclusion (c) below.

Theorem 5.1. *Let $(\mathcal{H}, (\cdot \,|\, \cdot))$ be a separable Hilbert space and let $A : \mathcal{D}(A) \to \mathcal{H}$ be a selfadjoint operator with spectral measure P_A. Let E denote the space $\mathcal{L}(\mathcal{D}(A), \mathcal{H}) \widehat{\otimes}_\tau \mathcal{D}(A)$ where the Hilbert space $\mathcal{D}(A)$ is endowed with the graph norm.*

Let $(\Gamma, \mathcal{E}, \mu)$ be a σ-finite measure space. Suppose that the measurable function $u : \mathbb{R} \times \Gamma \to \mathbb{C}$ has the property that for every $h \in \mathcal{D}(A)$, the function $u(\cdot, \gamma)$ is $P_A h$-integrable in $\mathcal{D}(A)$.

Let $f : \Gamma \to \mathcal{L}(\mathcal{D}(A), \mathcal{H})$ be a strongly μ-measurable in $\mathcal{L}_s(\mathcal{D}(A)\mathcal{H})$ function for which there exist positive μ-measurable functions α, β, v, on Γ, with the following properties:

(i) *$|u(\sigma, \gamma)| \le v(\gamma)$ for every $\sigma \in \mathbb{R}$ and $\gamma \in \Gamma$,*
(ii) *$\|f(\gamma)h\| \le \alpha(\gamma)\|Ah\| + \beta(\gamma)\|h\|$ for every $h \in \mathcal{D}(A)$, $\gamma \in \Gamma$, and*
(iii) *$\int_\Gamma v(\gamma)(\alpha(\gamma) + \beta(\gamma)) \, d\mu(\gamma) < \infty$.*

Then for each $h \in \mathcal{D}(A)$,

(a) *the function $\gamma \longmapsto f(\gamma) \int_S u(\sigma, \gamma) \, d(P_A h)(\sigma)$, $\gamma \in \Gamma$, is Bochner μ-integrable in $\mathcal{H}$ for each $S \in \mathcal{B}(\mathbb{R})$,*
(b) *the function $\gamma \longmapsto u(\sigma, \gamma) f(\gamma) g$, $\gamma \in \Gamma$, is Bochner μ-integrable in $\mathcal{H}$ for each $\sigma \in \mathbb{R}$ and $g \in \mathcal{D}(A)$, and*
(c) *the $\mathcal{L}(\mathcal{D}(A), \mathcal{H})$-valued function $\sigma \longmapsto \int_T u(\sigma, \gamma) f(\gamma) \, d\mu(\gamma)$, $\sigma \in \mathbb{R}$, is strongly measurable in $\mathcal{L}(\mathcal{D}(A), \mathcal{H})$ and $(P_A h)$-integrable in E with respect to the $\mathcal{D}(A)$-valued measure $P_A h$, for each set $T \in \mathcal{E}$.*

Moreover the equality

$$\int_S \left(\int_T u(\sigma, \gamma) f(\gamma) \, d\mu(\gamma) \right) d(P_A h)(\sigma)$$

$$= \int_T f(\gamma) \left(\int_S u(\sigma, \gamma) \, d(P_A h)(\sigma) \right) d\mu(\gamma) \qquad (5.14)$$

holds for every $S \in \mathcal{B}(\mathbb{R})$ and $T \in \mathcal{E}$.

Proof. Property (a) follows from the strong μ-measurability of f in $\mathcal{L}(\mathcal{D}(A), \mathcal{H})$ and the norm estimates

$$\int_\Gamma \left\| f(\gamma) \int_S u(\sigma, \gamma) \, d(P_A h)(\sigma) \right\| d\mu(\gamma)$$

$$\leq \int_\Gamma \alpha(\gamma) \left\| A \int_S u(\sigma, \gamma) \, d(P_A h)(\sigma) \right\| d\mu(\gamma)$$

$$+ \int_\Gamma \beta(\gamma) \left\| \int_S u(\sigma, \gamma) \, d(P_A h)(\sigma) \right\| d\mu(\gamma)$$

$$\leq \|P_A h\|_{sv(\mathcal{D}(A))} \int_\Gamma v(\gamma)(\alpha(\gamma) + \beta(\gamma)) \, d\mu(\gamma)$$

$$< \infty.$$

Here $\|P_A h\|_{sv(\mathcal{D}(A))}$ is the semivariation norm of the $\mathcal{D}(A)$-valued measure $P_A h$. The estimates (i)-(iii) also give Property (b).

Now let B be an element of the σ-algebra $\mathcal{B}(\mathbb{R}) \otimes \mathcal{E}$ with

$$B(\gamma) = \{\sigma \in \mathbb{R} : (\sigma, \gamma) \in B\}$$

its section at $\gamma \in \Gamma$. We check that the $\mathcal{L}(\mathcal{D}(A), \mathcal{H}) \otimes \mathcal{D}(A)$-valued function

$$\Phi_B : \gamma \longmapsto f(\gamma) \otimes \int_{B(\gamma)} u(\sigma, \gamma) \, d(P_A h)(\sigma), \quad \gamma \in \Gamma,$$

is integrable in the Banach space $E = \mathcal{L}(\mathcal{D}(A), \mathcal{H}) \widehat{\otimes}_\tau \mathcal{D}(A)$ and

$$\left\| \int_\Gamma \Phi_B(\gamma) \, d\mu(\gamma) \right\|_\tau \leq \|P_A h\|_{sv(\mathcal{D}(A))} \int_\Gamma v(\gamma)(\alpha(\gamma) + \beta(\gamma)) \, d\mu(\gamma).$$

Firstly, the inequality

$$\int_\Gamma v(\gamma) \|f(\gamma)^*\|_{\mathcal{L}(\mathcal{H}, \mathcal{D}(A))} \, d\mu(\gamma) = \int_\Gamma v(\gamma) \|f(\gamma)\|_{\mathcal{L}(\mathcal{D}(A), \mathcal{H})} \, d\mu(\gamma)$$

$$= \sup_{\|\psi\|_\infty \leq 1} \int_\Gamma v(\gamma) \|f(\gamma)\psi(\gamma)\|_\mathcal{H} \, d\mu(\gamma)$$

$$\leq \int_\Gamma v(\gamma)(\alpha(\gamma) + \beta(\gamma)) \, d\mu(\gamma)$$

follows from assumption (ii), where the supremum is taken over $\mathcal{D}(A)$-valued $\mathcal{E}$-simple functions ψ. It follows that

$$\int_\Gamma \|\Phi_B(\gamma)\|_\tau \, d\mu(\gamma) \leq \|P_A h\|_{sv(\mathcal{D}(A))} \int_\Gamma v(\gamma)(\alpha(\gamma) + \beta(\gamma)) \, d\mu(\gamma).$$

We are not assuming that f is strongly μ-measurable for the *uniform norm* of $\mathcal{L}(\mathcal{D}(A), \mathcal{H})$, so some caution is needed.

Now $k \circ \Phi_B : \Gamma \to \mathcal{L}(\mathcal{D}(A), \mathcal{H} \otimes \mathcal{H})$ defines a linear map $L_B : \mathcal{D}(A) \to L^1(\mu, \mathcal{H} \otimes \mathcal{H})$ by

$$L_B g = [k \circ \Phi_B g], \quad g \in \mathcal{D}(A).$$

The bounded linear operator L_B is the limit in $\mathcal{L}(\mathcal{D}(A), L^1(\mu, \mathcal{H} \widehat{\otimes}_\pi \mathcal{H}))$ of $(P_n \otimes I)L_B$, $n = 1, 2, \ldots$, for conditioning operators $P_n : L^1(\mu, \mathcal{H}) \to L^1(\mu, \mathcal{H})$, $n = 1, 2, \ldots$, on finitely generated σ-algebras. Then $(P_n \otimes I)L_B = [k \circ \Phi_{B,n}]$ for

$$\Phi_{B,n} : \gamma \longmapsto (P_n f)(\gamma) \otimes \int_{B(\gamma)} u(\sigma, \gamma) \, d(P_A h)(\sigma), \quad \gamma \in \Gamma,$$

and $P_n f$ is an $\mathcal{L}(\mathcal{D}(A), \mathcal{H})$-valued $\mathcal{E}$-simple function for each $n = 1, 2, \ldots$, so $\Phi_{B,n}$ has values in $\mathcal{L}_s(\mathcal{D}(A), \mathcal{H}) \otimes \mathcal{D}(A)$. Moreover,

$$\|(P_n \otimes I)L_B\|_{\mathcal{L}(\mathcal{D}(A), L^1(\mu, \mathcal{H} \widehat{\otimes}_\pi \mathcal{H}))} \leq \int_\Gamma \|\Phi_B\|_\tau \, d\mu,$$

$\int_B (P_n \otimes I)L_B \, d\mu \in \mathcal{L}(\mathcal{D}(A), \mathcal{H}) \otimes \mathcal{D}(A)$ for each $n = 1, 2 \ldots$, and

$$\lim_{n \to \infty} \int_C (P_n \otimes I)L_B \, d\mu = \int_C L_B \, d\mu \tag{5.15}$$

in $\mathcal{L}_s(\mathcal{D}(A), \mathcal{H} \widehat{\otimes}_\pi \mathcal{H})$ for each $C \in \mathcal{E}$.

We want to show that for each $C \in \mathcal{E}$, the linear map

$$g \longmapsto \int_C L_B g \, d\mu, \quad g \in \mathcal{D}(A),$$

belongs to the *uniform closure* kE of $k\big(\mathcal{L}(\mathcal{D}(A), \mathcal{H}) \otimes \mathcal{D}(A)\big)$ in the space $\mathcal{L}(\mathcal{D}(A), \mathcal{H} \widehat{\otimes}_\pi \mathcal{H})$ of linear operators. An appeal to Lemma 5.2 and equation (5.15) shows that this is true on any set $\Gamma_n = \{\|\Phi_B\|_\tau \leq n\}$ and because the vector measure $L_B \mu : \mathcal{E} \to \mathcal{L}_s(\mathcal{D}(A), \mathcal{H} \widehat{\otimes}_\pi \mathcal{H})$ has finite variation $V(L_B \mu)(\Gamma) = \int_\Gamma \|\Phi_B(\gamma)\|_\tau \, d\mu(\gamma)$ in the uniform operator norm of $\mathcal{L}(\mathcal{D}(A), \mathcal{H} \widehat{\otimes}_\pi \mathcal{H})$, we have $L_B \mu : \mathcal{E} \to kE$. The equality $L_B \mu = k \circ (\Phi_B . \mu)$ ensures that Φ_B is μ-integrable in E.

Because the Banach space E is bilinear admissible for $\mathcal{D}(A), \mathcal{H}$, the integral

$$\int_B \big(u(\sigma, \gamma) f(\gamma)\big) \otimes d\big((P_A h) \otimes \mu\big)(\sigma, \gamma) \in E$$

is uniquely defined by the scalar equation

$$\left(\int_B \big(u(\sigma, \gamma) f(\gamma)\big) \otimes d\big((P_A h) \otimes \mu\big)(\sigma, \gamma) \Big| u \otimes v \otimes w \right)$$

$$= \int_B \big(u(\sigma, \gamma)(f(\gamma) u | v)\big) \, d\big((P_A h | w) \otimes \mu\big)(\sigma, \gamma)$$

$$= \left(\int_\Gamma \Phi_B \, d\mu \Big| u \otimes v \otimes w \right)$$

for each $B \in \mathcal{B}(\mathbb{R}) \otimes \mathcal{E}$, $u, w \in \mathcal{D}(A)$ and $v \in \mathcal{H}$. Moreover, statement (c) holds and

$$\int_S \left(\int_T u(\sigma, \gamma) f(\gamma) \, d\mu(\gamma) \right) \otimes d(P_A h)(\sigma)$$

$$= \int_{S \times T} (u(\sigma, \gamma) f(\gamma)) \otimes d((P_A h) \otimes \mu)(\sigma, \gamma),$$

for all $S \in \mathcal{B}(\mathbb{R})$ and $T \in \mathcal{E}$. The equality (5.14) of the iterated integrals follows from the scalar version of Fubini's Theorem and the bilinear admissability of E. $\qquad \square$

Corollary 5.1. *Let* $(\mathcal{H}, (\cdot | \cdot))$ *be a separable Hilbert space and* $H_0 : \mathcal{D}(H_0) \to \mathcal{H}$ *be a selfadjoint operator with spectral measure* P_0. *Let*

$$E = \mathcal{L}(\mathcal{D}(H_0), \mathcal{H}) \widehat{\otimes}_\tau \mathcal{D}(H_0)$$

where the Hilbert space $\mathcal{D}(H_0)$ *is endowed with the graph norm.*

Suppose that $V : \mathcal{D}(V) \to \mathcal{H}$ *is a selfadjoint operator with dense domain* $\mathcal{D}(H_0) \subset \mathcal{D}(V)$. *We suppose that* $H = H_0 + V$ *is also a selfadjoint operator on* $\mathcal{D}(H_0)$, $\epsilon > 0$ *and the function* $f_\epsilon : \mathbb{R}_+ \times \mathbb{R} \to \mathcal{L}(\mathcal{D}(H_0), \mathcal{H})$ *is given by*

$$f_\epsilon(t, \sigma) = e^{itH} V e^{-i(\sigma - i\epsilon)t}, \quad t \in \mathbb{R}_+, \ \sigma \in \mathbb{R}.$$

Then for each $h \in \mathcal{D}(H_0)$,

(a) *the function* $t \longmapsto e^{-\epsilon t} e^{itH} V e^{-iH_0 t} g$, $t \in \mathbb{R}_+$, *is Bochner integrable in* $\mathcal{H}$ *for each* $g \in \mathcal{D}(H_0)$,

(b) *the function* $t \longmapsto f_\epsilon(t, \sigma) g$, $t \in \mathbb{R}_+$, *is Bochner* μ-integrable in $\mathcal{H}$ for *each* $\sigma \in \mathbb{R}$ *and* $g \in \mathcal{D}(H_0)$, *and*

(c) *the* $\mathcal{L}(\mathcal{D}(H_0), \mathcal{H})$-valued function $\sigma \longmapsto \int_T f_\epsilon(t, \sigma) \, dt$, $\sigma \in \mathbb{R}$, *is strongly measurable in* $\mathcal{L}(\mathcal{D}(H_0), \mathcal{H})$ *and* $(P_0 h)$-integrable in E with respect to *the* $\mathcal{D}(H_0)$-valued measure $P_0 h$, *for each set* $T \in \mathcal{B}(\mathbb{R}_+)$.

Moreover the equality

$$\int_S \left(\int_T f_\epsilon(t, \sigma) \, dt \right) d(P_0 h)(\sigma) = \int_T e^{-\epsilon t} e^{itH} V e^{-iH_0 t} P_0(S) h \, dt$$

holds for every $S \in \mathcal{B}(\mathbb{R})$ *and* $T \in \mathcal{B}(\mathbb{R}_+)$.

Proof. Statements (a) and (b) follow directly from the fact that e^{-itH} and e^{-itH_0} are continuous unitary semigroups on $\mathcal{H}$ and $\mathcal{D}(H_0)$. The selfadjoint operator V is closed and has domain containing $\mathcal{D}(H_0)$. By the Closed Graph Theorem it is bounded in the graph norm of the Hilbert space $\mathcal{D}(H_0)$. Conditions (i)-(iii) of Theorem 5.1 are satisfied for the function $u(t, \sigma) = e^{-\epsilon t} e^{-i\sigma t}$ and $f : t \longmapsto e^{itH} V$, $t \in \mathbb{R}_+$, $\sigma \in \mathbb{R}$, where we take $v(t) = e^{-\epsilon t}$, $t \in \mathbb{R}_+$, and α, β are constants. $\qquad \square$

5.5 Application to the Lippmann-Schwinger equations

The 'Lippmann-Schwinger equations' (5.9) and (5.10) for the scattering solutions $\psi(\omega, \lambda)$ of the stationary Schrödinger equation (5.2), are valid for stationary scattering theory under rather stringent conditions.

For example, in the stationary Schrödinger equation (5.2) in $\mathbb{R}^d$, suppose that V is a uniformly bounded potential for which there exist $C > 0$ and $\rho > d$ such that

$$|V(x)| \leq C(1 + |x|)^{-\rho}, \quad x \in \mathbb{R}^d. \tag{5.16}$$

Let $\mathbb{S}^{d-1} = \{x \in \mathbb{R}^d : |x| = 1\}$ be the unit sphere in $\mathbb{R}^d$. Then for any $\lambda > 0$ and $\omega \in \mathbb{S}^{d-1}$, equation (5.2) has a unique solution $\psi(\,\cdot\,; \omega, \lambda)$ with asymptotics

$$\psi(x; \omega, \lambda) = e^{i\sqrt{\lambda}\langle \omega, x \rangle} + a \frac{e^{i\sqrt{\lambda}|x|}}{|x|^{(d-1)/2}} + o(|x|^{-(d-1)/2})$$

as $|x| \to \infty$ in $\mathbb{R}^d$. The solution $\psi(\,\cdot\,; \omega, \lambda)$ is associated with the incoming plane wave

$$\psi_0(x; \omega, \lambda) = e^{i\sqrt{\lambda}\langle \omega, x \rangle}, \quad x \in \mathbb{R}^d.$$

Then equations (5.9) and (5.10) are satisfied and if we define

$$\psi^-(x, \xi) = \psi(x; \omega, \lambda), \quad \psi_+(x; \xi) = \overline{\psi(x; -\omega, \lambda)}$$

for $\xi = \sqrt{\lambda}\omega$ and $x \in \mathbb{R}^d$, then the transformations

$$(\mathcal{F}^\pm f)(\xi) = (2\pi)^{-d/2} \int_{\mathbb{R}^d} \overline{\psi^\pm(x; \xi)} f(x) \, dx, \quad f \in L^2(\mathbb{R}^d),$$

define isometries from the absolutely continuous subspace $\mathcal{H}^{(a)}$ of $H = -\Delta + V$ onto $L^2(\mathbb{R}^d)$ such that the wave operators (5.1) have the representation

$$W_\pm = (\mathcal{F}^\pm)^* \mathcal{F}_0$$

with respect to the Fourier transform $\mathcal{F}_0$ on $L^2(\mathbb{R}^d)$ [137, equation (1.23)] and

$$\mathcal{F}^\pm H = |\xi|^2 \mathcal{F}^\pm.$$

The analysis can be pushed through to $\rho > (d + 1)/2$ in the bound (5.16) by utilising an appropriate function space for asymptotic estimates [137, Section 3.5] and to the case of short-range potentials $\rho > 1$ by averaging over $\omega \in \mathbb{S}^{d-1}$ [137, Section 2.3].

Under the conditions for which Theorem 4.4 is valid, from equations (5.5) and (5.7), for each $h \in \mathcal{D}(H_0)$, we have

$$W_- F_0(B)h = F_0(B)h - \lim_{\epsilon \downarrow 0} \int_B R(\lambda - i\epsilon)V \, d(F_0 h)(\lambda), \qquad (5.17)$$

$$W_- F_0(B)h = F_0(B)h - \lim_{\epsilon \downarrow 0} \int_B R_0(\lambda - i\epsilon)VW_- \, d(F_0 h)(\lambda), \quad (5.18)$$

for every Borel subset B of $\mathbb{R}_+ = [0, \infty)$, so equations (5.17) and (5.18) are equalities between $\mathcal{H}$-valued measures. The $\mathcal{H}$-valued measures $B \longmapsto W_- F_0(B)h$ and $B \longmapsto F_0(B)h$ are absolutely continuous on $\mathbb{R}_+$. Roughly speaking,

$$\frac{d(F_0 \psi_0(\,\cdot\,; \omega, \lambda'))}{d\lambda} = \delta(\lambda - \lambda')\psi_0(\,\cdot\,; \omega, \lambda'),$$

$$\frac{d(W_- F_0 \psi_0(\,\cdot\,; \omega, \lambda'))}{d\lambda} = \delta(\lambda - \lambda')\psi(\,\cdot\,; \omega, \lambda'),$$

although the plane wave $\psi_0(\,\cdot\,; \omega, \lambda')$ is not an element of $\mathcal{H}$ and the measures are not differentiable in $\mathcal{H}$, but it is clear that equations (5.5) and (5.7) are the integrated versions of the 'Lippmann-Schwinger equations' (5.9) and (5.10), valid under very general conditions.

Theorem 5.2. *Let $H_0 : \mathcal{D}(H_0) \to \mathcal{H}$ be a positive selfadjoint operator in the Hilbert space $\mathcal{H}$ with absolutely continuous spectrum. Let $V : \mathcal{D}(V) \to \mathcal{H}$ be a symmetric operator such that $\mathcal{D}(H_0) \subseteq \mathcal{D}(V)$ and $H = H_0 + V$ is selfadjoint on $\mathcal{D}(H_0)$. Suppose also that the limits (5.1) exist. Let F_0 be the spectral measure of H_0 and F the spectral measure of H and $R_0(z) = (H_0 - z)^{-1}$, $R(z) = (H - z)^{-1}$ the corresponding resolvents.*

Then for each $\epsilon > 0$, the two $\mathcal{L}(\mathcal{D}(A), \mathcal{H})$-valued functions $\lambda \longmapsto R(\lambda - i\epsilon)V$ and $\lambda \longmapsto R_0(\lambda - i\epsilon)VW_-$, $\lambda \in \mathbb{R}$, are $(F_0\psi)$-integrable in the Banach space $\mathcal{L}(\mathcal{D}(H_0), \mathcal{H}) \widehat{\otimes}_\tau \mathcal{D}(H_0)$ and the equalities

$$W_- \psi - \psi = -\lim_{\epsilon \downarrow 0} \int_\mathbb{R} R(\lambda - i\epsilon)V \, d(F_0\psi)(\lambda)$$

$$= -\lim_{\epsilon \downarrow 0} \int_\mathbb{R} R_0(\lambda - i\epsilon)VW_- \, d(F_0\psi)(\lambda).$$

holds for each $\psi \in \mathcal{D}(H_0)$.

Proof. We first show how the first equation of (5.5) follows from Theorem 5.1. According to equations (5.11) and (5.12), we can write

$$W_- \psi - \psi = \lim_{\epsilon \downarrow 0} i \int_0^\infty e^{itH} V \left(\int_\mathbb{R} e^{-i(\lambda - i\epsilon)t} \, d(F_0\psi)(\lambda) \right) dt \qquad (5.19)$$

for every $\psi \in \mathcal{D}(H_0)$. The function $u : \mathbb{R} \times \mathbb{R}_+ \to \mathbb{C}$ defined by

$$u(\lambda, t) = ie^{-i(\lambda - i\epsilon)t}, \quad \lambda \in \mathbb{R}, \ t \geq 0,$$

is a uniformly bounded continuous function on $\mathbb{R} \times \mathbb{R}_+$, so $u(\cdot, t)$ is $(F_0\psi)$-integrable and $F_0(H_0\psi)$-integrable in $\mathcal{H}$ for each $t > 0$.

Let $B(t)\psi = e^{itH}V\psi$ for every $t \in \mathbb{R}$ and $\psi \in \mathcal{D}(H_0)$ and $v(t) = e^{-\epsilon t}$, $t \geq 0$. Then the $\mathcal{L}(\mathcal{D}(H_0), \mathcal{H})$-valued function $t \longmapsto B(t)$, $t \in \mathbb{R}_+$, is strongly continuous in $\mathcal{L}_s(\mathcal{D}(H_0), \mathcal{H})$ and $|u(\lambda, t)| \leq v(t)$ for all $\lambda \in \mathbb{R}$, $t \in \mathbb{R}_+$.

By assumption the domain $\mathcal{D}(V)$ of the symmetric operator V contains the domain $\mathcal{D}(H_0)$ of the selfadjoint operator H_0, so by the Closed Graph Theorem, there exists $c > 0$ such that

$$\|V\psi\|^2 \leq c^2(\|H_0\psi\|^2 + \|\psi\|^2)$$

for every $\psi \in \mathcal{D}(H_0)$, so also

$$\|B(t)\psi\|^2 \leq c^2(\|H_0\psi\|^2 + \|\psi\|^2)$$

for every $\psi \in \mathcal{D}(H_0)$, because e^{itH} is a unitary operator for every $t \geq 0$. Condition (ii) of Theorem 5.1 is easily verified.

Now for each $t \geq 0$, Borel subset S of $\mathbb{R}$ and $\psi \in \mathcal{D}(H_0)$, the equality

$$B(t) \int_S u(\lambda, t) \, d(F_0\psi)(\lambda) = ie^{-\epsilon t} e^{itH} V e^{-itH_0} F_0(S)\psi$$

holds, so the function $t \longmapsto B(t) \int_S u(\lambda, t) \, d(F_0\psi)(\lambda)$, $t \geq 0$, is continuous in $\mathcal{H}$ and

$$\int_0^\infty \left\| B(t) \int_S u(\lambda, t) \, d(F_0\psi)(\lambda) \right\| dt \leq c\|\psi\|_{\mathcal{D}(H_0)}/\epsilon,$$

so it is Bochner integrable in $\mathcal{H}$, verifying conclusion (1) of Theorem 5.1. Similarly, the function $t \longmapsto u(\lambda, t)B(t)\psi$, $t > 0$, is Bochner integrable in $\mathcal{H}$ for each $\psi \in \mathcal{D}(H_0)$ and $\lambda \in \mathbb{R}$, because $u(\lambda, t)B(t)\psi = ie^{-\epsilon t} e^{it(H-\lambda)}V\psi$ and

$$\int_0^\infty u(\lambda, t)B(t)\psi \, dt = -R(\lambda - i\epsilon)V\psi,$$

so conclusion (2) of Theorem 5.1 is also verified directly.

It follows from equation (5.19) and Theorem 5.1, that for each $\epsilon > 0$, the $\mathcal{L}(\mathcal{D}(H_0), \mathcal{H})$-valued function $\lambda \longmapsto R(\lambda - i\epsilon)V$, $\lambda \in \mathbb{R}$, is $(F_0\psi)$-integrable in $\mathcal{L}(\mathcal{D}(H_0), \mathcal{H}) \widehat{\otimes}_\tau \mathcal{D}(H_0)$ and the equalities

$$W_-\psi - \psi = \lim_{\epsilon \downarrow 0} i \int_0^\infty e^{itH} V \left(\int_{\mathbb{R}} e^{-i(\lambda - i\epsilon)t} \, d(F_0\psi)(\lambda) \right) dt$$

$$= -\lim_{\epsilon \downarrow 0} \int_{\mathbb{R}} R(\lambda - i\epsilon)V \, d(F_0\psi)(\lambda)$$

hold for each $\psi \in \mathcal{D}(H_0)$.

Under our assumptions, we can also verify that

$$W_-\psi - \psi = \lim_{\epsilon\downarrow 0} i \int_0^\infty e^{itH_0} VW_- \left(\int_{\mathbb{R}} e^{-i(\lambda - i\epsilon)t} \, d(F_0\psi)(\lambda) \right) \, dt \qquad (5.20)$$

for every $\psi \in \mathcal{D}(H_0)$ [7, (20)].

The intertwining property $HW_- = W_-H_0$ and the assumption that $\mathcal{D}(H) = \mathcal{D}(H_0) \subseteq \mathcal{D}(V)$ ensures that $W_-\mathcal{D}(H_0) \subset \mathcal{D}(V)$, so that the right-hand side of equation (5.20) makes sense and $\lambda \longmapsto R_0(\lambda - i\epsilon)VW_-$, $\lambda \in \mathbb{R}$, is actually an $\mathcal{L}(\mathcal{D}(H_0), \mathcal{H})$-valued function.

Now if we define $B(t)\psi = e^{itH_0}VW_-\psi$ for every $\psi \in \mathcal{D}(H_0)$ and $t \geq 0$, and leave the functions u, v as above, then

$$\int_0^\infty u(\lambda, t)B(t)\psi \, dt = -R_0(\lambda - i\epsilon)VW_-\psi,$$

and an application of Theorem 5.1 ensures that $\mathcal{L}(\mathcal{D}(H_0), \mathcal{H})$-valued function $\lambda \longmapsto R_0(\lambda - i\epsilon)VW_-$, $\lambda \in \mathbb{R}$, is $(F_0\psi)$-integrable in $\mathcal{L}_s(\mathcal{D}(H_0), \mathcal{H})\widehat{\otimes}_\tau \mathcal{D}(H_0)$ and

$$W_-\psi - \psi = -\lim_{\epsilon\downarrow 0} \int_{\mathbb{R}} R_0(\lambda - i\epsilon)VW_- \, d(F_0\psi)(\lambda).$$

$\square$

Chapter 6

Random evolutions

The Feynman-Kac formula is used in the theory of stochastic processes to represent solutions of the heat equation with a source or sink term and finds widespread applications in mathematical finance and mathematical physics. I. Kluvánek pointed out in [81] that the Feynman-Kac formula may also be viewed as a means to represent the superposition of two general evolutions and is readily interpreted in terms of integration with respect to operator valued measures.

6.1 Evolution processes

The concepts of direct interest in modelling of evolving quantum systems are the dynamical group $t \longmapsto e^{-itH}$, $t \in \mathbb{R}$, representing the evolution of quantum states and the spectral measure Q representing the observation of the position of the system in the underlying classical configuration space Σ of the system. These lead to the operator valued set functions described in [64].

Suppose that Ω is a nonempty set. A family $\mathcal{S}$ of subsets of Ω is called a *semi-algebra* of sets if $\Omega \in \mathcal{S}$ and for every $A \in \mathcal{S}$ and $B \in \mathcal{S}$, there exist a positive integer n and pairwise disjoint set $X_j \in \mathcal{S}$, $j = 0, 1, \ldots, n$, such that

$$A \cap B = X_0, \quad A \setminus B = \bigcup_{j=1}^{n} X_j.$$

and the union $\bigcup_{j=0}^{k} X_j$ belongs to $\mathcal{S}$ for every $k = 0, 1, \ldots, n$. The example of a semi-algebra to keep in mind is the collection of all subintervals $[a, b)$, $0 \le a < b \le 1$, of $[0, 1)$.

Let $(\Sigma, \mathcal{E})$ be a measurable space. For each $s \geq 0$, suppose that $\mathcal{S}_s$ is a semi-algebra of subsets of a nonempty set Ω such that $\mathcal{S}_s \subseteq \mathcal{S}_t$ for every $0 \leq s < t$. For every $s \geq 0$, there are given functions $X_s : \Omega \to \Sigma$ with the property that $X_s^{-1}(B) \in \mathcal{S}_t$ for all $0 \leq s \leq t$ and $B \in \mathcal{E}$. It follows that the cylinder sets

$$E = \{X_{t_1} \in B_1, \ldots, X_{t_n} \in B_n\}$$

$$:= \{\omega \in \Omega : X_{t_1}(\omega) \in B_1, \ldots, X_{t_n}(\omega) \in B_n\} \tag{6.1}$$

$$= X_{t_1}^{-1}(B_1) \cap \cdots \cap X_{t_n}^{-1}(B_n)$$

belong to $\mathcal{S}_t$ for all $0 \leq t_1 < \ldots < t_n \leq t$ and $B_1, \ldots, B_n \in \mathcal{E}$.

Let $\mathcal{X}$ be a Banach space. The vector space of all continuous linear operators $T : \mathcal{X} \to \mathcal{X}$ is denoted by $\mathcal{L}(\mathcal{X})$. It is equipped with the uniform operator topology, but measures with values in $\mathcal{L}(\mathcal{X})$ of practical interest are only σ-additive for the strong operator topology.

A *semigroup* S of operators acting on $\mathcal{X}$ is a map $S : [0, \infty) \to \mathcal{L}(\mathcal{X})$ such that $S(0) = Id_{\mathcal{X}}$, the identity map on $\mathcal{X}$ and $S(t + s) = S(t)S(s)$ for all $s, t \geq 0$. The semigroup S represents the evolution of state vectors belonging to $\mathcal{X}$. If $\lim_{t \to 0+} S(t)x = x$ for every $x \in \mathcal{X}$, then S is called a C_0-semigroup of operators.

An $\mathcal{L}(\mathcal{X})$-valued *spectral measure* Q on $\mathcal{E}$ is a map $Q : \mathcal{E} \to \mathcal{L}(\mathcal{X})$ that is σ-additive in the strong operator topology and satisfies $Q(\Sigma) = Id_{\mathcal{X}}$ and $Q(A \cap B) = Q(A)Q(B)$ for all $A, B \in \mathcal{E}$. In quantum theory, Q is typically multiplication by characteristic functions associated with the position observables, but the spectral measures associated with momentum operators also appear.

Now suppose that $M_t : \mathcal{S}_t \to \mathcal{L}(\mathcal{X})$ is an additive operator valued set function for each $t \geq 0$. The system

$$(\Omega, \langle \mathcal{S}_t \rangle_{t \geq 0}, \langle M_t \rangle_{t \geq 0}; \langle X_t \rangle_{t \geq 0}) \tag{6.2}$$

is called a *time homogeneous Markov evolution process* if there exists a $\mathcal{L}(\mathcal{X})$-valued spectral measure Q on $\mathcal{E}$ and a semigroup S of operators acting on $\mathcal{X}$ such that for each $t \geq 0$, the operator $M_t(E) \in \mathcal{L}(\mathcal{X})$ is given by

$$M_t(E) = S(t - t_n)Q(B_n)S(t_n - t_{n-1}) \cdots Q(B_1)S(t_1) \tag{6.3}$$

for every cylinder set $E \in \mathcal{S}_t$ of the form (6.1) and the process is called an (S, Q)-*process*. The basic ingredients are the semigroup S describing the evolution of states and the spectral measure Q describing observation of states represented by vectors in X. An explicit proof that formula (6.13) actually does define an additive set function has been written, for example, in [85, Proposition 7.1].

In many significant examples the family $\mathcal{S}_t$ is a σ-*algebra* of subsets of a nonempty set Ω and $M_t : \mathcal{S}_t \to \mathcal{L}(\mathcal{X})$ is σ-*additive* for the strong operator topology of $\mathcal{L}(\mathcal{X})$, that is, the sum

$$M_t \left(\bigcup_{j=1}^{\infty} E_j \right) x = \sum_{j=1}^{\infty} M_t(E_j)x, \quad x \in \mathcal{X},$$

converges in the norm of $\mathcal{X}$ for all pairwise disjoint $E_j \in \mathcal{S}_t$, $j = 1, 2, \ldots$. Then we say that the process (6.2) is σ-*additive*. It is well-known that σ-additivity fails spectacularly for the unitary group $t \longmapsto e^{it\Delta}$.

Example 6.1. Let $(\Sigma, \mathcal{B})$ be a measurable space. Suppose that

$$(\Omega, \mathcal{S}, \langle P^x \rangle_{x \in \Sigma}; \langle X_t \rangle_{t \geq 0})$$

is a Markov process with $P^x(X_0 = x) = 1$ for each $x \in \Sigma$ [28, p. 6]. Then the *transition functions* $p_t(x, dy)$ are given by $p_t(x, B) = P^x(X_t \in B)$ for $x \in \Sigma$, $B \in \mathcal{B}$ and $t \geq 0$. For any signed measure $\mu : \mathcal{B} \to \mathbb{R}$ and $t \geq 0$, the measure $S(t)\mu : \mathcal{B} \to \mathbb{R}$ is given by

$$(S(t)\mu)(B) = \int_{\Sigma} p_t(x, B) \, d\mu(x), \quad \text{for } B \in \mathcal{B},$$

and the spectral measure Q is given by $Q(B)\mu = \chi_B.\mu$, $B \in \mathcal{B}$. Then $M_t(E)\mu$ defined by equation (6.13), is the measure

$$B \longmapsto \int_{\Sigma} P^x(\{X_t \in B\} \cap E) \, d\mu(x),$$

$B \in \mathcal{B}$, so M_t is an operator valued measure acting on the space $\mathcal{X}$ of signed measures with the total variation norm. In general, the semigroup S is not a C_0-semigroup on $\mathcal{X}$. For a *Feller process* [28, p. 50], S is a weak*-continuous semigroup of operators on the space of signed Borel measures $M(\Sigma) = C_0(\Sigma)'$.

For the case of Brownian motion in $\mathbb{R}^d$ considered previously in Section 4.3, we have

$$p_t(x, dy) = \frac{e^{-\frac{|x-y|^2}{2t}}}{(2\pi t)^{d/2}} \, dy, \quad t > 0, \; x \in \mathbb{R}^d.$$

Example 6.2. Let $(\Sigma, \mathcal{E}, \mu)$ be a measure space and let $T_t : \Sigma \to \Sigma$, $t \in \mathbb{R}$, be a group of measure preserving maps. Then $S(t) : f \longmapsto f \circ T_{-t}$ for $f \in$

$L^2(\mu)$ and $t \in \mathbb{R}$, defines a continuous unitary group S of linear operators on $L^2(\mu)$. Observe that if δ_x is the unit point mass at $x \in \Sigma$, then

$$\delta_x \circ T_{-t} = \delta_x \circ T_t^{-1} = \delta_{T_t x}.$$

Suppose that the spectral measure Q is defined by $Q(B)f = \chi_B . f$ for $B \in \mathcal{B}$ and $f \in L^2(\mu)$. The measure M_t defined by equation (6.13) is given by $M_t = S(t)(Q \circ \sigma^{-1})$. Here $\sigma : x \longmapsto T_t x$, for $t \geq 0$ and $x \in \Sigma$. The process $\langle X_t \rangle_{t \geq 0}$ is given by the evaluation maps $X_t(\omega) = \omega(t)$ for $t \geq 0$ and $\omega \in \Omega = \sigma(\Sigma)$.

If we take Σ to be the phase space of a system in classical mechanics with μ the Liouville measure, then the dynamical flow $t \longmapsto T_t x$, $t \in \mathbb{R}$, $x \in \Sigma$, of the system satisfies the assumptions above. The operator valued path space measure M_t is concentrated on the sample space Ω of all classical paths in the time interval $[0, t]$.

The viewpoint of operator valued measures of the process (6.2) deals with the essential mathematical representations of the concepts of interest to physics: the *dynamical group* represented by S and *observations*, represented by the spectral measure Q. It is due to I. Kluvánek [81–85].

There are two results relevant to the proof of Theorem 7.1 in the next chapter for dominated semigroups. The *modulus semigroup* is the smallest semigroup of operators such that $|S(t)u| \leq |S|(t)u$ for all $u \in L_+^2(\mu)$. If the modulus semigroup exists, then S is called a *dominated* semigroup of operators on $L^2(\mu)$.

Theorem 6.1 ([71, Theorem 3.1]). *Let $(\Sigma, \mathcal{E}, \mu)$ be a σ-finite measure space. For a semigroup S of operators on $L^2(\mu)$ and $Q(B)f = \chi_B f$, $B \in \mathcal{E}$, $f \in L^2(\mu)$, the operator valued, additive set function $M_t : \mathcal{S}_t \to \mathcal{L}(L^2(\mu))$ of the Markov evolution process (6.2) has uniformly bounded range on the algebra $\mathcal{S}_t$ generated by the cylinder sets (6.1) for every $t \geq 0$ if and only if S is dominated. In this case, the modulus semigroup $|S|$ of S exists and*

$$(|S|(t)u, v) = |(M_t u, v)|(\Omega)$$

for every $t > 0$, $u, v \in L^2(\mu)$ with $u \geq 0$ μ-a.e. and $v \geq 0$ μ-a.e.

Theorem 6.2 ([64]). *Suppose that (6.2) is a σ-additive Markov evolution process on the Banach space $\mathcal{X}$ such that X is progressively measurable, that is, $(\omega, s) \longmapsto X_s(\omega)$, $0 \leq s \leq t$, is jointly $(\mathcal{S}_t \otimes \mathcal{B}([0, t]))$-measurable for every $t > 0$.*

For any uniformly bounded $\mathcal{E}$-measurable function $V : \Sigma \to \mathbb{C}$, the operator valued map

$$S_V : t \longmapsto \int_\Omega \exp\left[\int_0^t V \circ X_s \, ds\right] dM_t, \quad t \geq 0$$

is a semigroup of operators on $\mathcal{X}$. If S is a C_0-semigroup on $\mathcal{X}$ with $S(t) = e^{tA}$, $t \geq 0$, then $S_V(t) = e^{t(A+Q(V))}$, $t \geq 0$.

Proof. If $x \in \mathcal{X}$ and we let $u(0) = x$ and

$$u(t) = \int_\Omega e^{\int_0^t V \circ X_r \, dr} \, d(M_t x), \quad t > 0,$$

then an application of the Fubini-Tonelli Theorem gives

$$\int_0^t \int_\Omega V \circ X_s . e^{\int_0^s V \circ X_r \, dr} \, d(M_t x) \, ds = \int_\Omega \left(\int_0^t V \circ X_s e^{\int_0^s V \circ X_r \, dr} \, ds\right) d(M_t x).$$

Applying formula (6.3) to the left-hand side and the Fundamental Theorem of Calculus to the right-hand side of this equation gives

$$\int_0^t S(t-s)Q(V)u(s) \, ds = \int_\Omega \left(e^{\int_0^t V \circ X_r \, dr} - 1\right) d(M_t x) = u(t) - S(t)x.$$

The expression $u(t) = S(t)x + \int_0^t S(t-s)Q(V)u(s) \, ds$ is a form of Duhamel equation and by iterating it, we obtain the Dyson series expansion

$$e^{t(A+Q(V))}x = e^{tA}x + \sum_{n=1}^\infty \int_0^t \int_0^{s_k} \cdots \int_0^{s_2} \left[e^{(t-s_k)A} B e^{(s_k - s_{k-1})A} \cdots B e^{s_1 A} x\right] ds_1 \cdots ds_k$$

for the infinitesimal generator A of S [76, Theorem IX.2.1], that is, we obtain the Feynman-Kac formula

$$e^{t(A+Q(V))} = \int_\Omega e^{\int_0^t V \circ X_r \, dr} \, dM_t, \quad t > 0. \tag{6.4}$$

Here $Q(V) = \int_\Sigma V \, dQ$ is a bounded linear operator on $\mathcal{X}$. The identity (6.4) first appeared in [81] and an exposition in greater detail appears in [64]. $\square$

We shall always assume that the dominated semigroup S is associated with a progressively measurable process (6.2) with $\mathcal{X} = L^2(\mu)$—the case in all examples of practical interest. In effect, this is a mild regularity assumption on S weaker than that of [119] and [97]. On the other hand, considerable effort is devoted to establishing this fact in the theory of Dirichlet forms [48]. We look more closely at this regularity assumption in the next section.

6.2 Measurable functions

A necessary and sufficient condition for the progressive measurability for metric space valued random processes was given by K.-L. Chung and J. Doob [27, Proposition 33] using the notion of *separable Borel measurability*.

As pointed out in the proof of a similar result in [34, Théorème IV.30] for real valued processes, separable Borel measurability is the same notion as *strong measurability* familiar from vector integration [38, Definition II.1]. In the context of Banach spaces, Pettis's Measurability Theorem [38, Theorem II.2] yields the equivalence of strong measurability and separable Borel measurability. We first clarify the measurability situation for metric space valued functions.

A function $f : \Omega \to \Sigma$ from a measurable space $(\Omega, \mathcal{S})$ with values in a metric space (Σ, d) is called $\mathcal{S}$-*measurable* if $f^{-1}(B) \in \mathcal{S}$ for every Borel subset B of Σ. An $\mathcal{S}$-measurable function $f : \Omega \to \Sigma$ with finitely many values is called $\mathcal{S}$-*simple*. A function $f : \Omega \to \Sigma$ is $\mathcal{S}$-simple if and only if there exists a finite partition $\{\Omega_1, \ldots, \Omega_n\}$, $n \in \mathbb{N}$, into $\mathcal{S}$-measurable sets such that $f(\Omega_j) = \{\sigma_j\}$ for $\sigma_j \in \Sigma$, $j = 1, \ldots, n$.

Let T be a Hausdorff topological space and $A \subset T$. Then $[A]$ denotes the set of all elements in T which are the limit of some sequence of elements of A. A set $A \subset T$ is called *sequentially closed* if $A = [A]$. The *sequential closure* $\overline{A}^s$, of a set $A \subset T$, is the smallest sequentially closed subset of T which contains A, that is, the intersection of all sequentially closed subsets of T containing A. For a metric space (Σ, d) and a nonempty set Ω, the product topology of $T = \Sigma^\Omega = \{f | f : \Omega \to \Sigma\}$ is called the topology of *pointwise convergence*.

Proposition 6.1. *Let $\mathcal{B}$ be the Borel σ-algebra of a metric space (Σ, d) and let $f : \Omega \to \Sigma$ be a function from a measurable space $(\Omega, \mathcal{S})$ into Σ. Then the following conditions are equivalent.*

a) *f is the uniform limit on Ω of a sequence of countably valued $\mathcal{S}$-measurable Σ-valued functions.*

b) *f is Borel measurable with separable range.*

c) *f belongs to the pointwise sequential closure of $\mathcal{S}$-simple Σ-valued functions on Ω.*

d) *f is the pointwise limit on Ω of a sequence of $\mathcal{S}$-simple Σ-valued functions.*

Proof. The argument is omitted in the proof of [34, Théorème IV.30], so we give it here. Condition a) implies condition b), because countably valued functions f_n, $n = 1, 2, \ldots$, are necessarily $(\mathcal{S}\text{-}\mathcal{B})$-measurable and if $f_n \to f$ pointwise as $n \to \infty$, then $f(\Omega) \subset \overline{\cup_n f_n(\Omega)}$ and $\overline{\cup_n f_n(\Omega)}$ is separable.

To see that b) implies a), let $\{S_{n,k}\}_k$ be a partition of $\overline{f(\Omega)}$ into disjoint Borel sets with diameter less than $1/n$: if $\{x_k\}_{k \in \mathbb{N}}$ is a dense subset of $\overline{f(\Omega)}$ and

$$B_{n,k} = \{x \in \Sigma : d(x, x_k) < 1/(2n)\}, \quad k = 1, 2, \ldots,$$

then $S_{n,k} = (B_{n,k} \cap \overline{f(\Omega)}) \setminus (\cup_{j<k} B_{n,j})$ for $k = 1, 2, \ldots$, gives one such partition for each n. For $f^{-1}(S_{n,k})$ nonempty, we choose $\omega_{n,k} \in f^{-1}(S_{n,k})$ and $f_n(\omega) = f(\omega_{n,k})$ for $\omega \in f^{-1}(S_{n,k})$, $k = 1, 2, \ldots$. Then f_n is countably valued for each $n = 1, 2, \ldots$ and we have $d(f_n(\omega), f(\omega)) < 1/n$ for every $\omega \in \Omega$ because $\{S_{n,k}\}_k$ covers $f(\Omega)$.

Let $S(\mathcal{S}, \Sigma)$ be the collection of all $\mathcal{S}$-simple Σ-valued functions. Then in the notation above, $[S(\mathcal{S}, \Sigma)] \subset \overline{S(\mathcal{S}, \Sigma)}^s$. Clearly a countably valued $\mathcal{S}$-measurable function belongs to $[S(\mathcal{S}, \Sigma)]$, so a) implies c). If $f \in [[S(\mathcal{S}, \Sigma)]]$, then c) holds. The pointwise sequential limit of elements of $[[S(\mathcal{S}, \Sigma)]]$ belongs to $[[S(\mathcal{S}, \Sigma)]]$ too, and so $[[S(\mathcal{S}, \Sigma)]]$ is sequentially closed and we have $[[S(\mathcal{S}, \Sigma)]] = \overline{S(\mathcal{S}, \Sigma)}^s$, that is, c) implies a).

Finally, that d) implies c) is clear. To show that a) implies d), the countably valued functions $\{f_n\}_{n=1}^\infty$ converging uniformly to f can be truncated to $\mathcal{S}$-simple Σ-valued functions $\{s_n\}_{n=1}^\infty$ converging pointwise to f by embedding $f(\Omega)$ into $[0,1]^\mathbb{N}$ as in [26, Lemma I.1.2]. $\qquad\square$

If any of the conditions a)-c) of Proposition 6.1 are satisfied, then f is said to be *strongly measurable*. If $(\Omega, \mathcal{S}, \mu)$ is a σ-finite measure space, then f is said to be *strongly μ-measurable* if the conditions above hold off a set of measure zero. For a strongly μ-measurable function with values in a complete metric space (Σ, d), there exists a set Ω_0 of full μ-measure such that $f(\Omega_0)$ is contained in a σ-compact set, because when μ is a finite measure, $\mu \circ f^{-1}$ is a regular Borel measure on Σ [46, 423B (a), 432B].

Remark 6.1. a) Suppose that $(\Omega, \mathcal{S}, \mu)$ is a σ-finite measure space and $f : \Omega \to \Sigma$ is a Borel measurable function with values in a Souslin space Σ [46, 423A]. A similar line of reasoning ensures that there exists a set Ω_0 of full μ-measure such that $f(\Omega_0)$ is contained in the countable union of compact metrisable sets [46, 432B, 423D(c)] and f is the limit on Ω_0 of a sequence of $\mathcal{S}$-simple functions.

b) Suppose that $\mathcal{F}$ is a family of uniformly continuous functions $\xi : \Sigma \to \mathbb{R}$ separating *the completion* of $f(\Omega)$ in the metric d. In the case that $f(\Omega)$

is separable, for $f : \Omega \to \Sigma$ to be Borel measurable it suffices to check that the scalar valued function $\xi \circ f$ is Borel measurable for each $\xi \in \mathcal{F}$, as in Pettis's Measurability Theorem mentioned above, see [46, 423B (a), 423F (b)].

c) In general, the sequential closure $\overline{A}^s$ of a subset A of a Hausdorff topological space T may also be defined recursively by setting $A_0 = A$ and proceeding as follows. Let ω_1 be the smallest uncountable ordinal. Suppose that $0 < \alpha < \omega_1$ and that A_β has been defined for all ordinals β satisfying $0 \le \beta < \alpha$. Define $A_\alpha = \left[\bigcup_{0 \le \beta < \alpha} A_\beta \right]$. Then $\overline{A}^s = \bigcup_{0 \le \alpha < \omega_1} A_\alpha$. For the special case of $T = \Sigma^\Omega$ with (Σ, d) a metric space and A the collection of $\mathcal{S}$-simple Σ-valued functions, Proposition 6.1 ensures the remarkable property that $\overline{A}^s = [[A]]$, that is, the pointwise limit of strongly measurable functions is strongly measurable, so ensuring that the recursion stops at most at the second step: $[[A]] = [A]$.

6.3   Progressive measurability

The only impediment to this simple argument used in the proof of Theorem 6.2 is the question of *measurability*, so that $s \longmapsto V \circ X_s$, $s \ge 0$, is Lebesgue measurable almost everywhere and $\int_0^t V \circ X_s \, ds$ is $\sigma(\mathcal{S}_t)$-measurable for each $t > 0$. In order to apply the Fubini-Tonelli Theorem, it is enough to require that the random process X is *progressively measurable*, that is, the mapping

$$(s, \omega) \longmapsto X_s(\omega), \quad 0 \le s \le t, \ \omega \in \Omega,$$

is $(\mathcal{B}([0, t]) \otimes \sigma(\mathcal{S}_t))$-measurable into $(\Sigma, \mathcal{B})$ for each $t > 0$.

If the real valued process X is right-continuous, then it is necessarily progressively measurable [28, Theorem 1, p. 38]—this covers most cases of practical interest. However, the Feynman-Kac formula (6.4) suggests that we ought to be able to choose a sample space Ω and random process X that is progressively measurable just under weak assumptions about the measurable space $(\Sigma, \mathcal{B})$. In the proof of the Cwikel-Lieb-Rozenbljum inequality for dominated semigroups in [67], the assumption that the underlying process is progressively measurable appears as an extraneous condition in relation to the rest of the proof, in which a Fubini-type argument is again a central feature. Here we examine conditions where progressive measurability can be achieved simply by modifying the process X.

Recall that a topological space Σ is called *Polish* if it is separable and its topology can be defined from a metric under which Σ is complete.

The following statement is the main result of the present section.

Theorem 6.3. *Suppose that $\mathcal{B}$ is the Borel σ-algebra of a Polish space Σ, S is a C_0-semigroup of injective operators acting on a separable Banach space $\mathcal{X}$ and $Q : \mathcal{B} \to \mathcal{L}(\mathcal{X})$ is a spectral measure also acting on $\mathcal{X}$. Additionally, suppose that for each $t > 0$, formula (6.3) defines a σ-additive measure $M_t : \sigma(\mathcal{S}_t) \to \mathcal{L}(\mathcal{X})$ with respect to the random process X and sample space Ω.*

Then there exists a progressively measurable process $\tilde{X}_s : \Omega \to \Sigma$, $s \geq 0$, such that $\{X_s \neq \tilde{X}_s\}$ is M_t-null for all $0 \leq s \leq t$ and $t > 0$.

The progressive measurability is with respect to the given filtration $\sigma(\mathcal{S}_t)$, $t \geq 0$. It turns out that $\sigma(\mathcal{S}_t)$ is also generated by $\tilde{X}_s$, $0 \leq s \leq t$. In [100], an elementary proof of the existence of progressively measurable modification $\tilde{X}$ of a *measurable* adapted process X is given. There it is assumed that X is adapted to a given filtration $\mathcal{F} = \langle \mathcal{F}_t \rangle_{t \geq 0}$ and $\tilde{X}$ is proved to be progressively measurable with respect to the *same* filtration $\mathcal{F}$. A different proof appears in [34, Théorème IV.30 (b)] and the result may be viewed as complementary to Theorem 6.3. The proof of Theorem 6.3 above appeals to the separability of $L^2(M_t)$ and the assumption that S is strongly continuous.

Under the assumptions of Theorem 6.3, we shall construct a modification $\tilde{X}_s : \Omega \to \Sigma$, $s \geq 0$, of X such that the mapping

$$(s, \omega) \longmapsto \tilde{X}_s(\omega), \quad 0 \leq s \leq t, \ \omega \in \Omega,$$

is $(\mathcal{B}([0,t]) \otimes \sigma(\mathcal{S}_t))$-measurable into $(\Sigma, \mathcal{B})$ for each $t > 0$.

Proof. Because $\mathcal{B}$ is the Borel σ-algebra of a Polish space Σ, there exists a countable subfamily $\mathcal{A}$ of $\mathcal{B}$ such that $\sigma(\mathcal{A}) = \mathcal{B}$ [46, 424B (a)]. The algebra generated by $\mathcal{A}$ is also countable, so we may assume that $\mathcal{A}$ is already a countable algebra of subsets of Σ. The algebra $\mathcal{S}^0$ of subsets of Ω generated by $\{X_r \in A\}$ for $r \geq 0$ rational and $A \in \mathcal{A}$ is also countable. The σ-algebra generated by the random variables X_s, $s \geq 0$, is denoted by $\mathcal{S}$. For each $t > 0$, the operator valued measure $M_t : \sigma(\mathcal{S}_t) \to \mathcal{L}(\mathcal{X})$ measures the σ-algebra $\sigma(\mathcal{S}_t)$ generated by the collection of all cylinder sets (6.1).

Step 1. For each $t \geq 0$, the linear subspace $\mathrm{span}\{[\chi_E]_{M_t} : E \in \mathcal{S}^0 \cap \mathcal{S}_t\}$ is dense in $L^2(M_t)$.

Because $\mathrm{span}\{[\chi_E]_{M_t} : E \in \mathcal{S}_t\}$ is dense in $L^1(M_t)$ by [80, Theorem of Extension] and the topology of $L^2(M_t)$ is weaker than the topology of

$L^1(M_t)$ on $\{[f]_{M_t} : \|f\|_\infty \le 1\}$, it is enough to show that span$\{[\chi_E]_{M_t} :$ $E \in \mathcal{S}^0 \cap \mathcal{S}_t\}$ is $L^1(M_t)$-dense in span$\{[\chi_E]_{M_t} : E \in \mathcal{S}_t\}$.

Let $t > 0$ and $E = \{X_s \in A\} \in \mathcal{S}_t$ for $0 \le s \le t$ and $A \in \mathcal{B}$. Suppose that $E_n = \{X_{s_n} \in A_n\} \in \mathcal{S}^0 \cap \mathcal{S}_t$ for $n = 1, 2, \ldots$, where $s_n \to s$ and $\sup_{\|\xi\|_{\mathcal{X}'} \le 1} |\langle Qx, \xi \rangle|(A\Delta A_n) \to 0$ as $n \to \infty$ for each $x \in \mathcal{X}$. Because $\mathcal{A}$ is a countable algebra of subsets of Σ for which $\sigma(\mathcal{A}) = \mathcal{B}$, such sets $A_n \in \mathcal{A}$, $n = 1, 2, \ldots$, exist by [116, Proposition 2]. This observation also handles the case $t = 0$.

Appealing to the assumption that the semigroup S in formula (6.3) defining the operator valued measure M_t is a C_0-semigroup and Q is a spectral measure, it follows that

$$M_t(E_n \cap F)x \to M_t(E \cap F)x \qquad (6.5)$$

in the norm of $\mathcal{X}$ as $n \to \infty$ for each $x \in \mathcal{X}$ and each $F \in \mathcal{S}_t$. Because $\{[\chi_F]_{M_t} : F \in \mathcal{S}_t\}$ is $L^1(M_t)$-dense in $\{[\chi_F]_{M_t} : F \in \sigma(\mathcal{S}_t)\}$ [80, Theorem of Extension], the convergence (6.5) is also valid for $F \in \sigma(\mathcal{S}_t)$. Appealing to the Hahn decomposition of a real valued measure, we also have $|\langle M_t x, \xi \rangle|(E_n \cap F) \to |\langle M_t x, \xi \rangle|(E \cap F)$ as $n \to \infty$ for each $x \in \mathcal{X}$, $\xi \in \mathcal{X}'$ and $F \in \sigma(\mathcal{S}_t)$.

The locally convex Hausdorff topology of $L^1(m)$ for an $\mathcal{L}(\mathcal{X})$-valued measure m is defined by the family of seminorms

$$p_x[m] : [f]_m \longmapsto \sup_{\|\xi\|_{\mathcal{X}'} \le 1} \int_\Omega |f| \, d|\langle mx, \xi \rangle|, \quad [f]_m \in L^1(m), \qquad (6.6)$$

for $x \in \mathcal{X}$ [86, p. 24].

A result of Bartle-Dunford-Schwartz [86, Theorem II.1.1] gives finite positive measures $\mu_{x,t}$ such that $\mu_{x,t}(E) \le \sup_{\|\xi\|_{\mathcal{X}'} \le 1} |\langle M_t x, \xi \rangle|(E)$ for each $x \in \mathcal{X}$ and $E \in \sigma(\mathcal{S}_t)$ and $\lim_{\mu_{x,t}(E) \to 0} \sup_{\|\xi\|_{\mathcal{X}'} \le 1} |\langle M_t x, \xi \rangle|(E) = 0$. A glance at the construction of the measure $\mu_{x,t}$ in [86, Theorem II.1.1] and dominated convergence ensures that $\mu_{x,t}(E_n \cap F) \to \mu_{x,t}(E \cap F)$ as $n \to \infty$ for each $x \in \mathcal{X}$ and $F \in \sigma(\mathcal{S}_t)$, that is, $\chi_{E_n} \to \chi_E$ weakly in $L^1(\mu_{x,t})$ as $n \to \infty$, for each $x \in \mathcal{X}$. By the Hahn-Banach Theorem, χ_E belongs to the norm closure in $L^1(\mu_{x,t})$ of the balanced convex hull of $\{[\chi_F]_{\mu_{x,t}} : F \in \mathcal{S}^0 \cap \mathcal{S}_t\}$ for each $x \in \mathcal{X}$ and so, by [86, Theorem III.2.2], also to the closure in $L^1(M_t)$ of the balanced convex hull of $\{[\chi_F]_{M_t} : F \in \mathcal{S}^0 \cap \mathcal{S}_t\}$. Because the same argument applies to the intersection of finitely many cylinder sets like E, it follows that span$\{[\chi_F]_{M_t} : F \in \mathcal{S}^0 \cap \mathcal{S}_t\}$ is $L^1(M_t)$-dense in span$\{[\chi_F]_{M_t} : F \in \mathcal{S}_t\}$.

The measurable space $(\Sigma, \mathcal{B})$ is isomorphic to either $(C, \mathcal{P}C)$ for some countable set C or $(\mathbb{R}, \mathcal{B}(\mathbb{R}))$ [46, 424C]. It suffices to prove the result for

$\Sigma = \mathbb{R}$. The countable case is handled in a similar manner. Under the measure space isomorphism, Borel measurability for an $\mathbb{R}$-valued function guarantees Borel measurability for the corresponding Σ-valued function f and Proposition 6.1 ensures that $f : \Omega \to \Sigma$ is strongly measurable.

The space $L^0(M_t) = \{[f]_{M_t} | f : \Omega \to \mathbb{R} \text{ is } M_t\text{-measurable}\}$ is given the weakest topology for which the map

$$[f]_{M_t} \longmapsto \left[\frac{|f|}{1 + |f|} \right]_{M_t}, \quad [f]_{M_t} \in L^0(M_t),$$

is continuous in $L^1(M_t)$. Because $\mathcal{X}$ is separable, for some countable dense subset $\mathcal{D}$ of the unit ball of $\mathcal{X}$, the topology of $L^1(M_t)$ is determined by the countable family of seminorms $p_x[M_t]$, $x \in \mathcal{D}$, defined by equation (6.6). It follows that both $L^1(M_t)$ and $L^0(M_t)$ are metrisable and complete.

Step 2. The mapping $s \longmapsto [X_s]_{M_t}$, $0 \le s \le t$, is strongly Borel measurable in $L^0(M_t)$ for each $t \ge 0$.

As mentioned above, the limit of a sequence of strongly measurable functions is strongly measurable. Let $X_s^{(n)} = \chi_{\{|X_s| \le n\}} X_s$ for $s \ge 0$ and $n = 1, 2, \ldots$. Because $L^1(M_t)$ embeds in $L^0(M_t)$ and for each $0 \le s \le t$, $[X_s^{(n)}]_{M_t} \to [X_s]_{M_t}$ M_t-a.e. and in $L^0(M_t)$ as $n \to \infty$, it suffices to prove that mapping $s \longmapsto [X_s^{(n)}]_{M_t}$, $0 \le s \le t$, is strongly Borel measurable in $L^2(M_t)$ for each $t \ge 0$. By Step 1, $L^2(M_t)$ is separable because we may take the span over the rationals to obtain our countable dense set. The separability of $\mathcal{X}$ ensures that $L^2(M_t)$ is metrisable. The uniform boundedness principle ensures that $L^2(M_t)$ is sequentially complete, so $L^2(M_t)$ is a Polish space as well.

For each $n = 1, 2, \ldots$, let $\mathcal{Y}_n$ denote the closure in $L^2(M_t)$ of the linear span of the collection $\Xi_n = \{[X_s^{(n)}]_{M_t} : 0 \le s \le t\}$ of random variables. For each bounded $\sigma(\mathcal{S}_t)$-measurable function $\varphi : \Omega \to \mathbb{C}$, $u \in \mathcal{X}$ and $v \in \mathcal{X}'$, the continuous linear functional $\xi[\varphi, u, v] : L^2(M_t) \to \mathbb{C}$ is defined by

$$\langle f, \xi[\varphi, u, v] \rangle = \int_\Omega f\varphi \, d|\langle M_t u, v \rangle|, \quad f \in L^2(M_t).$$

Then the collection $\{\xi[\varphi, u, v] : \varphi \in \Xi_n, u \in \mathcal{X}, v \in \mathcal{X}'\}$ of continuous linear functionals on $L^2(M_t)$ clearly separates the closed subspace $\mathcal{Y}_n$ of $L^2(M_t)$.

The closed linear subspace $\mathcal{Y}_n$ is itself a Polish space, so according to [46, 423B (a), 423F (b)], the Borel σ-algebra of $\mathcal{Y}_n$ is generated by the continuous linear functionals $y \longmapsto \langle y, \xi[\varphi, u, v] \rangle$, $y \in \mathcal{Y}_n$, as φ ranges over

Ξ_n and $u \in \mathcal{X}$ and $v \in \mathcal{X}'$. Because S is a C_0-semigroup, inspection of formula (6.3) shows that the scalar valued function

$$s \longmapsto \int_E X_s^{(n)} X_r^{(n)} \, d\langle M_t u, v \rangle, \quad 0 \le s \le t, \tag{6.7}$$

is *continuous* for each $0 \le r \le t$, $u \in \mathcal{X}$, $v \in \mathcal{X}'$ and $E \in \mathcal{S}_t$. The collection of sets $E \in \sigma(\mathcal{S}_t)$ for which (6.7) is Borel measurable is a monotone class, so it is the whole σ-algebra $\sigma(\mathcal{S}_t)$. An appeal to the Hahn Decomposition Theorem for real valued measures establishes the Borel measurability of the scalar valued function $s \longmapsto \langle X_s^{(n)}, \xi[\varphi, u, v] \rangle$, $0 \le s \le t$, for each $\varphi \in \Xi_n$, $u \in \mathcal{X}$, $v \in \mathcal{X}'$, so that the $\mathcal{Y}_n$-valued function $s \longmapsto [X_s^{(n)}]_{M_t}$, $0 \le s \le t$, is Borel measurable. From the separability of $L^2(M_t)$ and Proposition 6.1, we see that $s \longmapsto [X_s^{(n)}]_{M_t}$, $0 \le s \le t$, is strongly Borel measurable in $L^2(M_t)$ for each $t \ge 0$. Taking the limit as $n \to \infty$, it follows that $s \longmapsto [X_s]_{M_t}$, $0 \le s \le t$, is strongly Borel measurable in $L^0(M_t)$.

Step 3. Let $T > 0$. There exists a process $\hat{X}_s : \Omega \to \Sigma$, $0 \le s \le T$, such that

  i) $\{X_s \ne \hat{X}_s\}$ is M_t-null for all $0 \le s \le T$,

  ii) $\hat{X}_s$ is $\sigma(\mathcal{S}_s)$-measurable for each $0 \le s \le T$, and

  iii) the mapping $(s, \omega) \longmapsto \hat{X}_s(\omega)$, $0 \le s \le t$, $\omega \in \Omega$, is strongly $(\mathcal{B}([0,t]) \otimes \sigma(\mathcal{S}_t))$-measurable for each $0 < t \le T$.

Let $T > 0$. The proof of [27, Proposition 33] applies verbatim in the proof of Step 3 with convergence in probability replaced by convergence in the space $L^0(M_T)$, see also [34, Théorème IV.30]. For convenience, the elementary construction is reproduced here.

By Proposition 6.1 and Step 2, the closure C of the set $\{[X_s]_{M_T} : 0 \le s \le T\}$ in $L^0(M_T)$ is separable. Let $\{S_{n,k}\}_k$ be a partition of C into disjoint Borel sets with diameter less than $1/n^2$ with the $(n+1)$-th partition a refinement of the n-th. Define $A_{n,k,j} \subset [0,T]$ by

$$A_{n,k,j} = \{t \in [0,T] : [X_t]_{M_T} \in S_{n,k}, \; j2^{-n} < t \le (j+1)2^{-n}\}.$$

Then $\{A_{n,k,j}\}_{k,j}$ is a partition of $[0,T]$ for each $n = 1, 2, \dots$. For nonempty $A_{n,k,j}$, choose $t_{n,k,j} \in A_{n,k,j}$ recursively so that $t_{n,k,j} \in \{t_{n+1,\ell,m}\}$ and define $\phi_n : [0,T] \to [0,T]$ by

$$\phi_n = \sum_{k,j} t_{n,k,j} \chi_{A_{n,k,j}}.$$

Then ϕ_n is Borel measurable and

$$X_{\phi_n(t)} = X_t, \quad \text{if } t = t_{m,k,j} \text{ and } n \geq m,$$
$$|\phi_n(t) - t| < 2^{-n}, \quad \text{if } t \in [0, T].$$

For each $t \in [0, T]$, we have $X_{\phi_n(t)}(\omega) \to X_t(\omega)$ as $n \to \infty$ for M_T-almost all $\omega \in \Omega$.

Let σ be any element of Σ. Then the function $\hat{X}$ defined for each $t \in [0, T]$ by

$$\hat{X}_t(\omega) = \begin{cases} \lim_{n \to \infty} X_{\phi_n(t)}(\omega), & \text{for all } \omega \in \Omega \text{ for which the limit exists,} \\ \sigma, & \text{otherwise,} \end{cases}$$

has the required properties.

If $t > s$, then the formula $M_t(E) = S(t - s)M_s(E)$ holds for every $E \in \sigma(\mathcal{S}_t)$. Inspection of equation (6.3) shows that the equality is valid for any cylinder set (6.1). Equality on $\sigma(\mathcal{S}_t)$ follows from the σ-additivity of M_t and M_s in the strong operator topology and the continuity of the linear operator $S(t - s)$.

Because the bounded linear operator $S(r) : \mathcal{X} \to \mathcal{X}$ is assumed to be injective for each $r > 0$, the operator valued measures $M_t \upharpoonright \sigma(\mathcal{S}_s)$ and M_s are mutually absolutely continuous on the σ-algebras $\sigma(\mathcal{S}_s)$. Hence, choosing an unbounded sequence of times $0 < T_n < T_{n+1}$, $n = 1, 2, \ldots$, the process above may be applied on each time interval $[T_n, T_{n+1})$, $n = 1, 2, \ldots$, so that the set $\{X_s \neq \tilde{X}_s^{(n)}\}$ is $M_{T_{n+1}}$-null for all $T_n \leq s < T_{n+1}$. Then for $T_n < t \leq T_{N+1}$, the set $\{X_s \neq \tilde{X}_s^{(n)}\}$ is M_t-null for all $T_n \leq s \leq t$. The required progressively measurable process is obtained by setting $T_0 = 0$ and

$$\tilde{X}_s = \sum_{n=0}^{\infty} \tilde{X}_s^{(n)} \chi_{[T_n, T_{n+1})}(s), \quad s \geq 0,$$

in the case $\Sigma = \mathbb{R}$ and the image of $\tilde{X}$ under the Borel isomorphism in the case of a general Polish space Σ. $\qquad\Box$

Remark 6.2. a) If $(\Omega, \mathcal{F}, P)$ is a general probability measure space and the process X_t, $t \geq 0$, is adapted to some filtration $\{\mathcal{F}_t\}_{t \geq 0}$, then it requires more work to show that there exists a progressively measurable modification $\tilde{X}_t$, $t \geq 0$, adapted to the same filtration $\{\mathcal{F}_t\}_{t \geq 0}$, see [34, Théorème IV.30], [74] and [100] on this point. Step 2 above is a crucial feature leading to the conclusion of Theorem 6.3 in the present setting, where we have appealed to the Markov property of M_t, $t \geq 0$, and the assumption that S is a C_0-semigroup of operators acting on the separable Banach space $\mathcal{X}$. As pointed out in [27, Proposition 32] for the case of a general metric space (Σ, d) in the probabilistic setting, a strongly progressively measurable modification exists only if the mapping $s \longmapsto [X_s]_{M_t}$, $0 \leq s \leq t$, is strongly Borel measurable in the space $L^0(M_t, \Sigma)$ of M_t-measurable Σ-valued functions for each $t \geq 0$.

b) In the probabilistic setting, a *Feller semigroup* is a sub-Markov C_0-semigroup on the space $C_0(\Sigma)$ of continuous functions vanishing at infinity. The topological space Σ is assumed to be a locally compact Hausdorff space with a countable base, so it is σ-compact and Polish. A slight modification of the proof above also applies to the weak*-continuous semigroup dual to a Feller semigroup. In this setting, a much stronger property than Theorem 6.3 obtains using martingale theory: there exists a modification $\tilde{X}$ of the process which is right continuous with left limits [28, Theorem 2.6].

c) For a general metric space (Σ, d) rather than a Polish space and $t > 0$, the mapping $s \longmapsto [X_s]_{M_t}$, $0 \leq s \leq t$, is strongly Borel measurable in the space $L^0(M_t, \Sigma)$ of M_t-measurable Σ-valued functions provided that

$$\lim_{s \to r} \sup_{0 \leq s \leq t} \sup_{\|\xi\|_{\mathcal{X}'} \leq 1} |\langle M_t x, \xi \rangle|(\{d(X_s, X_r) > \epsilon\}) = 0$$

for each $x \in \mathcal{X}$, and $0 \leq r \leq t$ (stochastic continuity), because then the mapping $s \longmapsto [X_s]_{M_t}$, $0 \leq s \leq t$, is actually a *continuous* map from the compact set $[0, t]$ into the metrisable space $L^0(M_t, \Sigma)$ and so it is strongly Borel measurable in $L^0(M_t, \Sigma)$.

Moreover, a function from $[0, t]$ into $L^0(M_t, \Sigma)$ is strongly λ-measurable in $L^0(M_t, \Sigma)$ with respect to Lebesgue measure λ if and only if it is Lusin λ-measurable [46, 411M, 418E-J]. If we only require that the subset

$$\{(\omega, s) : X_s(\omega) \neq \tilde{X}_s(\omega)\}$$

of $\Omega \times [0, t]$ be $(M_t \otimes \lambda)$-null, as is sufficient for the validity of the Feynman-Kac formula (6.4), then *stochastic quasi-continuity* is a necessary and sufficient condition for the existence of the strongly progressively measurable process $\tilde{X}$, see [37] for the probabilistic case. As Step 2 of the proof above demonstrates, the process X of Theorem 6.3 is automatically stochastically quasi-continuous when (Σ, d) is separable and complete.

d) Theorem 1 also applies to the case in which $(\Sigma, \mathcal{B})$ is a *standard Borel space* [46, 424A], that is, $\mathcal{B}$ is the Borel σ-algebra of some Polish topology on Σ. The common spaces of distributions are standard Borel spaces [125, pp. 112-117]. When Σ is a Souslin space [46, 423A], the spectral measure Q is supported by the countable union $\Sigma_Q = \cup_n \Sigma_n$ of compact metrisable subsets Σ_n, $n = 1, 2, \ldots$, of Σ [46, 432B, 423D(c)], which is a standard

Borel space by [125, Corollary 2, p. 102]. Then a version of Theorem 1 applies to Σ_Q-valued processes.

e) If we omit the assumption that Σ is a Polish space (or even a Souslin space), then for any bounded measurable function $V : \Sigma \to \mathbb{R}$, a real valued progressively measurable process $\widetilde{V \circ X}_s$, $s \geq 0$, may be obtained as above, so that the Feynman-Kac formula becomes

$$e^{t(A+Q(V))} = \int_\Omega e^{\int_0^t \widetilde{V \circ X}_s \, ds} \, dM_t. \tag{6.8}$$

Equation (6.8) suffices as an ingredient in the proof of the general CLR-inequality for dominated semigroups in the next chapter. The real valued random process $t \longmapsto e^{\int_0^t \widetilde{V \circ X}_s \, ds}$, $t \geq 0$, is called a *multiplicative functional* in the probability literature [28, p. 358]. The process $\widetilde{V \circ X}$ ought to be considered as a junior grade *renormalisation* if the process X is itself essentially wild.

6.4 Operator bilinear integration

In order to treat random evolutions in the next section and to obtain a generalisation of the Feynman-Kac formula, Theorem 6.2, relevant to random evolutions, we need to integrate operator valued functions with respect to operator valued measures.

Let Y be a Banach space, $(\Omega, \mathcal{S})$ a measurable space, and $M : \mathcal{S} \to \mathcal{L}(Y)$ an operator valued measure, by which we mean that M is σ-additive for the strong operator topology. Suppose that X is another Banach space and τ is a completely separated norm tensor product topology on $X \otimes Y$. Let I_X be the identity operator on X. The tensor product $I_X \otimes T : X \otimes Y \to X \otimes Y$ of the identity map and a continuous linear operator $T : Y \to Y$ need not. be continuous for the topology τ, so this is encompassed in the conditions below.

For each $A \in \mathcal{S}$, we denote the linear map $I_X \otimes [M(A)] : X \otimes Y \to X \otimes Y$ by $M_X(A)$. An operator valued measure whose range is an equicontinuous family of linear operators is called an *equicontinuous* operator valued measure. If there exists $C > 0$ such that $\|M_X(A)\phi\|_\tau \leq C\|\phi\|_\tau$ for every $\phi \in X \otimes Y$ and $A \in \mathcal{S}$, then we say that the Banach space X is (M, τ)-*admissible*.

Lemma 6.1. *Let X be an (M, τ)-admissible Banach space. Then M_X extends uniquely to an equicontinuous operator valued measure acting on $X \widehat{\otimes}_\tau Y$.*

Proof. Because X is (M, τ)-admissible, $\{M_X(A) : A \in \mathcal{S}\}$ is an equicontinuous family of operators acting on the normed space $X \otimes_\tau Y$. For each $A \in \mathcal{S}$, the unique continuous linear extension of $M_X(A)$ to all of $X \widehat{\otimes}_\tau Y$ is denoted by $\hat{M}_X(A)$. Then $\{\hat{M}_X(A) : A \in \mathcal{S}\}$ is an equicontinuous family of operators acting in $\mathcal{L}(X \widehat{\otimes}_\tau Y)$.

By property (T1) of the norm tensor product topology τ (see Section 1.4), the $(X \otimes Y)$-valued set function $M_X a$ is σ-additive for all elements a belonging to the dense subset $X \otimes Y$ of $X \widehat{\otimes}_\tau Y$; equicontinuity ensures that $\hat{M}_X a$ is σ-additive for all $a \in X \widehat{\otimes}_\tau Y$. $\square$

The uniquely defined operator valued measure of the above statement is also denoted by $M_X : \mathcal{S} \to \mathcal{L}(X \widehat{\otimes}_\tau Y)$.

Definition 6.1. Let $(\Omega, \mathcal{S})$ be a measurable space and X, Y Banach spaces. Suppose that τ is a completely separated norm tensor product topology on $X \otimes Y$. Let $M : \mathcal{S} \to \mathcal{L}(Y)$ be an operator valued measure.

An operator valued function $\Phi : \Omega \to \mathcal{L}(X)$ is said to be *M-integrable* in $\mathcal{L}(X \widehat{\otimes}_\tau Y)$, if for each $A \in \mathcal{T}$, there exists an operator $[\Phi \otimes M](A) \in \mathcal{L}(X \widehat{\otimes}_\tau Y)$ such that for every $x \in X$ and $y \in Y$, the X-valued function $\Phi x : \omega \mapsto \Phi(\omega)x$, $\omega \in \Omega$, is integrable in $X \widehat{\otimes}_\tau Y$ with respect to the Y-valued measure $My : A \mapsto M(A)y$ in the sense of Definition 2.2 and the equality

$$[\Phi \otimes M](A)(x \otimes y) = \int_A [\Phi x] \otimes d[My]$$

holds for every $A \in \mathcal{S}$. The operator $[\Phi \otimes M](A)$ is also written as $\int_A \Phi \otimes dM$. Sometimes, we write $M(\Phi)$ for the definite integral $[\Phi \otimes M](\Omega)$.

If the space X is not (M, τ)-admissible, then it may happen that the only operator valued function integrable with respect to M is the function equal to zero almost everywhere.

Let $(\Omega, \mathcal{S})$ be a measurable space and X, Y Banach spaces. Suppose that τ is a completely separated norm tensor product topology on $X \otimes Y$. We check that $\Phi \otimes M$ is σ-additive.

Lemma 6.2. *Let $M : \mathcal{S} \to \mathcal{L}(Y)$ be an operator valued measure and let the operator valued function $\Phi : \Omega \to \mathcal{L}(X)$ be M-integrable. The set function $A \mapsto [\Phi \otimes M](A), A \in \mathcal{S}$, is σ-additive in $\mathcal{L}(X \widehat{\otimes}_\tau Y)$.*

Proof. For each $a \in X \widehat{\otimes}_\tau Y$, there exists $a_n \in X \otimes Y$, $n = 1, 2, \ldots$, such that $\|a - a_n\|_\tau \to 0$. Then $[\Phi \otimes M](A)a_n \to [\Phi \otimes M](A)a$ for each $A \in \mathcal{S}$ because $[\Phi \otimes M](A) \in \mathcal{L}(X \widehat{\otimes}_\tau Y)$. The Vitali-Hahn-Saks theorem ensures that $[\Phi \otimes M]a$ is σ-additive in $X \widehat{\otimes}_\tau Y$. $\qquad \square$

It is useful to have a condition interpreting Definition 6.1 in terms of approximation by simple functions.

Proposition 6.2. *Let $M : \mathcal{S} \to \mathcal{L}(Y)$ be an operator valued measure and $\Phi : \Omega \to \mathcal{L}(X)$, an operator valued function. Suppose that for every nonzero element $\eta = \sum_{j=1}^{n} x_j \otimes y_j$ of $X \otimes Y$, $n = 1, 2, \ldots$, there exist X-valued $\mathcal{S}$-simple functions $s_{j,k} : \Omega \to X$, $k = 1, 2, \ldots$, and $j = 1, \ldots, n$, for which there exists $C > 0$ independent of η and $\{s_{j,k}\}$ such that,*

(i) *for each $j = 1, \ldots, n$, $s_{j,k} \to \Phi x_j$ $(M y_j)$-a.e. as $k \to \infty$,*

$$\lim_{k,l \to \infty} \left\| \int_A s_{j,k}(\omega) \otimes d[M y_j](\omega) - \int_A s_{j,l}(\omega) \otimes d[M y_j](\omega) \right\|_\tau = 0$$

for each $A \in \mathcal{S}$, and

(ii) $\left\| \sum_{j=1}^{n} \int_A s_{j,k}(\omega) \otimes d[M y_j](\omega) \right\|_\tau \leq \|\eta\|_\tau$ *for all $A \in \mathcal{S}$ and $k = 1, 2, \ldots$.*

Then the operator valued function Φ is M-integrable in $\mathcal{L}(X \widehat{\otimes}_\tau Y)$ and

$$[\Phi \otimes M](A)\eta = \lim_{k \to \infty} \sum_{j=1}^{n} \int_A s_{j,k}(\omega) \otimes d[M y_j](\omega). \qquad (6.9)$$

If for dense sets of $x \in X$ and $y \in Y$, the Y-valued measure My has σ-finite X-semivariation in $X \otimes_\tau Y$ on the set $\{\Phi x \neq 0\}$, then conditions (i) *and* (ii) *are also necessary. Moreover, the convergence in* (6.9) *is uniform for all $A \in \mathcal{S}$.*

Proof. If condition (i) holds, then by Definition 2.2, $\Phi x_j : \Omega \to X$ is $(M y_j)$-integrable for each $j = 1, 2, \ldots, n$. Condition (ii) ensures that $p\left(\sum_{j=1}^{n} ([\Phi x_j] \otimes [M y_j])(A) \right) \leq q(\eta)$ for all $A \in \mathcal{S}$. For every $x \in X$ and $y \in Y$, the X-valued function Φx is My-integrable in $X \widehat{\otimes}_\tau Y$, and the map $(x, y) \mapsto [\Phi x] \otimes [My]$, $x \in X, y \in Y$, is bilinear. Hence, the linear map, $(\Phi \otimes M)(A) : X \otimes Y \to X \widehat{\otimes}_\tau Y$ defined by

$$[\Phi \otimes M](A)\eta = \sum_{j=1}^{n} ([\Phi x_j] \otimes [M y_j])(A),$$

for each η as above, satisfies $p([\Phi \otimes M](A)\eta) \le q(\eta)$, for each $A \in \mathcal{S}$ and $\eta \in X \otimes Y$. Therefore, $[\Phi \otimes m](A)$ is the restriction to $X \otimes Y$ of a continuous linear operator, also denoted by $[\Phi \otimes m](A)$, from $X \widehat{\otimes}_\tau Y$ to $X \widehat{\otimes}_\tau Y$. The collection of linear maps $[\Phi \otimes M](A)$, $A \in \mathcal{S}$, is equicontinuous. According to Definition 6.1, the function $\Phi : \Omega \to \mathcal{L}(X)$ is M-integrable in $\mathcal{L}(X \widehat{\otimes}_\tau Y)$.

Suppose now that Φ is M-integrable, X, Y and $(X \widehat{\otimes}_\tau Y, p)$ are Banach spaces, and for dense sets $X_0 \subseteq X$ and $Y_0 \subseteq Y$, My has σ-finite X-semivariation in $X \widehat{\otimes}_\tau Y$ on $\{\Phi x \ne 0\}$ for each $x \in X_0, y \in Y_0$. By Lemma 6.2 and the condition that τ is normed, $\Phi \otimes M$ is σ-additive in the strong operator topology of $\mathcal{L}(X \widehat{\otimes}_\tau Y)$, so the range of the measure $[\Phi \otimes M]\xi$ is necessarily bounded in $X \widehat{\otimes}_\tau Y$ for each $\xi \in X \widehat{\otimes}_\tau Y$. By the uniform boundedness principle, the family of operators $[\Phi \otimes m](A)$, $A \in \mathcal{S}$, is bounded in the uniform operator norm of $\mathcal{L}(X \widehat{\otimes}_\tau Y)$.

Let $C > \sup_{A \in \mathcal{S}} \|[\Phi \otimes M](A)\|_{\mathcal{L}(X \widehat{\otimes}_\tau Y)}$. Let $\eta \in X_0 \otimes Y_0$ be a vector of the form $\sum_{j=1}^n x_j \otimes y_j$ with $n = 1, 2, \dots$. According to Definition 2.2 and Theorem 2.1, there exist simple functions $s_{j,k}$ such that for the norm p of $X \widehat{\otimes}_\tau Y$, (i) is satisfied, uniformly for all $A \in \mathcal{S}$. We can chop the simple functions $s_{j,k}$ off on the set $\{\Phi x_j \ne 0\}$, if necessary, in order to apply Theorem 2.1 as stated. Hence, $[\Phi \otimes M](A)\eta = \lim_{k \to \infty} \sum_{j=1}^n \int_A s_{j,k} \otimes d[My_j]$, uniformly for $A \in \mathcal{S}$. We can extract a subsequence $\{s'_{j,k}\}_{k=1}^\infty$ of $\{s_{j,k}\}_{k=1}^\infty$, so that for every $k = 1, 2, \dots$ and $A \in \mathcal{S}$, we have $p(\sum_{j=1}^n \int_A s'_{j,k} \otimes d[My_j]) \le Cp(\eta)$. Because $X_0 \otimes Y_0$ is dense in $X \otimes_\tau Y$ by condition (T3) of a tensor product topology, it follows that conditions (i) and (ii) are satisfied for the norm p of $X \widehat{\otimes}_\tau Y$ and for $q = Cp$, for *every* $\eta \in X \otimes Y$. $\qquad\square$

Proposition 6.3. *Let Y be a Banach, $M : \mathcal{S} \to \mathcal{L}(Y)$ an operator valued measure and suppose that X is an (M, τ)-admissible Banach space for the completely separated norm tensor product topology τ on $X \otimes Y$.*

Let $\phi : \Omega \to \mathbb{C}$ be a scalar valued function and define the function $\Phi : \Omega \to \mathcal{L}(X)$ by $\Phi = \phi I$. The following conditions are equivalent.

(i) *ϕ is M-integrable in $\mathcal{L}(Y)$ and $I_X \otimes ([\phi.M](A)) \in \mathcal{L}(X \otimes_\tau Y)$ for each $A \in \mathcal{S}$.*

(ii) *ϕ is M_X-integrable in $\mathcal{L}(X \otimes_\tau Y)$.*

(iii) *Φ is M-integrable in $\mathcal{L}(X \widehat{\otimes}_\tau Y)$.*

If any of the conditions holds, then on $X \otimes Y$, the equalities

$$[\Phi \otimes M](A) = [\phi.M_X](A) = I_X \otimes ([\phi.M](A))$$

hold for every $A \in S$. If $\phi : \Omega \to \mathbb{C}$ is bounded and S-measurable, then Φ is M-integrable in $\mathcal{L}(X \widehat{\otimes}_\tau Y)$.

Proof. To show that (i) implies (ii), we can find simple functions $\phi_k \in$ **sim**(S), $k = 1, 2, \ldots$, converging everywhere to ϕ, so that as $k \to \infty$, $[\phi_k.(My)](A) \to [\phi.(My)](A)$ in Y, for each $A \in S$ and $y \in Y$. Then for each $x \in X$, $x \otimes ([\phi_k.(My)](A)) \to x \otimes ([\phi.(My)](A))$ in $X \otimes_\tau Y$ as $k \to \infty$, because property (T3) of a tensor product topology ensures that $x \otimes Y$ is isomorphic to Y in the case that $x \neq 0$; if $x = 0$, we get the zero vector. From the definition of M_X, for each $x \in X$, we have

$$\lim_{k \to \infty} \int_A \phi_k \, d[M_X(x \otimes y)] = x \otimes \int_A \phi \, d[My] = \left(I_X \otimes \int_A \phi \, dM \right)(x \otimes y).$$

By assumption, $I_X \otimes ([\phi.M](A)) \in \mathcal{L}(X \otimes_\tau Y)$ for each $A \in S$, so ϕ is M_X-integrable in $\mathcal{L}(X \otimes_\tau Y)$ and $[\phi.M_X](A) = I_X \otimes ([\phi.M](A))$ for every $A \in S$.

Now suppose that (ii) is true, and find simple functions $\phi_k \in$ **sim**(S), $k = 1, 2, \ldots$, converging everywhere to ϕ, so that for each $A \in S$, $x \in X$ and $y \in Y$, we have $[\phi_k.M_X](A)(x \otimes y) \to [\phi.M_X](A)(x \otimes y)$ in $X \otimes_\tau Y$, as $k \to \infty$. Then,

$$\lim_{k \to \infty} \int_A [\phi_k x] \otimes d[My] = \lim_{k \to \infty} \int_A \phi_k \, d[M_X(x \otimes y)] = \left(\int_A \phi \, dM_X \right)(x \otimes y).$$

A glance at Definition 2.2 verifies that Φx is My-integrable and

$$\int_A [\Phi x] \otimes d[My]) = \left(\int_A \phi \, dM_X \right)(x \otimes y).$$

Now $[\phi.M_X](A) \in \mathcal{L}(X \widehat{\otimes}_\tau Y)$, so it has a unique continuous linear extension to all of $X \widehat{\otimes}_\tau Y$, denoted by the same symbol. From Definition 6.1, the function Φ is M-integrable and $\Phi \otimes M = \phi.M_X$.

Suppose that (iii) holds, y is an element of Y and $x \in X$ is a nonzero vector. By the Hahn-Banach theorem, there exists $x' \in X'$ such that $\langle x, x' \rangle = 1$. The assumption that Φ is M-integrable means that Φx is My-integrable and $[(\Phi \otimes M)(A)](x \otimes y) = ([\Phi x] \otimes [My])(A)$, for each $A \in S$.

Proposition 2.2 ensures that $\langle \Phi x, x' \rangle = \langle \phi x, x' \rangle = \phi$ is My-integrable in Y and

$$\left\langle \int_A \phi \, d[My], y' \right\rangle = \left\langle \int_A \langle \Phi x, x' \rangle \, d[My], y' \right\rangle = \left\langle \int_A [\Phi x] \otimes d[My], x' \otimes y' \right\rangle.$$

The vector $x \in X$ is nonzero, so $J_x : y \mapsto x \otimes y$, $y \in Y$, is an isomorphism of Y onto $x \otimes Y$ by property (T3) of the tensor product topology τ. A calculation shows that for each $y \in Y$, we have

$$\int_A \phi \, d[My] = (x' \otimes I_Y) \circ (\Phi \otimes M)(A) \circ J_x y.$$

The right-hand side of this equation is an element of Y because Y is sequentially complete. Here, the continuous linear extension to $X \widehat{\otimes}_\tau Y$ of the map $x' \otimes I_Y : u \otimes y \mapsto \langle u, x' \rangle y$, for $u \in X$ and $y \in Y$, has been denoted by the same symbol. Consequently, ϕ is M-integrable in $\mathcal{L}(Y)$ and

$$\int_A \phi\, dM = (x' \otimes I_Y) \circ (\Phi \otimes M)(A) \circ J_x \in \mathcal{L}(Y), \qquad \text{for all } A \in \mathcal{S}.$$

Moreover, it is readily verified that $[\Phi \otimes M](A)(x \otimes y) = \big(I_X \otimes ([\phi.M](A))\big)(x \otimes y)$ for all $x \in X$ and $y \in Y$, so $I_X \otimes ([\phi.M](A))$ necessarily belongs to $\mathcal{L}(X \otimes_\tau Y)$.

If ϕ is bounded and $\mathcal{S}$-measurable, then it is the uniform limit of $\mathcal{S}$-simple functions ϕ_k, $k = 1, 2, \ldots$. Let $\Phi_k = \phi_k I_X$ for every $k = 1, 2, \ldots$. The assumption that X is (M, τ)-admissible means that the range of M_X is contained in an equicontinuous set in $\mathcal{L}(X \otimes_\tau Y)$, so there exists an equicontinuous subset of $\mathcal{L}(X \otimes_\tau Y)$ containing the ranges of each of the measures $\phi_k.M_X = \Phi_k \otimes M$, $k = 1, 2, \ldots$. The limit measure $\phi.M_X = \Phi \otimes M$ therefore takes its values in $\mathcal{L}(X \widehat{\otimes}_\tau Y)$. $\qquad \square$

Definition 6.2. Let X, Y and $X \widehat{\otimes}_\tau Y$ be Banach spaces. Let $M : \mathcal{S} \to \mathcal{L}(Y)$ be an additive set function. We say that M has *finite $\mathcal{L}(X)$-semivariation in $\mathcal{L}(X \otimes_\tau Y)$* if

(i) $A \otimes M(E) \in \mathcal{L}(X \otimes_\tau Y)$ for each $A \in \mathcal{L}(X)$ and $E \in \mathcal{S}$, and
(ii) there exists $C > 0$ such that $\| \sum_{j=1}^n A_j \otimes M(E_j) \|_{\mathcal{L}(X \otimes_\tau Y)} \leq C$, for all $A_j \in \mathcal{L}(X)$ with $\|A_j\| \leq 1$ and pairwise disjoint $E_j \in \mathcal{S}$, $j = 1, \ldots, n$ and $n = 1, 2, \ldots$.

Let $\beta_{\mathcal{L}(X)}(M)(E)$ be the smallest such number C as the sets E_j above range over subsets of $E \in \mathcal{S}$. Then the set function $\beta_{\mathcal{L}(X)}(M)$ is called the $\mathcal{L}(X)$-semivariation of M in $\mathcal{L}(X \otimes_\tau Y)$. It follows from the property (T1) of the norm tensor product topology τ in Section 1.4, that if $X \neq 0$, then there exists $K > 0$, such that for every additive set function $M : \mathcal{S} \to \mathcal{L}(Y)$, the semivariation $\|M\|$ of M in the operator norm is bounded by $K \beta_{\mathcal{L}(X)}(M)$.

If the $\mathcal{L}(X)$-semivariation of M in $\mathcal{L}(X \otimes_\tau Y)$ is continuous and $X \neq 0$, then the fact mentioned above shows that M is σ-additive for the uniform operator topology of $\mathcal{L}(Y)$—a condition which is rarely satisfied for operator valued measures arising in applications. It is therefore a useful observation that the following result does not require the $\mathcal{L}(X)$-semivariation in $\mathcal{L}(X \otimes_\tau Y)$ to be continuous.

Theorem 6.4. *Let X, Y and $X \widehat{\otimes}_\tau Y$ be Banach spaces. Let $M : \mathcal{S} \to \mathcal{L}(Y)$ be an operator valued measure such that M has finite $\mathcal{L}(X)$-semivariation $\beta_{\mathcal{L}(X)}(M)$ in $\mathcal{L}(X \otimes_\tau Y)$, and for each $y \in Y$, My has the continuous X-semivariation in $X \otimes_\tau Y$. Let $\Phi : \Omega \to \mathcal{L}(X)$ be a function such that,*

(a) *for each $x \in X$ and $y \in Y$, $\Phi x : \Omega \to X$ is strongly My-measurable;*
(b) *there exists $C > 0$ such that for each $x \in X$ and $y \in Y$, the bound $\|\Phi(\omega)x\|_X \le C\|x\|_X$ holds for My-almost all $\omega \in \Omega$.*

Then Φ is M-integrable in $\mathcal{L}(X \widehat{\otimes}_\tau Y)$. If $\|\Phi\|_\infty$ denotes the smallest number C satisfying (b)*, then $\|(\Phi \otimes M)(\Omega)\| \le \|\Phi\|_\infty \beta_{\mathcal{L}(X)}(M)(\Omega)$.*

Proof. We verify the conditions of Proposition 6.2 to show that Φ is M-integrable. Any element ϕ of $X \otimes Y$ may be written in the form $\phi = \sum_{j=1}^n x_j \otimes y_j$, with $\{x_j\}$ a linearly independent set. Each function Φx_j is My_j-integrable by Theorem 2.2, so there exist X-valued $\mathcal{S}$-simple functions $\{s_{j,k}\}$ such that condition (i) of Proposition 6.2 holds. We need to modify the approximating sequence to ensure condition (ii) of that proposition is also valid.

Let X_0 be the linear span of $\{x_j\}$ and equip it with the norm of X. For each $\omega \in \Omega$, define the linear map $\tilde{\Phi}_k(\omega) : X_0 \to X$ by setting $\tilde{\Phi}_k(\omega)x_j = s_{j,k}(\omega)$. Because we have assumed that $\{x_j\}$ is a linearly independent set, $\tilde{\Phi}_k$ is well-defined for each $k = 1, 2, \ldots$. The space X_0 is finite dimensional, so for each $j = 1, \ldots, n$, the operators $\Phi_k(\omega)$ converge to $\Phi(\omega) \upharpoonright X_0$ in the uniform operator topology of $\mathcal{L}(X_0, X)$ as $k \to \infty$, for My_j-almost all $\omega \in \Omega$.

Let $\epsilon > 0$. The set

$$A_k = \bigcap_{l=k}^\infty \{\omega \in \Omega : \|\tilde{\Phi}_l(\omega)x\|_X \le (\|\Phi\|_\infty + \epsilon)\|x\|_X \text{ for all } x \in X_0\}$$

belongs to the σ-algebra $\mathcal{S}$, for $\tilde{\Phi}_l$ is an $\mathcal{L}(X_0, X)$-valued $\mathcal{S}$-simple function for every $l = 1, 2, \ldots$. The pointwise convergence of $\{\tilde{\Phi}_k\}$ in the uniform operator topology of $\mathcal{L}(X_0, X)$ guarantees that $\cup_{k=1}^\infty A_k$ is a set of full My_j-measure for each $j = 1, \ldots, n$.

Let $P : X \to X_0$ be any norm one projection and define $\Phi_k(\omega) = \chi_{A_k}(\omega)\tilde{\Phi}_k(\omega) \circ P$, for all $k = 1, 2, \ldots$ and $\omega \in \Omega$. Then $\|\Phi_k(\omega)\|_{\mathcal{L}(X)} \le \|\Phi\|_\infty + \epsilon$ for all $k = 1, 2, \ldots$ and $\omega \in \Omega$. Moreover, Φ_k is a family of $\mathcal{L}(X)$-valued $\mathcal{S}$-simple functions such that the X-valued $\mathcal{S}$-simple functions $\Phi_k x_j$, for $k = 1, 2, \ldots$ and $j = 1, \ldots, n$ satisfy condition (i) of Proposition 6.2. The setwise convergence of the indefinite integrals follows from the argument of Vitali's convergence theorem; see the proof of Lemma 2.1. By the finiteness of the $\mathcal{L}(X)$-semivariation $\beta_{\mathcal{L}(X)}(M)$ of M, for every $A \in \mathcal{S}$ we have

$$\left\| \sum_{j=1}^{n} \int_A [\Phi_k x_j] \otimes d[M y_j] \right\|_{\tau} \leq (\|\Phi\|_\infty + \epsilon)\beta_{\mathcal{L}(X)}(M)(\Omega)\|\phi\|_{\tau},$$

for all $k = 1, 2, \ldots$. Hence, condition (ii) of Proposition 6.2 also holds, so Φ is M-integrable and (6.9) holds. Because ϵ is any positive number and ϕ is any element of $X \otimes Y$, the bound $\|(\Phi \otimes M)(\Omega)\| \leq \|\Phi\|_\infty \beta_{\mathcal{L}(X)}(M)(\Omega)$ is valid. $\qquad \square$

The following result is a slight modification of the bounded convergence theorem of [11, Theorem 7]. Again, it is significant that we only assume that the operator valued measure M has *pointwise* continuous X-semivariation in $X \otimes_\tau Y$.

Theorem 6.5 (Bounded Convergence Theorem). *Let X, Y and $X \widehat{\otimes}_\tau Y$ be Banach spaces. Let $M : \mathcal{S} \to \mathcal{L}(Y)$ be an operator valued measure such that M has finite $\mathcal{L}(X)$-semivariation $\beta_{\mathcal{L}(X)}(M)$ in $\mathcal{L}(X \otimes_\tau Y)$, and for each $y \in Y$, (My) has continuous X-semivariation in $X \otimes_\tau Y$. Let $\Phi_k : \Omega \to \mathcal{L}(X)$, $k = 1, 2, \ldots$, be functions such that*

(a) *for every $x \in X$ and $y \in Y$, the function $\Phi_k x : \Omega \to X$ is (My)-measurable;*

(b) *there exists a positive number C with the property that for every $x \in X$ and $y \in Y$, and $k = 1, 2, \ldots$, the bound $\|\Phi_k(\omega)x\|_X \leq C\|x\|_X$ holds, for (My)-almost all $\omega \in \Omega$.*

If for each $x \in X$ and $y \in Y$, $\Phi_k(\omega)x \to \Phi(\omega)x$ as $k \to \infty$, for $((My))$-almost every $\omega \in \Omega$, then Φ is M-integrable and

$$\int_A \Phi_k \otimes dM \to \int_A \Phi \otimes dM$$

in the strong operator topology of $\mathcal{L}(X \widehat{\otimes}_\tau Y)$ as $k \to \infty$, uniformly for $A \in \mathcal{S}$.

Proof. We already know by Theorem 6.4 that Φ is M-integrable in $\mathcal{L}(X \widehat{\otimes}_\tau Y)$ and that the bound $\| [\Phi_k \otimes M](A)\|_{\mathcal{L}(X \widehat{\otimes}_\tau Y)} \leq C\beta_{\mathcal{L}(X)}(M)(\Omega)$ holds, for all $k = 1, 2, \ldots$ and $A \in \mathcal{S}$. The vector valued bounded convergence theorem, Theorem 2.2, tells us that for all $\phi \in X \otimes Y$, the vectors $[\Phi_k \otimes M](A)\phi, k = 1, 2, \ldots$, converge to $[\Phi \otimes M](A)\phi$ in $X \widehat{\otimes}_\tau Y$, as

$k \to \infty$, uniformly for $A \in \mathcal{S}$. The equicontinuity for the values of the indefinite integrals $\Phi_k \otimes M$, $k = 1, 2, \ldots$, guarantees the convergence for every $\phi \in X \widehat{\otimes}_\tau Y$. $\qquad\square$

We have the following analogues of Propositions 2.2 and 2.3, and Corollaries 2.1 and 2.2. The proofs are similar to the earlier ones, so they are omitted.

Proposition 6.4. *Let* $M : \mathcal{S} \to \mathcal{L}(Y)$ *be an* $\mathcal{L}(Y)$-*valued measure. If* $\Phi : \Omega \to \mathcal{L}(X)$ *is* M-*integrable in* $\mathcal{L}(X \widehat{\otimes}_\tau Y)$, *then for all* $x \in X$, $y \in Y$, $x' \in X'$ *and* $y' \in Y'$, *the* $\mathcal{L}(X)$-*valued function* Φ *is integrable with respect to the scalar measure* $\langle My, y' \rangle$, *the scalar valued function* $\langle \Phi x, x' \rangle$ *is integrable with respect to the* $\mathcal{L}(Y)$-*valued measure* M *and the following equalities hold for all* $A \in \mathcal{S}$:

$$\left\langle \int_A \Phi \, d\langle My, y' \rangle \, , \, x \otimes x' \right\rangle = \left\langle \int_A \langle \Phi x, x' \rangle \, dM \, , \, y \otimes y' \right\rangle$$
$$= \int_A \langle \Phi x, x' \rangle \, d\langle My, y' \rangle.$$

Corollary 6.1. *Let* $M : \mathcal{S} \to \mathcal{L}(Y)$ *be an* $\mathcal{L}(Y)$-*valued measure. If* $\Phi : \Omega \to \mathcal{L}(X)$ *is* M-*integrable in* $\mathcal{L}(X \widehat{\otimes}_\tau Y)$, *then* $\Phi \otimes M \ll M$.

Corollary 6.2. *Suppose that* $M : \mathcal{S} \to \mathcal{L}(Y)$ *is an* $\mathcal{L}(Y)$-*valued measure. If* $\Phi : \Omega \to \mathcal{L}(X)$ *is* M-*integrable in* $\mathcal{L}(X \widehat{\otimes}_\tau Y)$, *and* $f : \Omega \to \mathbb{C}$ *is a bounded* $\mathcal{S}$-*measurable function, then* $f\Phi$ *is* M-*integrable,* Φ *is* $f.M$-*integrable and the equalities* $(f\Phi) \otimes M = \Phi \otimes (f.M) = f.(\Phi \otimes M)$ *hold.*

Proposition 6.5. *Suppose that* X_j, Y_j, $j = 1, 2$ *are Banach spaces and* τ_1 *is a completely separated norm tensor product topology on* $X_1 \otimes Y_1$, *and* τ_2 *is a completely separated norm tensor product topology on* $X_2 \otimes Y_2$.

Let $M : \mathcal{S} \to \mathcal{L}(Y_1)$ *be a measure and suppose that* $S : X_1 \to X_2$ *and* $T : Y_1 \to Y_2$ *are continuous linear maps whose tensor product* $S \otimes T : X_1 \otimes_{\tau_1} Y_1 \to X_2 \otimes_{\tau_2} Y_2$ *is continuous. If* $\Phi : \Omega \to \mathcal{L}(X_1)$ *is* M-*integrable in* $\mathcal{L}(X_1 \widehat{\otimes}_{\tau_1} Y_1)$, *then* $S\Phi$ *is* TM-*integrable in* $\mathcal{L}(X_2 \widehat{\otimes}_{\tau_2} Y_2)$ *and*

$$(S \otimes T) \int_A \Phi \otimes dM = \int_A [S\Phi] \otimes d[TM],$$

for every $A \in \mathcal{S}$.

Similarly, if $M : \mathcal{S} \to \mathcal{L}(Y_2)$ *is a measure and* $\Phi : \Omega \to \mathcal{L}(X_2)$ *is* M-*integrable in* $\mathcal{L}(X_2 \widehat{\otimes}_{\tau_2} Y_2)$, *then the function* ΦS *is* MT-*integrable in the space* $\mathcal{L}(X_1 \widehat{\otimes}_{\tau_1} Y_1)$ *and the equality* $\left[\int_A \Phi \otimes dM \right] (S \otimes T) = \int_A [\Phi S] \otimes d[MT]$ *holds for every* $A \in \mathcal{S}$.

As shown in Section 1.5.2, operator valued measures taking their values in the Banach lattice of positive operators on an L^p-space have bounded $\mathcal{L}(X)$-semivariation for any Banach space X and the same holds true for operator valued measures dominated by a positive measure. The bounded (S, Q)-processes on L^p-spaces considered in Section 6.5 below have this property, so Theorem 1.10 and Theorem 6.6 below see use in the next section on random evolutions.

Let $(\Gamma, \mathcal{E}, \mu)$ be a σ-finite measure space and $1 \leq p \leq \infty$.

Theorem 6.6. *Suppose that M is positive and that N is dominated by M. Let $F : \Omega \to \mathcal{L}(X)$ be a function such that for each $x \in X$ and $y \in L^p(\mu)$, the function Fx is strongly (My)-measurable and $\|F\|_{\mathcal{L}(X)}$ is M-integrable in $\mathcal{L}(L^p(\mu))$. Then the function F is N-integrable in $\mathcal{L}(L^p(\mu; X))$ and the estimate*

$$\left\| \int_A F \otimes dN \right\|_{\mathcal{L}(L^p(\mu;X))} \leq \left\| \int_A \|F\|_{\mathcal{L}(X)} \, dM \right\|_{\mathcal{L}(L^p(\mu))}$$

holds for all $A \in \mathcal{S}$.

Proof. We establish the result for the special case $N = M \geq 0$ first. For $n \in \mathbb{N}$, let $g = \sum_{j=1}^{n} x_j \chi_{G_j}$ be an X-valued $\mathcal{E}$-simple function with $x_j \in X$ and G_j pairwise disjoint for $j = 1, \ldots, n$, such that $\|g\|_{L^p(\mu;X)} \leq 1$. Then, making use of Theorem 2.3, we have

$$\left\| \left(\int_A F(\omega) \otimes dM(\omega) \right) g \right\|_{L^p(\mu;X)}^p$$

$$= \left\| \sum_j \int_A [Fx_j](\omega) \otimes d[M\chi_{G_j}](\omega) \right\|_{L^p(\mu;X)}^p$$

$$= \int_\Gamma \left\| \left(\sum_j \int_A [Fx_j](\omega) \otimes d[M\chi_{G_j}](\omega) \right)(\gamma) \right\|_X^p d\mu(\gamma)$$

$$\leq \int_\Gamma \left(\sum_j \left\| \left(\int_A [Fx_j](\omega) \otimes d[M\chi_{G_j}](\omega) \right)(\gamma) \right\|_X \right)^p d\mu(\gamma)$$

$$\leq \int_\Gamma \left(\sum_j \left(\int_A \|F(\omega)x_j\|_X \, d[M\chi_{G_j}](\omega) \right)(\gamma) \right)^p d\mu(\gamma)$$

$$= \left\| \sum_j \int_A \|F(\omega)x_j\|_X \, d[M\chi_{G_j}](\omega) \right\|_{L^p(\mu)}^p .$$

By assumption, the nonnegative function $\|F\|_{\mathcal{L}(X)}$ is M-integrable in $\mathcal{L}(L^p(\mu))$, so the right-hand side is bounded by

$$\left\|\sum_j \int_A \|F(\omega)\|_{\mathcal{L}(X)} \|x_j\|_X \, d[M\chi_{G_j}](\omega)\right\|^p_{L^p(\mu)}$$

$$= \left\|\left(\int_A \|F(\omega)\|_{\mathcal{L}(X)} \, dM(\omega)\right)\left(\sum_j \|x_j\|_X \chi_{G_j}\right)\right\|^p_{L^p(\mu)}$$

$$\leq \left\|\int_A \|F(\omega)\|_{\mathcal{L}(X)} \, dM(\omega)\right\|^p_{\mathcal{L}(L^p(\mu))} \left\|\sum_j \|x_j\|_X \chi_{G_j}\right\|^p_{L^p(\mu)}$$

$$= \left\|\int_A \|F(\omega)\|_{\mathcal{L}(X)} \, dM(\omega)\right\|^p_{\mathcal{L}(L^p(\mu))} \|g\|^p_{L^p(\mu;X)}.$$

holding for all $A \in \mathcal{S}$. Since X-valued $\mathcal{E}$-simple functions are dense in $L^p(\mu; X)$ this establishes the required inequality and completes the proof for the positive case.

The inequality for the case where N is dominated by M follows analogously to the positive case taking note that, when $x \in X$ and $E \in \mathcal{E}$,

$$\left\|\left(\int_A [Fx](\omega) \otimes d[N\chi_E](\omega)\right)(\gamma)\right\|_X \leq \left(\int_A \|F(\omega)x\|_X \, d[M\chi_E](\omega)\right)(\gamma)$$

holds true for all $A \in \mathcal{S}$ and μ-almost every $\gamma \in \Gamma$ by Theorem 2.3. $\qquad\square$

6.5 Random evolutions

A *random evolution* $F_t : \Omega \to \mathcal{L}(\mathcal{X})$, $t \geq 0$, is a family of operator valued random variables acting on some Banach space $\mathcal{X}$ of states, whose evolution is influenced by some random changes in the environment. Random events are measured by a family of operator valued measures $M_t : \sigma(\mathcal{S}_t) \to \mathcal{L}(L^p(\mu))$, $t \geq 0$.

For example, suppose that a particle moves in a straight line with constant speed, until it suffers a random collision, after which it changes velocity, and again moves in a straight line with a new constant speed. The situation may be described, more abstractly, as an evolving system whose mode of evolution changes due to random changes in the environment. The notion of a 'random evolution' was introduced by R. Griego and R. Hersh [56] to provide a mathematical formulation of such a randomly influenced dynamical system (see [61] for a later survey, [77, Chapter 10] and [62] for elementary and historical accounts). A *random evolution* or more prosaically, a *multiplicative operator functional* is an operator valued function M satisfying a linear differential equation of the form

$$\frac{dM}{ds}(s,t) = -V(X_s)M(s,t), \quad 0 \le s < t. \tag{6.10}$$

The coefficient V is an operator valued function and X_s is a random variable for each $s \ge 0$. In the case that X_s is Markovian with respect to a family of probability measures P^x, the expected value $u(x,t) = P^x[M(0,t)]$ satisfies the equation

$$\frac{du}{dt}(x,t) = Gu(x,t) + V(x)u(x,t). \tag{6.11}$$

Here G denotes the generator of the Markov process $\langle X_s \rangle_{s \ge 0}$. This is a generalisation of the Feynman-Kac formula considered in Theorem 6.2 where V is now allowed to be an operator valued function.

For example, a finite state Markov chain is defined on the state space $\Sigma = \{1,\ldots,n\}$ by its generator, an $(n \times n)$ matrix $Q = \{q_{ij}\}$ with the property that $q_{ij} \ge 0$ for $i \neq j$ and $\sum_{j=1}^{n} q_{ij} = 0$ for each $i = 1,\ldots,n$.

If we define $q_i = -q_{ii}$ and $\Pi_{ij} = q_{ij}/q_i$ if $q_i \neq 0$ and zero otherwise, then there exists a Markov process X_t, $t \ge 0$, with probability measures P^i, $i = 1,\ldots,n$, such that $P^i(X_0 = j) = \delta_{ij}$. The sample path $t \mapsto X_t$ is piecewise constant with jumps at the instants $\tau_1 < \tau_2 < \ldots$, where $P^i(\tau_1 > t) = e^{-q_i t}$ and $P^i(X_{\tau_1} = j) = \Pi_{ij}$ for $1 \le i, j \le n$, $i \neq j$.

Given real numbers $v_1,\ldots,v_n$, there is a continuous group T_i of linear operators on the space $C_0(\mathbb{R})$ of continuous functions vanishing at infinity, defined by the formula

$$(T_i(t)f)(x) = f(x + v_i t), \quad x \in \mathbb{R}$$

for each $f \in C_0(\mathbb{R})$ and $t \in \mathbb{R}$. Then

$$M[0,t] = T_{X_0}(\tau_1)T_{X_{\tau_1}}(\tau_2 - \tau_1)\cdots T_{X_{\tau_{N(t)}}}(t - \tau_{N(t)}) \tag{6.12}$$

is a random evolution. The random integer $N(t)$ is defined by the formula $\tau_{N(t)} \le t < \tau_{N(t)+1}$. For smooth $f_i \in C_0(\mathbb{R})$, $i = 1,\ldots,n$, the expectations $u_i(t) = \mathbb{E}^i(M(0,t]f_i)$, $t \ge 0$, is the solution of the system of equations

$$\frac{\partial u_i}{\partial t} = v_i \frac{\partial u_i}{\partial x} + \sum_{j=1}^{n} q_{ij}u_j, \quad u_i(0) = f_i, \ i = 1,\ldots,n.$$

The point of departure in this chapter is to re-interpret the term 'random' so as to obtain the representation of solutions to a more general class of partial differential equations. The representation we are aiming for is one in which the expectation value above is replaced by the integral with respect to the operator valued measures M_t of an (S, Q)-process, associated with a semigroup S of bounded linear operators acting on an L^p-space and the spectral measure Q of multiplication by characteristic functions.

We want to obtain a representation of solutions u_t, $t \geq 0$, to initial value problems for certain partial differential equations in the form

$$u_t = \left(\int_\Omega F_t \otimes dM_t \right) u_0.$$

Here $F_t : \Omega \to \mathcal{L}(X)$ is an operator valued 'random variable'. It turns out that we need to define F_t in terms of the adjoint of a functional satisfying equation (6.10), but we retain the term *multiplicative operator functional*. In the example described by formula (6.12), instead we consider the multiplicative operator functional

$$F_t = T_{X_{\tau_{N(t)}}}(t - \tau_{N(t)}) \cdots T_{X_{\tau_1}}(\tau_2 - \tau_1) T_{X_0}(\tau_1)$$

in the natural ordering between increasing times and operator actions.

For the case of a scalar perturbation V, the multiplicative operator functional F_t is given by $\omega \mapsto \exp \left[\int_0^t V(\omega(s)) \, ds \right]$, $\omega \in \Omega$. Because of our more general setting, equation (6.11) can be solved in cases where G is not the generator of a probabilistic Markov process.

In this section, we apply the theory of bilinear integration developed in Section 6.4 to integration with respect to (S, Q, t)-set functions M_t in order to establish conditions for which the integral of a multiplicative operator functional F_t, $t \geq 0$, defines a C_0-semigroup $t \mapsto \int_\Omega F_t \otimes dM_t$, $t \geq 0$, acting on $L^p(\mu; X)$. The construction of the multiplicative operator function $\langle F_t \rangle_{t \geq 0}$ itself follows work of J. Hagood [58].

Now suppose that the system $(\Omega, \langle S_t \rangle_{t \geq 0}, \langle M_t \rangle_{t \geq 0}; \langle X_t \rangle_{t \geq 0})$ is a σ-additive *time homogeneous Markov evolution process* as in Section 6.1, that is, there exist a $\mathcal{L}(\mathcal{Y})$-valued spectral measure Q on $\mathcal{E}$ and a semigroup S of operators acting on $\mathcal{Y}$ such that for each $t \geq 0$, the operator $M_t(E) \in \mathcal{L}(\mathcal{Y})$ is given by

$$M_t(E) = S(t - t_n)Q(B_n)S(t_n - t_{n-1}) \cdots Q(B_1)S(t_1) \qquad (6.13)$$

for every cylinder set $E \in \mathcal{S}_t$ of the form (6.1) and the process is called an (S, Q)-*process*. The basic ingredients are the semigroup S describing the evolution of states and the spectral measure Q describing observation of states represented by vectors in $\mathcal{Y}$. For our purposes, attention is restricted to the Banach space $\mathcal{Y} = L^p(\mu)$ with $1 \leq p < \infty$ and μ a σ-finite measure.

Now let $\mathcal{X}$ be a Banach space. A *multiplicative operator functional* is a measurable mapping $F_t : \Omega \to \mathcal{L}(\mathcal{X})$, $t \geq 0$, such that a.e.,

(i) $t \mapsto F_t(\omega)$ is continuous for the weak operator topology

(ii) $F_{s+t}(\omega) = F_s(\theta_t \omega) F_t(\omega)$, $F_0(\omega) = Id_{\mathcal{X}}$.

Here $\theta_t : \Omega \to \Omega$ is a *shift map*: $X_{s+t}(\omega) = X_s(\theta_t \omega)$. If $\mathcal{X} = \mathbb{R}$ and $V : \Sigma \to \mathbb{R}$ is a suitable measurable function, $F_t = e^{- \int_0^t V \circ X_s \, ds}$ is an example. For a random evolution in the sense of probability theory, the operators in (ii) are usually written in the opposite order.

With the right idea of bilinear integration, it is straightforward that for an integrable multiplicative operator functional F_t, $t \geq 0$, the formula $R(t) = \int_\Omega F_t(\omega) \otimes dM_t(\omega)$, $t \geq 0$, gives a semigroup R of continuous linear operators acting on $L^p(\mu, \mathcal{X})$.

Proving that R is a C_0-semigroup requires additional assumptions [64, Chapter 5]. For example, $\sum_j A_j \otimes M_t(E_j)$ should form a bounded subset of $\mathcal{L}(L^p(\mu, \mathcal{X}))$ if $\|A_j\|_{\mathcal{L}(\mathcal{X})} \leq 1$ and (E_j) are pairwise disjoint, that is, M_t has *finite* $\mathcal{L}(\mathcal{X})$-*semivariation* in $\mathcal{L}(L^p(\mu, \mathcal{X}))$.

The boundedness properties are satisfied if M_t comes from a Markov process or according to Theorem 1.10, is dominated by a positive $\mathcal{L}(L^p(\mu))$-valued measure. In the case that Q is the spectral measure of multiplication by characteristic functions acting on $L^p(\mu)$, Theorem 6.1 shows that M_t is dominated by a positive $\mathcal{L}(L^p(\mu))$-valued measure if and only if S is a *dominated* semigroup of operators on $L^p(\mu)$. The following example is from [64, Theorem 5.3.3].

Example 6.3. Let $c_j : \mathbb{R} \to (0, \infty)$, $j = 1, 2, \ldots$, be continuous functions such that $\sup_{x,j} c_j(x) < \infty$ and $\inf_x c_j(x) > 0$. Suppose that $A = \{a_{jk}\}$ is an infinite matrix such that $\sup_{j \in \mathbb{N}} \sum_{k=1}^\infty |a_{jk}| < \infty$ and $\sup_{k \in \mathbb{N}} \sum_{j=1}^\infty |a_{jk}| < \infty$.

Then the solution to the equations

$$\frac{\partial u_j}{\partial t}(t, x) = c_j(x) \frac{\partial u_j}{\partial x}(t, x) + \sum_{k=1}^\infty a_{jk} u_k(t, x)$$

$$u_j(0, x) = f_j(x), \quad j = 1, 2, \ldots$$

with $\sum_{j=1}^\infty \|f_j\|_2^2 + \|f_j'\|_2^2 < \infty$ can be written as

$$u_j(t, x) = \left(\left[\int_\Omega F_t \otimes dM_t \right] f \right)_j (x).$$

Here $\Sigma = \mathbb{N}$, $\mathcal{X} = L^2(\mathbb{R})$, the multiplicative operator functional F_t is constructed from the operators $V_j : f \mapsto c_j f'$ and M_t is the (S, Q, t)-measure constructed from $S(t) = e^{At}$ on ℓ^2 and Q multiplication by characteristic functions. The sample space Ω consists of piecewise constant paths. Now $\{a_{jk}\}$ need not be the generator of a Markov process.

Chapter 7

The Cwikel-Lieb-Rozenbljum inequality for dominated semigroups

The subject of scattering theory touched upon in Chapter 5 dealt with the long time asymptotics of quantum systems. The asymptotics of quantum systems as $\hbar \to 0$ is termed the *semiclassical approximation*, where the leading term in the asymptotic approximation is usually given by a quantity obtained from Hamiltonian mechanics on classical phase space, as would be consistent with physical intuition. For example, the number of bound states in a quantum system should correlate with the volume of classical states with negative energy in the corresponding classical phase space. Volume in classical phase space is determined by Liouville measure.

An inequality establishes that this is the case in wide generality. Moreover, its proof appeals to the theory that has been outlined in the preceding chapters. The essential idea of the proof given in this chapter is a *tour de force* of classical functional analysis due to E. Lieb [90]. We can implement Lieb's proof under the rather weak assumption on the free Hamiltonian H_0 that e^{-tH_0}, $t > 0$, is a dominated semigroup of operators on $L^2(\mu)$, such as in the case that the Schrödinger operator is coupled to a magnetic field in $\mathbb{R}^{3n}$, $n = 1, 2, \ldots$. An appeal to our Feynman-Kac formula, Theorem 6.2, and our treatment of operator traces in Chapter 3 via bilinear integration serve to push Lieb's proof into the setting of general measure spaces and quantum field theory.

7.1 Asymptotic estimates for bound states

The Cwikel-Lieb-Rozenbljum or CLR inequality refers to an upper estimate for the number of negative eigenvalues (bound state energies) for a Schrödinger operator, proved by very different methods in [30, 90, 118].

In the case of the Laplacian operator $\Delta = \partial^2/\partial x_1^2 + \cdots + \partial^2/\partial x_d^2$ defined in $L^2(\mathbb{R}^d)$, the spectrum $\sigma(H_0)$ of selfadjoint operator $H_0 = -\Delta$ is $[0, \infty)$ and the semigroup $S(t) = e^{-tH_0}$, $t \geq 0$, is defined by the functional calculus for the selfadjoint operator H_0. Then for each $f \in L^2(\mathbb{R}^d)$ and $t > 0$,

$$(S(t)f)(x) = \frac{1}{(4\pi t)^{d/2}} \int_{\mathbb{R}^d} e^{-\frac{|x-y|^2}{4t}} f(y)\, dy, \quad f \in L^2(\mathbb{R}^d), \ x \in \mathbb{R}^d.$$

Suppose that $V : \Sigma \to \mathbb{R}$ is a measurable function with positive and negative parts $V_+ = V \vee 0$ and $V_- = (-V) \vee 0$. Let $N(H_0 + Q(V))$ denote the number of eigenvalues of $H_0 + Q(V)$ belonging to the half-line $(-\infty, 0]$. The selfadjoint operator of multiplication by V is denoted by $Q(V)$ and $H_0 + Q(V)$ is defined as a *form sum*. [115, Section VII.6]

The CLR inequality for the Schrödinger operator H_0 is

$$N(H_0 + Q(V)) \leq c_d \int_{\mathbb{R}^d} V_-(x)^{\frac{d}{2}}\, dx, \quad d = 3, 4, \ldots, \tag{7.1}$$

where the constant c_d depends on the dimension d but not the interaction potential V.

The significance of the bound (7.1) is that the asymptotic limit

$$\lim_{\lambda \to \infty} \frac{N(H_0 + \lambda V)}{\lambda^{\frac{d}{2}}} = (2\pi)^{-d} \sigma_d \int_{\mathbb{R}^d} |V(x)|^{\frac{d}{2}}\, dx \tag{7.2}$$

holds if $V \leq 0$ belongs to $L^{d/2}(\mathbb{R}^d)$, where σ_d is the volume of the unit ball in $\mathbb{R}^d$ [128, Theorem 10.7]. In classical phase space $\mathbb{R}^d \times \mathbb{R}^d$ with the Hamiltonian $H(p, x) = p^2 + V(x)$, the volume of $\{H \leq 0\} \subseteq \mathbb{R}^d \times \{V \leq 0\}$ is

$$\int_{\{H \leq 0\}} 1\, dpdx = \sigma_d \int_{\mathbb{R}^d} V_-^{\frac{d}{2}}(x)\, dx,$$

so in the classical limit with $\lambda = 1/\hbar^2$ and Planck's constant $h \to 0$ ($\hbar = h/(2\pi)$), the number $N(-\hbar^2\Delta + V) = N(-\Delta + \hbar^{-2}V)$ of bound states of a quantum system is given asymptotically by the corresponding volume of phase space with $H \leq 0$ divided by $h^d = (2\pi)^d\hbar^d$. The bound (7.1) replaces the asymptotic limit (7.2) by an inequality. Moreover, such a bound is needed to prove (7.2) in the stated generality. Similar asymptotic estimates are useful in considering the problem 'Can you hear the shape of a drum?', see [115, Section XIII.15].

The bound (7.1) admits the following generalisation which encompasses many quantum systems, see [97]. Let $(\Sigma, \mathcal{E}, \mu)$ be a σ-finite standard Borel space [46, 424A] with a given Lusin μ-filtration $\mathcal{F}$, so that $L^2(\mu)$ is separable as described in Subsection 3.4.1. Suppose that H_0 is a selfadjoint operator defined in $L^2(\mu)$ and with spectrum contained in $[0, \infty)$.

The semigroup $S(t) = e^{-tH_0}$, $t \geq 0$, is assumed to consist of absolute integral operators and have the property that there exists a (smallest) semigroup $|S|(t)$, $t \geq 0$, of (pointwise) positive operators such that $|S(t)f| \leq |S|(t)f$ μ-a.e. for every $t \geq 0$ and nonnegative $f \in L^2(\mu)$—we take the term *positive operator* to mean positive in the partial order of operators acting on the Banach lattice $L^2(\mu)$. To avoid confusion, we say that a selfadjoint operator T is *hermitian positive* if its spectrum $\sigma(T)$ is a nonnegative set of real numbers. Also, suppose that $\sup_{t>0} \||S|(t)\|_{\mathcal{L}(L^2(\mu))} < \infty$ and suppose that $(\epsilon I + H_0)^{-1}$ and $Q(V)(\epsilon I + H_0)^{-1}$ are compact linear operators for each (some) $\epsilon > 0$.

Theorem 7.1. *Let $G : [0, \infty) \to [0, \infty)$ be a convex function with*

$$\int_0^\infty e^{-z} G(z) \frac{dz}{z} = 1.$$

Let $m : E \mapsto \chi_E$, $E \in \mathcal{E}$, and suppose that $\int_E \langle |S|(t), dm \rangle < \infty$ for every $t > 0$ and every set E with finite μ-measure. Then

$$N(H_0 + Q(V)) \leq \int_0^\infty \frac{dt}{t} \int_\Sigma \langle |S|(t) G(tV_-), dm \rangle. \tag{7.3}$$

Some explanation of the bound (7.3) is in order. The precise definition of the bilinear integral $\int_E \langle T, dm \rangle$ used in this chapter is given in Definition 7.1 below. If the right-hand side of (7.3) is infinite, then there is nothing to prove.

In applications, the bilinear integral $\int_\Sigma \langle |S|(t) G(tV_-), dm \rangle$ is the *trace* of the trace class operator $|S|(t) G(tV_-)$ on $L^2(\mu)$, see Chapter 3, but we don't preclude the possibility that $\int_\Sigma \langle |S|(t) G(tV_-), dm \rangle = \infty$ for some $t > 0$.

For the operator $H_0 = -\Delta$ in $L^2(\mathbb{R}^d)$, $d = 3, 4, \ldots$, $S = |S|$ and for Lebesgue measure λ on $\mathbb{R}^d$, the integral

$$\int_E \langle S(t), dm \rangle = (4\pi t)^{-d/2} \lambda(E)$$

is finite for each set E with finite Lebesgue measure and each $t > 0$. Moreover, we can choose

$$G(z) = \frac{(z-a)_+}{\int_a^\infty (z-a) e^{-z} \frac{dz}{z}}$$

to obtain the CLR inequality

$$N(H_0 + Q(V)) \leq C(G) \int_{\mathbb{R}^d} V_-(x)^{\frac{d}{2}} \, dx$$

where

$$C(G) = \frac{\int_a^\infty (t-a)t^{-d/2-1}\,dt}{(2\pi)^{d/2}\int_a^\infty (z-a)e^{-z}\frac{dz}{z}}.$$

For $d = 3$, the optimal value for $C(G)$ is 0.1156 obtained when $a = 1/4$ [90].

The idea that the CLR inequality (7.1) applies to quantum systems dominated by a Markov process first appears in [119], where the dominating process is assumed to be determined by an *ultracontractive* semigroup $|S|$. The proof of a bound similar to (7.3) is achieved in [119] by a version of the Trotter product formula rather than the Feynman-Kac formula that is a feature of Lieb's proof [90]. The proof in [119] also applies in the situation when $|S|$ is not a contraction $L^1(\mu)$. Despite the title of paper [119], the proof of the bound (7.3) in the general setting is achieved with *path integration* in the operator setting espoused by I. Kluvánek, see Theorem 6.2 above.

The CLR inequality in the form (7.1) is equivalent to the *global Sobolev inequality* on the underlying space with dimension $d = 3, 4, \ldots$. The argument of Li and Yau [89] gives the first intimation and the connection is more fully explored in [91,122]. However, our main result, Theorem 7.1, belongs to the domain of *measure theory*. Additional geometric information may then be extracted from L^p-related bounds on the semigroup e^{-tH_0}, $t \geq 0$. The work of [119] suggests that the bound (7.3) may be related to a weak form of logarithmic Sobolev inequality for an abstract measure space, independent of dimension, see [57].

The operator version of the *Feynman-Kac formula* Theorem 6.2 is used to prove the bound (7.3) with few assumptions about the regularity of the dominating semigroup. In particular, $|S|(t)$, $t > 0$, need not be a semigroup associated with the transition functions $p_t(x, dy)$, $x \in \Sigma$, of a Markov process in Σ. For example, if $H_0 = -\Delta + c\delta$ is the Laplacian with a point interaction defined in $L^2(\mathbb{R}^3)$, then e^{-tH_0}, $t \geq 0$, is a semigroup of (pointwise) positive operators bounded on $L^p(\mathbb{R}^3)$ only for $3/2 < p < 3$ [2, equation (3.4)] and such a semigroup cannot have a family of integral kernels which are the transition functions of a Markov process.

Besides the operator version of the Feynman-Kac formula, a critical ingredient in the approach of this chapter to the proof of the CLR bound (7.3) is the use of the bilinear integral $\int_\Sigma \langle T, dm \rangle \in [0, \infty]$ for a positive operator T on $L^2(\mu)$ as developed in Chapter 3. A similar notion of bilinear integration features in the connection between stationary state and time-dependent scattering theory in Chapter 5.

If $(Tu, u) \geq 0$ for all $u \in L^2(\mu)$ and $\int_\Sigma \langle T, dm \rangle < \infty$, then according to Theorem 3.3 above, T is a trace class operators on $L^2(\mu)$ and

$$\int_\Sigma \langle T, dm \rangle = \text{tr}(T).$$

The space $\mathfrak{C}_1(\mathcal{E}, L^2(\mu))$ of absolute integral operators T for which

$$\int_\Sigma \langle |T|, dm \rangle < \infty$$

considered in Chapter 3 forms a *lattice ideal* in the space of regular operators on $L^2(\mu)$, whereas the space $\mathcal{C}_1(L^2(\mu))$ of trace class operators is an *operator ideal* in the space of continuous linear operators on $L^2(\mu)$, that is, if $T \in \mathfrak{C}_1(\mathcal{E}, L^2(\mu))$ and $U : L^2(\mu) \to L^2(\mu)$ is a continuous linear operator with $|Uf| \leq |Tf|$ μ-a.e. for all $f \in L^2(\mu)$ with $f \geq 0$ μ-a.e., then $U \in \mathfrak{C}_1(\mathcal{E}, L^2(\mu))$, but on the other hand, $AT, TA \in \mathcal{C}_1(L^2(\mu))$ for all $A \in \mathcal{L}(L^2(\mu))$ and $T \in \mathcal{C}_1(L^2(\mu))$. As mentioned above, the intersections $\mathfrak{C}_1(\mathcal{E}, L^2(\mu)) \cap \mathcal{P}_+$ and $\mathcal{C}_1(L^2(\mu)) \cap \mathcal{P}_+$ with the collection $\mathcal{P}_+$ of bounded selfadjoint operators with nonnegative spectra coincide.

The proof of Theorem 7.1 is achieved in two steps. First a generalisation of Lieb's inequality [90] to dominated semigroups is given. The Feynman-Kac formula with respect to the evolution process associated with the semigroup S and the interchange of the order of integration with respect to the bilinear integral with respect to m and Lebesgue measure are essential ingredients. The last step is a standard application of the Birman-Schwinger principle.

7.2 Lattice traces for positive operators

The trace integral $\int_\Sigma \langle T, dm \rangle = \int_\Sigma \tilde{k}(x, x) \, d\mu(x)$ for an absolute integral operator with integral kernel k was defined in Chapter 3 for operators $T \in \mathfrak{C}_1(\mathcal{E}, L^2(\mu))$, with $\mathcal{E}$ a given Lusin filtration and

$$\tilde{k} = \limsup_{n \to \infty} \mathbb{E}(k | \mathcal{E}_n \otimes \mathcal{E}_n)$$

on $\Sigma \times \Sigma$. As shown in Proposition 3.7, $\tilde{k}(x, x) = k(x, x)$ for μ-almost all $x \in \Sigma$ in the case that μ is a σ-finite radon measure on the Hausdorff topological space Σ, the kernel $k : \Sigma \times \Sigma \to \mathbb{C}$ is continuous and $\mathcal{E}$ is suitably associated with k. In particular, k is necessarily $(\vee \mathcal{E})^2$-measurable.

For a general positive operator $T : L^2_+(\mu) \to L^2_+(\mu)$, we find that the requisite analysis is facilitated by approximating T from below by elements of $\mathfrak{C}_1(\mathcal{E}, L^2(\mu))$ in a similar fashion to measure theory: for a nonnegative measurable function f and a semifinite measure μ, the equality

$$\int_\Sigma f \, d\mu = \sup \left\{ \int_\Sigma s \, d\mu : 0 \le s \le f \right\} \in [0, \infty]$$

holds with respect to the supremum over μ-integrable simple functions s [45, 213B].

Let $(\Sigma, \mathcal{B}, \mu)$ be a σ-finite measure space equipped with a Lusin μ-filtration $\mathcal{E} = \langle \mathcal{E}_n \rangle_{n \in \mathbb{N}}$. The closure in $\mathfrak{C}_1(\mathcal{E}, L^2(\mu))$ of the collection of all finite rank operators $T : L^2(\mu) \to L^2(\mu)$ with integral kernels of the form

$$k = \sum_{j=1}^n f_j \otimes g_j,$$

for $f_j \in L^2(\mu)$, $g_j \in L^2(\mu)$ for $j = 1, \dots, n$ and $n = 1, 2, \dots$, is denoted by $L^2(\mu) \widehat{\otimes}_\varepsilon L^2(\mu)$.

The map $J : \Sigma \to \Sigma \times \Sigma$ defined by $J(x) = (x, x)$, $x \in \Sigma$, maps Σ bijectively onto the diagonal of $\Sigma \times \Sigma$. The following statement is what we observed in Proposition 3.5 above.

Proposition 7.1. *If k is the integral kernel of an operator belonging to $L^2(\mu) \widehat{\otimes}_\varepsilon L^2(\mu)$, then $\lim_{n \to \infty} \mathbb{E}(f | \mathcal{E}_n \otimes \mathcal{E}_n) \circ J$ converges μ-almost everywhere in Σ and in $L^1(\mu)$.*

Definition 7.1. Let $(\Sigma, \mathcal{B}, \mu)$ be a σ-finite measure space equipped with a Lusin μ-filtration $\mathcal{E} = \langle \mathcal{E}_n \rangle_{n \in \mathbb{N}}$. For a (lattice) positive operator T on $L^2(\mu)$ and $B \in \mathcal{B}$,

$$\int_B \langle T, dm \rangle = \sup \left\{ \int_B \tilde{k}(x, x) \, d\mu(x) : 0 \le T_k \le T, \ T_k \in L^2(\mu) \widehat{\otimes}_\varepsilon L^2(\mu) \right\}. \tag{7.4}$$

Here k is the integral kernel of the operator $T_k \in L^2(\mu) \widehat{\otimes}_\varepsilon L^2(\mu)$. The supremum may be infinite.

According to Theorem 3.1, the space $\mathfrak{C}_1(\mathcal{E}, L^2(\mu))$ is a lattice ideal in the space $\mathcal{L}_r(L^2(\mu))$ of regular operators on $L^2(\mu)$ and there is a one-to-one correspondence between absolute integral operators $T_k \in \mathcal{L}_r(L^2(\mu))$ and their integral kernels k [96, Theorem 3.3.5]. Consequently, the linear map $T_k \longmapsto \tilde{k} \circ J$, $T_k \in L^2(\mu) \widehat{\otimes}_\varepsilon L^2(\mu)$, is a Banach lattice homomorphism into $L^1(\mu)$ and the supremum (7.4) is the limit of an upwards directed set of finite integrals. Hence, if $\int_\Sigma \langle T, dm \rangle < \infty$, then $B \longmapsto \int_B \langle T, dm \rangle$, $B \in \mathcal{B}$, is a finite measure whose density with respect to μ is

$$\sup\{\tilde{k} \circ J : 0 \le T_k \le T,\ T_k \in L^2(\mu) \widehat{\otimes}_\varepsilon L^2(\mu)\}$$

in the Dedekind complete Banach lattice $L^1(\mu)$. In particular, for $T_k \in L^2(\mu) \widehat{\otimes}_\varepsilon L^2(\mu)$, we have

$$\int_B \langle T_k, dm \rangle = \int_B \tilde{k}(x, x)\, d\mu(x), \quad B \in \mathcal{B}.$$

It is clear that for $0 \le T_1 \le T_2$ we have

$$\int_\Sigma \langle T_1, dm \rangle \le \int_\Sigma \langle T_2, dm \rangle. \tag{7.5}$$

As noted above, if T is trace class or the integral kernel k_T of T is continuous, then T belongs to $L^2(\mu) \widehat{\otimes}_\varepsilon L^2(\mu)$ and it follows that for a positive kernel operator $T : L^2(\mu) \to L^2(\mu)$, we have

$$\int_\Sigma \langle T, dm \rangle = \operatorname{tr}(T)$$

if T is a trace class operator. In the case that $T \in L^2(\mu) \widehat{\otimes}_\varepsilon L^2(\mu)$, then

$$\int_\Sigma \langle T, dm \rangle = \int_\Sigma k_T(x, x)\, d\mu(x)$$

if μ is a σ-finite Radon measure on the Hausdorff space Σ and k_T is continuous, provided that the partitions $\mathcal{P}^\mathcal{E}$ that determine our Lusin μ-filtration $\mathcal{E}$ consist of relatively compact subsets of Σ.

Note that the only operator $T_k \in L^2([0,1]) \widehat{\otimes}_\varepsilon L^2([0,1])$ such that $T_k \le Id$, the identity operator, is the zero operator, so $\int_0^1 \langle Id, dm \rangle = 0$. On the other hand, if $\Sigma = \{1, \ldots, n\}$ and μ is counting measure, then

$$\int_0^1 \langle Id, dm \rangle = n.$$

Because we are dealing with a σ-finite measure μ it is also worthwhile to look at local approximations, leading to Theorem 7.2 below, which is a refinement of Theorem 3.3.

For a nonempty subset Γ of Σ, the σ-algebra $\{B \cap \Gamma : B \in \mathcal{B}\}$ is denoted by $\mathcal{B} \cap \Gamma$. For a set $B \in \mathcal{B}$ with $\mu(B) > 0$, $\mu_B : E \longmapsto \mu(E \cap B)$, $E \in \mathcal{B}$, is the relative measure on B. The completion of $L^2(\mu_B) \otimes L^2(\mu_B)$ with respect to the norm

$$f \longmapsto \|f\|_{L^1(\mu_B \otimes \mu_B))} + \int_B M_{\mathcal{E}^2}(f \chi_{B \times B}) \circ J\, d\mu$$

is denoted by $L^2(\mu_B)\widetilde{\otimes}_\varepsilon L^2(\mu_B)$. The tilde is used here to emphasise that we are now dealing with the integral kernels of operators rather than the absolute integral operators themselves.

As in Proposition 3.5, the functions

$$\mathbb{E}(f\chi_{B\times B}|\mathcal{E}_n \otimes \mathcal{E}_n) \circ J, \quad n = 1, 2, \ldots,$$

converge μ_B-a.e. and in $L^1(\mu_B)$ for each $f \in L^2(\mu_B)\widetilde{\otimes}_\varepsilon L^2(\mu_B)$.

The following observations are worth noting.

a) By the argument of Proposition 3.4, the projective tensor product $L^2(\mu_B)\widehat{\otimes}_\pi L^2(\mu_B)$ embeds onto a dense subspace of $L^2(\mu_B)\widetilde{\otimes}_\varepsilon L^2(\mu_B)$ and

$$\lim_{n\to\infty} \mathbb{E}(k\chi_{B\times B}|\mathcal{E}_n \otimes \mathcal{E}_n)(x,x) = \sum_{j=1}^\infty \phi_j(x)\psi_j(x)$$

for μ-almost all $x \in B$, if $k\chi_{B\times B} = \sum_{j=1}^\infty \phi_j \otimes \psi_j$ $(\mu \otimes \mu)$-a.e. with $\sum_{j=1}^\infty \|\phi_j\|_2\|\psi_j\|_2 < \infty$.

b) According to Proposition 3.7, if μ is a Radon measure on Σ, K is a compact subset of Σ and $k : \Sigma \times \Sigma \to \mathbb{C}$ is continuous, then $k\chi_{K\times K} \in L^2(\mu_K)\widetilde{\otimes}_\varepsilon L^2(\mu_K)$ and

$$\lim_{n\to\infty} \mathbb{E}(k\chi_{K\times K}|\mathcal{E}_n \otimes \mathcal{E}_n)(x,x) = k(x,x)$$

for μ-almost all $x \in K$. There is an underlying technical assumption here that Σ possesses a Lusin filtration, so the Borel σ-algebra of Σ is countably generated.

The space $L^2(\mu)\widetilde{\otimes}_{\varepsilon,\sigma}L^2(\mu)$ consists of all $(\mu \otimes \mu)$-equivalence classes $[f]$ of functions $f : \Sigma \times \Sigma \to \mathbb{C}$ for which there exist a partition $\mathcal{U}$ of Σ by sets in $\cup\mathcal{P}^\mathcal{E}$ such that $[f\chi_{U\times U}] \in L^2(\mu_U)\widetilde{\otimes}_\varepsilon L^2(\mu_U)$ for every $U \in \mathcal{U}$, that is, functions belonging to $L^2(\mu)\widetilde{\otimes}_{\varepsilon,\sigma}L^2(\mu)$ are *locally traceable* with respect to the Lusin μ-filtration $\mathcal{E}$.

Proposition 7.2. *If $f \in L^2(\mu)\widetilde{\otimes}_{\varepsilon,\sigma}L^2(\mu)$, then $\lim_{n\to\infty} \mathbb{E}(f|\mathcal{E}_n \otimes \mathcal{E}_n) \circ J$ converges μ-almost everywhere in Σ.*

Proof. Let $f \in L^2(\mu)\widetilde{\otimes}_{\varepsilon,\sigma}L^2(\mu)$ and let $\mathcal{U}$ be a partition of Σ by sets in $\cup\mathcal{P}^\mathcal{E}$ such that $[f\chi_{U\times U}] \in L^2(\mu_U)\widetilde{\otimes}_\varepsilon L^2(\mu_U)$ for every $U \in \mathcal{U}$.

Then for each $U \in \mathcal{U}$, there exists $n_U = 1, 2, \ldots$, such that $U \in \mathcal{P}^\mathcal{E}_{n_U}$ and

$$\mathbb{E}(f|\mathcal{E}_n \otimes \mathcal{E}_n)(x,y) = \mathbb{E}(f\chi_{U\times U}|\mathcal{E}_n \otimes \mathcal{E}_n)(x,y), \quad n \geq n_U,$$

for every $x,y \in U$ because $U_n(x) \subset U$ and $U_n(y) \subset U$ for all $n \geq n_U$, so that

$$\lim_{n\to\infty} \mathbb{E}(f|\mathcal{E}_n \otimes \mathcal{E}_n) \circ J$$

converges μ-a.e. in U. Because $\mathcal{U}$ is a partition of Σ, $\lim_{n\to\infty} \mathbb{E}(f|\mathcal{E}_n\otimes\mathcal{E}_n)\circ J$ converges μ-almost everywhere in Σ. $\qquad\square$

In the following result, we obtain a necessary and sufficient condition for a hermitian positive operator with kernel k to be trace class directly in terms of the finiteness of the integral $\int_\Sigma \tilde{k}(x,x)\, d\mu(x)$ along the diagonal.

Theorem 7.2. *Let $T_k : L^2(\mu) \to L^2(\mu)$ be a hermitian positive integral operator with kernel k and let $\mathcal{E}$ be a Lusin μ-filtration for which $\mathbb{E}(|k||\mathcal{E}_n \otimes \mathcal{E}_n)$ has finite values for each $n = 1, 2, \ldots$. The operator T_k is trace class if and only if $k \in L^2(\mu)\widetilde{\otimes}_{\mathcal{E},\sigma}L^2(\mu)$ and $\int_\Sigma \tilde{k}(x,x)\, d\mu(x) < \infty$. If T_k is trace class, then the formula*

$$\mathrm{tr}(T_k) = \int_\Sigma \tilde{k}(x,x)\, d\mu(x) \qquad (7.6)$$

holds with respect to the integral kernel $\tilde{k}$ of the operator T_k defined by

$$\tilde{k} = \lim_{n\to\infty} \mathbb{E}(k|\mathcal{E}_n \otimes \mathcal{E}_n),$$

wherever the limit exists. Moreover,

$$\int_\Sigma M_{\mathcal{E}^2}(k)(x,x)\, d\mu(x) \le 4\mathrm{tr}(T_k).$$

Proof. As in Proposition 7.1, if $k \in L^2(\mu)\widetilde{\otimes}_{\mathcal{E},\sigma}L^2(\mu)$ then for disjoint $\Sigma_j \in \cup\mathcal{P}^{\mathcal{E}}$, $j = 1, 2, \ldots$, such that $\cup_j\Sigma_j$ has full measure, the function $f\chi_{\Sigma_j\times\Sigma_j}$ belongs to $L^2(\mu_{\Sigma_j})\widetilde{\otimes}_{\mathcal{E}}L^2(\mu_{\Sigma_j})$ and

$$\tilde{k}(x,x) = \lim_{j\to\infty} \widetilde{k\chi_{\Sigma_j\times\Sigma_j}}(x,x)$$

for μ-almost all $x \in \Sigma_j$ and every $j = 1, 2, \ldots$. Moreover, $k\chi_{\Sigma_j\times\Sigma_j}$ is the integral kernel of the hermitian positive operator $Q(\Sigma_j)T_kQ(\Sigma_j)$ for each $j = 1, 2, \ldots$. Applying Theorem 3.3, the operator $Q(\Sigma_j)T_kQ(\Sigma_j)$ is trace class and

$$\int_{\Sigma_j} M_{\mathcal{E}^2}(k\chi_{\Sigma_j\times\Sigma_j}) \circ J\, d\mu \le 4\mathrm{tr}(Q(\Sigma_j)T_kQ(\Sigma_j)).$$

For any finite subset I of $\mathbb{N}$, let $n_I = \max\{n_{\Sigma_j} : j \in I\}$, where $\Sigma_j \in \mathcal{P}_{n_{\Sigma_j}}$ and

$$\mathbb{E}(k|\mathcal{E}_n \otimes \mathcal{E}_n)(x,y) = \mathbb{E}(k\chi_{(\cup_{i\in I}\Sigma_i)\times(\cup_{i\in I}\Sigma_i)}|\mathcal{E}_n \otimes \mathcal{E}_n)(x,y), \quad n \ge n_I,$$

for every $x, y \in \cup_{i\in I}\Sigma_i$ because $U_n(x) \subset \cup_{i\in I}\Sigma_i$ and $U_n(y) \subset \cup_{i\in I}\Sigma_i$ for all $n \ge n_I$. In particular, for $j \in I$ and $x \in \Sigma_j$, $U_n(x) \subset \Sigma_j$ for all $n \ge n_I$ and

$$\mathbb{E}(k|\mathcal{E}_n \otimes \mathcal{E}_n)(x,x) = \frac{\int_{U_n(x)\times U_n(x)} k\, d(\mu \otimes \mu)}{\mu(U_n(x))^2}.$$

Consequently, $\mathbb{E}(k|\mathcal{E}_n \otimes \mathcal{E}_n)(x, x) = \mathbb{E}(k\chi_{B \times B}|\mathcal{E}_n \otimes \mathcal{E}_n)(x, x)$ for $x \in B = \cup_{i \in I} \Sigma_i$ and $\mathrm{tr}(Q(B)T_k Q(B)) = \int_B \widetilde{k\chi_{B \times B}}(x, x)\, d\mu(x)$ by equation (7.6) and the bound

$$\mathrm{tr}(Q(B)T_k Q(B)) \leq \int_\Sigma \tilde{k}(x, x)\, d\mu(x)$$

follows. The noncommutative Fatou lemma shows that T_k is trace class. The rest of the proof follows that of Theorem 3.3. $\square$

Proposition 7.3. *Let $T : L^2(\mu) \to L^2(\mu)$ be a positive operator. For any uniformly bounded nonnegative μ-measurable functions V_1, V_2, the equalities*

$$\int_\Sigma \langle Q(V_2)TQ(V_1), dm \rangle = \int_\Sigma \langle Q(V_1 V_2)T, dm \rangle = \int_\Sigma \langle TQ(V_1 V_2), dm \rangle$$

of extended real numbers hold.

For any essentially bounded μ-measurable function $V \geq 0$,

$$\int_\Sigma \langle Q(V)T, dm \rangle \leq \|V\|_\infty \int_\Sigma \langle T, dm \rangle \in [0, \infty]. \tag{7.7}$$

Proof. We will first prove the inequality

$$\int_\Sigma \langle Q(V_1 V_2)T, dm \rangle \leq \int_\Sigma \langle Q(V_2)TQ(V_1), dm \rangle \tag{7.8}$$

of extended real numbers. The equalities are established by a similar argument.

Let $\epsilon > 0$, $V_j^{(\epsilon)} = V_j \vee \epsilon$, for $j = 1, 2$, and let λ be a positive number such that $\lambda < \int_\Sigma \langle Q(V_1^{(\epsilon)} V_2^{(\epsilon)})T, dm \rangle$. Then there exists $T_k \in L^2(\mu) \widehat{\otimes}_\epsilon L^2(\mu)$ such that $T_k \leq Q(V_1^{(\epsilon)} V_2^{(\epsilon)})T$ and

$$\lambda < \int_\Sigma \tilde{k} \circ J\, d\mu \leq \int_\Sigma \langle Q(V_1^{(\epsilon)} V_2^{(\epsilon)})T, dm \rangle.$$

Then $Q((V_1^{(\epsilon)})^{-1})T_k Q(V_1^{(\epsilon)}) \in L^2(\mu) \widehat{\otimes}_\epsilon L^2(\mu)$, the inequality

$$Q((V_1^{(\epsilon)})^{-1})T_k Q(V_1^{(\epsilon)}) \leq Q(V_2^{(\epsilon)})TQ(V_1^{(\epsilon)})$$

holds and $\int_\Sigma \langle Q((V_1^{(\epsilon)})^{-1})T_k Q(V_1^{(\epsilon)}), dm \rangle = \int_\Sigma \tilde{k} \circ J\, d\mu$. Consequently, $\lambda < \int_\Sigma \langle Q(V_2^{(\epsilon)})TQ(V_1^{(\epsilon)}), dm \rangle$, so that

$$\int_\Sigma \langle Q(V_1^{(\epsilon)} V_2^{(\epsilon)})T, dm \rangle \leq \int_\Sigma \langle Q(V_2^{(\epsilon)})TQ(V_1^{(\epsilon)}), dm \rangle.$$

If $\sup_{\epsilon > 0} \int_\Sigma \langle Q(V_2^{(\epsilon)})TQ(V_1^{(\epsilon)}), dm \rangle < \infty$, then the inequality (7.8) obtains in the limit by monotone convergence. The other inequalities are proved in a similar fashion.

A similar argument shows that for $\lambda < \|V\|_\infty^{-1} \int_\Sigma \langle Q(V^{(\epsilon)})T, dm \rangle$ there exists $T_k \in L^2(\mu) \widehat{\otimes}_\varepsilon L^2(\mu)$ such that $T_k \leq \|V\|_\infty^{-1} Q(V^{(\epsilon)})T$ and

$$\lambda < \int_\Sigma \tilde{k} \circ J \, d\mu \leq \|V\|_\infty^{-1} \int_\Sigma \langle Q(V^{(\epsilon)})T, dm \rangle.$$

Then $Q((V^{(\epsilon)})^{-1})T_k \in L^2(\mu) \widehat{\otimes}_\varepsilon L^2(\mu)$, the inequality

$$\|V\|_\infty Q((V^{(\epsilon)})^{-1})T_k \leq T$$

holds and $\|V\|_\infty \int_\Sigma \langle Q((V^{(\epsilon)})^{-1})T_k, dm \rangle \geq \int_\Sigma \tilde{k} \circ J \, d\mu > \lambda$. Consequently, $\lambda < \int_\Sigma \langle T, dm \rangle$, so that

$$\|V\|_\infty^{-1} \int_\Sigma \langle Q(V^{(\epsilon)})T, dm \rangle \leq \int_\Sigma \langle T, dm \rangle$$

for all $\epsilon > 0$. If $\int_\Sigma \langle T, dm \rangle < \infty$, monotone convergence ensures that $\sup_{\epsilon > 0} \int_\Sigma \langle Q(V^{(\epsilon)})T, dm \rangle = \int_\Sigma \langle Q(V)T, dm \rangle$, which establishes the bound (7.7). $\square$

It is well known that if T is a trace class operator on a Hilbert space $\mathcal{H}$ and B is any bounded linear operator on $\mathcal{H}$ then BT and TB are also trace class operators ($\mathcal{C}_1(\mathcal{H})$ is an *operator ideal*) and

$$\mathrm{tr}(BT) = \mathrm{tr}(TB).$$

By contrast, the space $\mathfrak{C}_1(\mathcal{E}, L^2(\mu))$ is a *lattice ideal* in $\mathcal{L}_r(L^2(\mu))$. For $T \in \mathfrak{C}_1(\mathcal{E}, L^2(\mu))$ and $B \in \mathcal{L}(L^2(\mu))$, the operator BT may not even be a kernel operator, see [20, Section 4.6.2], but we have the following trace property.

Proposition 7.4. *Let* $T_j : L^2(\mu) \to L^2(\mu)$, $j = 1, 2$, *be positive kernel operators. Then the equalities*

$$\int_\Sigma \langle T_1 T_2, dm \rangle = \int_\Sigma \langle T_2 T_1, dm \rangle$$

of extend real numbers hold.

Proof. We can find a partition Σ_n, $n = 1, 2, \ldots$, of Σ into sets with finite μ-measure. Let $\mu_n = \mu \upharpoonright (\cup_{\ell \leq n} \Sigma_n) \cap \mathcal{B}$. Let $f^*(x, y) = f(y, x)$ for a function of two variables $x, y \in \Sigma$.

An appeal to [60, Theorem 10.7] shows that the kernel of $T_1 T_2$ is

$$k_1 * k_2 : (x, y) \longmapsto \int_\Sigma k_1(x, z) k_2(z, y) \, d\mu(z)$$

for $(\mu \otimes \mu)$-almost all $(x, y) \in \Sigma \times \Sigma$ and

$$\int_\Sigma \int_\Sigma \int_\Sigma u(x)k_1(x,z)k_2(z,y)v(y)\,d\mu(y)d\mu(z)d\mu(x) < \infty$$

for every $u, v \in L^2_+(\mu)$. Suppose that $f, g \in L^\infty_+(\mu \otimes \mu)$ and $f \le k_1$ and $g \le k_2$. Then T_f and T_g are Hilbert-Schmidt operators on $L^2(\Sigma_n)$ so T_{f*g} is a trace class operator such that

$$\int_{\Sigma_n} \langle T_f T_g, dm \rangle = \|fg^*\|_{L^1(\mu_n \otimes \mu_n)} \le \|k_1.k_2^*\|_{L^1(\mu_n \otimes \mu_n)} = \|k_2.k_1^*\|_{L^1(\mu_n \otimes \mu_n)},$$

$$\int_{\Sigma_n} \langle T_f T_g, dm \rangle \le \int_\Sigma \langle T_1 T_2, dm \rangle$$

for $\mu(\Sigma_n) < \infty$ and $\|k_1.k_2^*\|_{L^1(\mu \otimes \mu)} = \sup\{\|fg^*\|_{L^1(\mu \otimes \mu)}\}$. It follows that $\|k_1.k_2^*\|_{L^1(\mu \otimes \mu)} \le \int_\Sigma \langle T_1 T_2, dm \rangle$.

Moreover, if $T_f \in L^2(\mu) \widehat{\otimes}_\varepsilon L^2(\mu)$ and $0 \le f \le k_1 * k_2$, then $\int_\Sigma \langle T_f, dm \rangle \le \|k_1.k_2^*\|_{L^1(\mu \otimes \mu)}$. The same argument applies to $T_2 T_1$, so

$$\int_\Sigma \langle T_1 T_2, dm \rangle = \int_\Sigma \langle T_2 T_1, dm \rangle = \|k_1.k_2^*\|_{L^1(\mu \otimes \mu)} \in [0, \infty].$$

□

We also note that a bilinear version of the Fubini-Tonelli Theorem holds. Let $(\Xi, \mathcal{B}, \nu)$ be a σ-finite measure space. For any function $f : \Xi \to \mathcal{L}_+(L^2(\mu))$ such that $\int_\Xi \int_\Sigma \langle f(\xi), dm \rangle \, d\nu(\xi) < \infty$, we say that f is $(m \otimes \nu)$-integrable if for each $u, v \in L^2(\mu)$, the scalar function $(fu, v) : \xi \longmapsto (f(\xi)u, v)$ is ν-integrable and there exists $T \in \mathfrak{C}_1(\mathcal{E}, L^2(\mu))$ such that

$$\int_\Xi \int_\Sigma \langle f(\xi), dm \rangle \, d\nu(\xi) = \int_\Sigma \langle T, dm \rangle \qquad (7.9)$$

$$\int_\Xi (f(\xi)u, v) \, d\nu = (Tu, v) \qquad (7.10)$$

for all $u, v \in L^2(\mu)$. Then we set

$$\int_{\Sigma \times \Xi} \langle f, d(m \otimes \nu) \rangle = \int_\Sigma \langle T, dm \rangle.$$

Because $\mathfrak{C}_1(\mathcal{E}, L^2(\mu))$ is a lattice ideal, for each $A \in \mathcal{B}$ there exists a positive operator $\int_A f \, d\nu \in \mathfrak{C}_1(\mathcal{E}, L^2(\mu))$ such that

$$\left(\left(\int_A f \, d\nu \right) u, v \right) = \int_A (f(\xi)u, v) \, d\nu \le (Tu, v)$$

for all $u, v \in L^2(\mu)_+$.

Remark 7.1. For each $u, v \in L^2(\mu)$, the tensor product $u \otimes v$, and $T \longmapsto \int_\Sigma \langle T, dm \rangle$ are continuous functionals on $\mathfrak{C}_1(\mathcal{E}, L^2(\mu))$, so it is natural to assume that both equalities (7.9) and (7.10) hold.

The following statement is a consequence of the definitions.

Proposition 7.5. *Let* $f : \Xi \to \mathcal{L}_+(L^2(\mu))$ *be a positive operator valued function such that f is $(m \otimes \nu)$-integrable. Suppose that $f(\xi) \in L^2(\mu) \widehat{\otimes}_{\mathcal{E}} L^2(\mu)$ for ν-almost all $\xi \in \Xi$.*

Then the scalar valued function $\xi \longmapsto \int_\Sigma \langle f(\xi), dm \rangle$ is ν-integrable and the equalities

$$\int_{\Sigma \times \Xi} \langle f, d(m \otimes \nu) \rangle = \int_\Sigma \left\langle \int_\Xi f \, d\nu, dm \right\rangle \tag{7.11}$$

$$= \int_\Xi \int_\Sigma \langle f(\xi), dm \rangle \, d\nu(\xi) \tag{7.12}$$

hold. Moreover, $\int_\Sigma \langle \int_A f \, d\nu, dm \rangle = \int_A \int_\Sigma \langle f(\xi), dm \rangle \, d\nu(\xi)$ for every $A \in \mathcal{B}$.

Proof. Equation (7.11) is the definition of $\int_{\Sigma \times \Xi} \langle f, d(m \otimes \nu) \rangle$ and (7.12) is a reformulation of assumption (7.9). By assumption, for ν-almost all $\xi \in \Xi$, we can find a martingale $\mathcal{F}_\xi$ with respect to the filtration $\{\mathcal{E}_n\}$ and a regularisation $k_\xi(x, y)$, $x, y \in \Sigma$, of the kernel associated with $f(\xi)$ such that

$$\left(\left(\int_A f \, d\nu \right) u, v \right) = \int_A \int_\Sigma \int_\Sigma k_\xi(x, y) u(x) \overline{v}(y) \, d\mu(x) d\mu(y) d\nu(\xi)$$

for all $A \in \mathcal{B}$ and $u, v \in L^2(\mu)$. Then for each $A \in \mathcal{B}$, we have

$$\int_\Sigma \left\langle \int_A f \, d\nu, dm \right\rangle = \int_A \int_\Sigma k_\xi(x, x) \, d\mu(x) d\nu(\xi)$$

$$= \int_A \int_\Sigma \langle f(\xi), dm \rangle \, d\nu(\xi)$$

by the scalar Fubini-Tonelli Theorem. $\qquad \square$

The following result follows from the observation in Proposition 3.1 that $\mathfrak{C}_1(\mathcal{E}, L^2(\mu))$ is a lattice ideal and an application of monotone convergence.

Proposition 7.6. *Let $M : \mathcal{B} \to \mathcal{L}_+(L^2(\mu))$ be a positive operator valued measure on a measurable space $(\Xi, \mathcal{B})$. If $\int_\Sigma \langle M(\Xi), dm \rangle < \infty$, then the set function $\langle M, m \rangle : A \mapsto \int_\Sigma \langle M(A), dm \rangle$, $A \in \mathcal{B}$, is a finite measure such that*

$$\int_\Sigma \langle M(A), dm \rangle \le \int_\Sigma \langle M(\Xi), dm \rangle, \quad A \in \mathcal{B},$$

and

$$\int_\Sigma \langle M(f), dm \rangle = \int_\Xi f \, d\langle M, m \rangle \le \|f\|_\infty \int_\Sigma \langle M(\Xi), dm \rangle$$

for all $\mathcal{B}$-measurable functions $f : \Xi \to [0, \infty]$.

7.3   The CLR inequality for dominated semigroups

Proof of Theorem 7.1. Step 1. (*Lieb's inequality*) If $\mathcal{X}$ denotes the (S, Q)-process (6.2), then according to Theorem 6.1, the dominating $(|S|, Q)$-process

$$|\mathcal{X}| = (\Omega, \langle \mathcal{S}_t \rangle_{t \geq 0}, \langle P_t \rangle_{t \geq 0}; \langle X_t \rangle_{t \geq 0})$$

with $P_t(\Omega) = |S|(t)$, $t \geq 0$, is σ-additive. According to Theorem 6.3 and the assumption that $(\Sigma, \mathcal{B})$ is a standard Borel space, we may assume from the outset that X is a progressively measurable process.

Also, by an appeal to Theorem 6.1 and the assumption that $S(t)$ is an absolute integral operator for each $t > 0$, the bounded linear operator $|S|(t)$ is the supremum in $\mathcal{L}_r(L^2(\mu))$ of an increasing family of absolute integral operators for each $t > 0$, directed by the cylindrical σ-algebras $\sigma(\langle X_s \rangle_{s \in J})$ as the finite subsets $J \subset [0, t]$ increase by inclusion. Because the absolute integral operators form a *band* in the Banach lattice $\mathcal{L}_r(L^2(\mu))$ of regular linear operators on $L^2(\mu)$ [96, Theorem 3.3.6], it follows that the positive linear operator $|S|(t)$ is a kernel operator on $L^2(\mu)$ for each $t > 0$.

For a bounded $\mathcal{E}$-measurable function $U \geq 0$, suppose that the limit

$$K_U = \lim_{\epsilon \to 0+} U^{\frac{1}{2}}(\epsilon I + H_0)^{-1} U^{\frac{1}{2}}$$

exists in the strong operator topology. Then K_U is selfadjoint and $\varphi(K_U)$ is defined by the functional calculus for selfadjoint operators for any bounded measurable function φ. Let $f : [0, \infty) \to [0, \infty)$ be a lower semicontinuous function with $f(0) = 0$ and set

$$F(x) = \int_0^\infty e^{-t} f(xt) \frac{dt}{t}, \quad x \geq 0. \tag{7.13}$$

We now show that $\int_\Sigma \langle F(K_U), dm \rangle \in [0, \infty]$ and

$$\int_\Sigma \langle F(K_U), dm \rangle \leq \int_0^\infty \frac{dt}{t} \int_\Sigma \left\langle \int_\Omega f\left(\int_0^t U(X_s)\, ds \right) dP_t, dm \right\rangle. \tag{7.14}$$

In the case that F is a uniformly bounded function on $[0, \infty)$ and $F(K_U)$ is a trace class operator, then an application of Proposition 3.3 ensures that the left-hand side of the equation is finite and

$$\int_\Sigma \langle F(K_U), dm \rangle = \text{tr}(F(K_U)).$$

To prove the bound (7.14), suppose first that U is a bounded measurable function and let

$$K_U(\epsilon) = U^{\frac{1}{2}}(\epsilon I + H_0)^{-1} U^{\frac{1}{2}}, \quad \epsilon > 0.$$

Then $K_U(\epsilon)$ is a compact selfadjoint linear operator because under our assumptions, $(\epsilon I + H_0)^{-1}$ is a compact operator for each $\epsilon > 0$. By the spectral theorem for selfadjoint operators, $K_U(\epsilon)$ has a countable set of real eigenvalues whose only limit point is zero.

For each $\lambda > 0$, the operator

$$A_U(\lambda, \epsilon) = U^{\frac{1}{2}}(\epsilon I + \lambda U + H_0)^{-1}U^{\frac{1}{2}}$$

has the property that

$$
\begin{aligned}
A_U(\lambda, \epsilon) &= K_U(\epsilon)(I + \lambda K_U(\epsilon))^{-1} \\
&= Q(U^{\frac{1}{2}})\left(\int_0^\infty e^{-\epsilon t}e^{-t(H_0+\lambda U)}\,dt\right)Q(U^{\frac{1}{2}}) \\
&= Q(U^{\frac{1}{2}})\left(\int_0^\infty e^{-\epsilon t}\left(\int_\Omega e^{-\lambda \int_0^t U\circ X_s\,ds}\,dM_t\right)dt\right)Q(U^{\frac{1}{2}}),
\end{aligned}
$$

by the Feynman-Kac formula (6.2). Hence, the expression

$$F(K_U(\epsilon)) = Q(U^{\frac{1}{2}})\left(\int_0^\infty e^{-\epsilon t}\left(\int_\Omega g\left(\int_0^t U\circ X_s\,ds\right)dM_t\right)dt\right)Q(U^{\frac{1}{2}}) \tag{7.15}$$

is valid for $g : x \mapsto e^{-\lambda x}$, $x \geq 0$, and $F : x \mapsto x/(1 + \lambda x)$, $x \geq 0$, so we take

$$F(x) = x\int_0^\infty e^{-t}g(xt)\,dt = \int_0^\infty e^{-t}f(xt)\frac{dt}{t}$$

for the function $f : x \mapsto xg(x)$, $x \geq 0$.

By the Stone-Weierstrass Theorem, the collection $\mathcal{K}$ of all linear combinations of functions $x \mapsto e^{-\lambda x}$, $x \geq 0$, for $\lambda > 0$, is dense in the space $C_0([0,\infty))$ of continuous functions on $[0,\infty)$ vanishing at infinity. The equality (7.15) is valid in the case that U is a bounded μ-measurable function and $f(x) = xg(x)$, $x \geq 0$, for $g \in C_0([0,\infty))$ because the left-hand side is continuous in the uniform operator topology and the uniform norm on g by the functional calculus for selfadjoint operators. The right-hand side is bounded by $C\|g\|_\infty\|U\|_\infty/\epsilon$ with $C = \sup_{t>0}\|P_t(\Omega)\|_{\mathcal{L}(L^2(\mu))}$, which we have assumed to be finite.

Next we show that for the functions $g : x \mapsto e^{-\lambda x}$, $x \geq 0$, and $f : x \mapsto xg(x)$, $x \geq 0$, the equality

$$
\begin{aligned}
\frac{1}{t}\int_\Sigma &\left\langle \int_\Omega f\left(\int_0^t W\circ X_s\,ds\right)dP_t, dm\right\rangle \\
&= \int_\Sigma\left\langle Q(W^{\frac{1}{2}})\left(\int_\Omega g\left(\int_0^t W\circ X_s\,ds\right)dP_t\right)Q(W^{\frac{1}{2}}), dm\right\rangle \tag{7.16}
\end{aligned}
$$

holds for every $t > 0$ and every nonnegative bounded measurable function W vanishing off a set of finite μ-measure. The right-hand side is finite because

$$\int_\Sigma \left\langle Q(W^{\frac{1}{2}}) \left(\int_\Omega g \left(\int_0^t W \circ X_s \, ds \right) dP_t \right) Q(W^{\frac{1}{2}}), dm \right\rangle$$

$$\leq \int_\Sigma \left\langle Q(W^{\frac{1}{2}}) P_t(\Omega) Q(W^{\frac{1}{2}}), dm \right\rangle$$

$$= \int_\Sigma \langle Q(W) P_t(\Omega), dm \rangle < \infty$$

by appealing to the monotone property (7.5) and Proposition 7.3. Also, for any $t > 0$ and bounded $\mathcal{S}_t$-measurable function $\varphi : \Omega \to \mathbb{C}$, the operator $\int_\Omega \varphi \, dP_t$ is necessarily an absolute integral operator because

$$Q(E) \left| \int_\Omega \varphi \, dP_t \right| \leq \|\varphi\|_\infty Q(E) P_t(\Omega)$$

for every set $E \in \mathcal{E}$ with finite μ-measure.

To prove the equality (7.16) for $g : x \mapsto e^{-\lambda x}$, $x \geq 0$, we first look at the inner integral on the left-hand side of the equation. By an application of Fubini's Theorem and the Markov property of the process $|\mathcal{X}|$, we have

$$\int_\Omega f \left(\int_0^t W \circ X_s \, ds \right) dP_t$$

$$= \int_\Omega \left(\int_0^t W \circ X_s \, ds \right) g \left(\int_0^t W \circ X_r \, dr \right) dP_t$$

$$= \int_0^t \int_\Omega W \circ X_s . g \left(\int_0^t W \circ X_r \, dr \right) dP_t ds$$

$$= \int_0^t P_{t-s} \left(g \left(\int_0^{t-s} W \circ X_r \, dr \right) \right)$$

$$\times \int_\Omega W \circ X_s . g \left(\int_0^s W \circ X_r \, dr \right) dP_s ds$$

$$= \int_0^t P_{t-s} \left(g \left(\int_0^{t-s} W \circ X_r \, dr \right) \right)$$

$$\times Q(W) \left(\int_\Omega g \left(\int_0^s W \circ X_r \, dr \right) dP_s \right) ds.$$

Applying the bilinear integral with respect to m to the integrand, for each

$0 \le s \le t$, we have

$$\int_\Sigma \left\langle P_{t-s}\left(g\left(\int_0^{t-s} W \circ X_r\, dr\right)\right) Q(W) \int_\Omega g\left(\int_0^s W \circ X_r\, dr\right) dP_s, dm \right\rangle$$

$$= \int_\Sigma \left\langle Q(W) P_s\left(g\left(\int_0^s W \circ X_r\, dr\right)\right) \right.$$

$$\left. \times P_{t-s}\left(g\left(\int_0^{t-s} W \circ X_r\, dr\right)\right), dm \right\rangle$$

$$= \int_\Sigma \left\langle Q(W) \int_\Omega g\left(\int_0^t W \circ X_s\, ds\right) dP_t, dm \right\rangle$$

by Proposition 7.4 and the semigroup property of the map

$$t \longmapsto \int_\Omega g\left(\int_0^t W \circ X_s\, ds\right) dP_t, \quad t \ge 0,$$

mentioned in Theorem 6.2, so an appeal to the Fubini-Tonelli Theorem, Proposition 7.5, with respect to $dm \otimes ds$ and Proposition 7.3 gives

$$\frac{1}{t}\int_\Sigma \left\langle \left(\int_\Omega f\left(\int_0^t W \circ X_s\, ds\right) dP_t\right), dm \right\rangle$$

$$= \int_\Sigma \left\langle \frac{1}{t}\int_0^t P_{t-s}\left(g\left(\int_0^{t-s} W \circ X_r\, dr\right)\right.\right.$$

$$\left.\left. \times Q(W) \int_\Omega g\left(\int_0^s W \circ X_r\, dr\right) dP_s ds, dm \right\rangle$$

$$= \frac{1}{t}\int_0^t \int_\Sigma \left\langle Q(W) \int_\Omega g\left(\int_0^t W \circ X_r\, dr\right) dP_t, dm \right\rangle ds$$

$$= \int_\Sigma \left\langle Q(W) \int_\Omega g\left(\int_0^t W \circ X_s\, ds\right) dP_t, dm \right\rangle$$

$$= \int_\Sigma \left\langle Q(W^{\frac{1}{2}}) \left(\int_\Omega g\left(\int_0^t W \circ X_s\, ds\right) dP_t\right) Q(W^{\frac{1}{2}}), dm \right\rangle.$$

Appealing to the Fubini-Tonelli Theorem again for $dm \otimes dt$, it follows that

$$\int_\Sigma \langle F(K_W(\epsilon)), dm \rangle$$

$$= \int_\Sigma \left\langle Q(W^{\frac{1}{2}}) \left(\int_0^\infty e^{-\epsilon t}\left(\int_\Omega g\left(\int_0^t W \circ X_s\, ds\right) dM_t\right) dt\right) Q(W^{\frac{1}{2}}), dm \right\rangle$$

$$\le \int_\Sigma \left\langle Q(W^{\frac{1}{2}}) \left(\int_0^\infty e^{-\epsilon t}\left(\int_\Omega g\left(\int_0^t W \circ X_s\, ds\right) dP_t\right) dt\right) Q(W^{\frac{1}{2}}), dm \right\rangle$$

$$= \int_0^\infty e^{-\epsilon t} \int_\Sigma \left\langle Q(W^{\frac{1}{2}}) \left(\int_\Omega g\left(\int_0^t W \circ X_s\, ds\right) dP_t\right) Q(W^{\frac{1}{2}}), dm \right\rangle dt$$

$$= \int_0^\infty \frac{e^{-\epsilon t}}{t} \int_\Sigma \left\langle \left(\int_\Omega f \left(\int_0^t W \circ X_s \, ds \right) dP_t \right), dm \right\rangle dt. \tag{7.17}$$

The inequality (7.17) is also valid if W is replaced by a bounded and μ-measurable function U and $f(x) = xg(x)$, $x \geq 0$, for $g \in C_0([0, \infty))$. To prove this, let $\Phi_t = \int_0^t W \circ X_s \, ds$. According to Proposition 7.6 and the assumption that $\int_\Sigma \langle W P_t(\Omega), dm \rangle < \infty$, the measures $\langle W (P_t \circ \Phi_t^{-1}), m \rangle$ and $\langle id.(P_t \circ \Phi_t^{-1}), m \rangle$ are finite Borel measures on $[0, \infty)$. The equality (7.16) is valid for all $g \in \mathcal{K}$ and so for all $g \in C_0([0, \infty))$ by uniform convergence. By monotone convergence, equality (7.16) is valid for all uniformly bounded and continuous functions $g : [0, \infty) \to \mathbb{C}$.

Replacing W by any bounded and measurable function $U \geq 0$, there is nothing to prove if the right-hand side of (7.17) is infinite, so suppose that $f : [0, \infty) \to [0, \infty)$ is a lower semicontinuous function with $f(0) = 0$ and

$$\int_0^\infty \frac{e^{-\epsilon t}}{t} \int_\Sigma \left\langle \left(\int_\Omega f \left(\int_0^t U \circ X_s \, ds \right) dP_t \right), dm \right\rangle dt < \infty.$$

Then for any $g \in C_0([0, \infty))$ such that $0 \leq \tilde{f}(x) \leq f(x)$, with $\tilde{f}(x) = xg(x)$ for $x \in [0, \infty)$, Proposition 7.6 ensures that

$$\int_0^\infty \frac{e^{-\epsilon t}}{t} \int_\Sigma \left\langle \left(\int_\Omega \tilde{f} \left(\int_0^t U \circ X_s \, ds \right) dP_t \right), dm \right\rangle dt < \infty.$$

Let $\tilde{F}$ be the function corresponding to $\tilde{f}$ according to formula (7.13).

Now let Σ_n, $n = 1, 2 \ldots$, be increasing sets with finite μ-measure such that $\Sigma = \cup_{n=1}^\infty \Sigma_n$ and put $U_n = U\chi_{\Sigma_n}$ for $n = 1, 2, \ldots$. According to formula (7.13), the function $\tilde{F}(x) = \int_0^\infty e^{-\frac{t}{x}} \tilde{f}(t) \frac{dt}{t}$, is increasing. The equality $\sup_n \int_\Sigma \left\langle \tilde{F}(K_{U_n}(\epsilon)), dm \right\rangle = \int_\Sigma \left\langle \tilde{F}(K_U(\epsilon)), dm \right\rangle$ follows from Theorem 3.3 and the monotonicity of the spectrum.

The integrals

$$\int_0^\infty \frac{e^{-\epsilon t}}{t} \int_\Sigma \left\langle \left(\int_\Omega \tilde{f} \left(\int_0^t U_n \circ X_s \, ds \right) dP_t \right), dm \right\rangle dt$$

converge to $\int_0^\infty \frac{e^{-\epsilon t}}{t} \int_\Sigma \left\langle \left(\int_\Omega \tilde{f} \left(\int_0^t U \circ X_s \, ds \right) dP_t \right), dm \right\rangle dt$ as $n \to \infty$ by Proposition 7.6 and dominated convergence. Then

$$\int_\Sigma \left\langle \tilde{F}(K_U(\epsilon)), dm \right\rangle$$
$$\leq \int_0^\infty \frac{e^{-\epsilon t}}{t} \int_\Sigma \left\langle \left(\int_\Omega \tilde{f} \left(\int_0^t U \circ X_s \, ds \right) dP_t \right), dm \right\rangle dt.$$

An appeal to Proposition 7.6 ensures that for $\Phi_t = \int_0^t U \circ X_s \, ds$, $t > 0$, the set function

$$A \longmapsto \int_0^\infty \frac{e^{-\epsilon t}}{t} \int_\Sigma \left\langle \left(\int_A f \, d(P_t \circ \Phi_t^{-1}) \right), dm \right\rangle dt$$

is a finite Borel measure on $[0, \infty)$, so finally, taking the supremum over all $0 \leq \tilde{f} \leq f$ gives the inequality (7.17) with W replaced by U.

Taking $\epsilon \to 0+$, monotone convergence now gives the inequality (7.14).

Step 2. (*Birman-Schwinger Principle*) Let $F : [0, \infty) \to [0, \infty)$ be a strictly increasing function and let $\{\mu_k\}$ denote the set of eigenvalues of $F(K_U)$. Because K_U is selfadjoint and F is nonnegative, it follows that $\mu_k \geq 0$. According to the Birman-Schwinger Principle [128, Theorem 8.1], for $U = V_-$, we have

$$N(H_0 + Q(V)) \leq \sum_{\{k : \mu_k \geq F(1)\}} 1$$

$$\leq \frac{1}{F(1)} \sum_{\{k : \mu_k \geq F(1)\}} \mu_k$$

$$\leq \frac{1}{F(1)} \mathrm{tr}(F(K_U))$$

$$= \frac{1}{F(1)} \int_\Sigma \langle F(K_U), dm \rangle$$

by Theorem 3.3. Then

$$N(H_0 + Q(V)) \leq F(1)^{-1} \int_\Sigma \langle F(K_U), dm \rangle$$

$$\leq F(1)^{-1} \int_0^\infty \frac{dt}{t} \int_\Sigma \left\langle \int_\Omega f \left(\int_0^t U(X_s) \, ds \right) dP_t, dm \right\rangle$$

by Lieb's inequality (7.14) when F is given by formula (7.13), f is lower semicontinuous and $F(1)$ is finite.

If f is convex, then by Jensen's inequality

$$f \left(\int_0^t U(X_s) \, ds \right) \leq t^{-1} \int_0^t f(tU(X_s)) \, ds,$$

so that

$$\int_0^\infty \frac{dt}{t} \int_\Sigma \left\langle \int_\Omega f \left(\int_0^t U(X_s) \, ds \right) dP_t, dm \right\rangle$$

$$\leq \int_0^\infty \frac{dt}{t^2} \int_\Sigma \left\langle \int_\Omega \left(\int_0^t f(tU(X_s)) \, ds \right) dP_t, dm \right\rangle$$

$$= \int_0^\infty \frac{dt}{t} \int_\Sigma \left\langle \int_0^t \left(\int_\Omega f(tU(X_s)) \, dP_t \right) \frac{ds}{t}, dm \right\rangle.$$

According to the Markov property (6.3) for the $(|S|, Q)$-process $|\mathcal{X}|$, the right-hand side is equal to

$$\int_0^\infty \frac{dt}{t} \int_\Sigma \left\langle \int_0^t P_{t-s}(\Omega)Q(f(tU))P_s(\Omega) \frac{ds}{t}, dm \right\rangle$$

$$= \int_0^\infty \frac{dt}{t} \int_0^t \int_\Sigma \langle P_t(\Omega)Q(f(tU)), dm \rangle \frac{ds}{t}$$

$$= \int_0^\infty \frac{dt}{t} \int_\Sigma \langle P_t(\Omega)Q(f(tU)), dm \rangle .$$

The convex function $f : [0, \infty) \to [0, \infty)$ is necessarily continuous, so if $F(1)$ is finite, then $f(0) = 0$. Consequently, the bound (7.3) is valid if we take $G = F(1)^{-1}f$.

The bound is valid for any V satisfying the stated conditions, for then $H_0 + Q(V)$ is defined by a form sum, so by replacing V by $V\chi_{\{|V| \leq n\}} + \epsilon I$ and H_0 by $H_0 + \epsilon I$, if necessary, and taking the limit as $n \to \infty$ and $\epsilon \to 0+$, the bound (7.3) is obtained. $\qquad\square$

Example 7.1. Let $\mathbf{P}$ be the momentum operator $-id/dx$ acting in $L^2(\mathbb{R})$. With $H(\mathbf{A}) = \frac{1}{2}(\mathbf{P} - \mathbf{A})^2$ and $S(t) = e^{-tH(\mathbf{A})}$, $t \geq 0$, the σ-additive evolution process

$$(\Omega, \langle \mathcal{S}_t \rangle_{t \geq 0}, \langle M_t^{\mathbf{A}} \rangle_{t \geq 0}; \langle X_t \rangle_{t \geq 0})$$

corresponds to a quantum particle on the line subject to a magnetic vector potential $\mathbf{A}$. Then $(\Omega, \langle \mathcal{S}_t \rangle_{t \geq 0}, \langle M_t \rangle_{t \geq 0}; \langle X_t \rangle_{t \geq 0})$ is the dominating process with $M_t(\Omega) = e^{-\frac{1}{2}t\mathbf{P}^2}$, $t \geq 0$. According to Theorem 7.1, the bound

$$N(H(\mathbf{A}) + Q(V)) \leq \int_0^\infty \frac{dt}{t} \int_\Sigma \langle e^{-\frac{1}{2}t\mathbf{P}^2} G(tV_-), dm \rangle$$

holds for any $\mathbf{A} : \mathbb{R} \to \mathbb{R}$ such that $\mathbf{A}$ and $\mathbf{A}'$ belong to $L_{\text{loc}}^2(\mathbb{R})$ and $V \in L^1(\mathbb{R}) + L^p(\mathbb{R})$ with $1 < p < \infty$ [128, Theorem 8.2, Theorem 15.5].

Other examples of dominated semigroups for which the CLR inequality is valid may be found in [119], [97].

Chapter 8

Linear operator equations

The analysis of the equation $AX - XB = Y$ for linear operators A, B, X and Y acting in a Hilbert space $\mathcal{H}$ has many applications in operator theory, differential equations and quantum physics, see [14] for a relaxed discussion with numerous examples.

Starting with the case of scalars, the equation $ax - xb = y$ has a unique solution provided that $a \neq b$. For the case of diagonal matrices $A = \mathrm{diag}(\lambda_1, \ldots, \lambda_n)$ and $B = \mathrm{diag}(\mu_1, \ldots, \mu_n)$, for any matrix $Y = \{y_{ij}\}_{i,j=1}^n$ there exists a unique solution X of the equation $AX - XB = Y$ if and only if $\lambda_i - \mu_j \neq 0$ for $i, j = 1, \ldots, n$ and then the solution $X = \{x_{ij}\}_{i,j=1}^n$ is given by

$$x_{ij} = \frac{y_{ij}}{\lambda_i - \mu_j}, \quad i, j = 1, \ldots, n.$$

The operator version is called the *Sylvester-Rosenblum Theorem* in [14], although earlier versions are due to M. Krein and Yu. Daletskii [14, p. 1]. For a continuous linear operator A on a Banach space $\mathcal{X}$, the *spectrum* $\sigma(A)$ of A is the set of all $\lambda \in \mathbb{C}$ for which $\lambda I - A$ is not invertible.

Theorem 8.1 (Sylvester-Rosenblum Theorem). *Let $\mathcal{X}$ be a Banach space and let A and B be continuous linear operators on $\mathcal{X}$ for which $\sigma(A) \cap \sigma(B) = \emptyset$. Then for each operator $Y \in \mathcal{L}(\mathcal{X})$, the equation $AX - XB = Y$ has a unique solution $X \in \mathcal{L}(\mathcal{X})$.*

As a taster for applications of the Sylvester-Rosenblum Theorem, suppose that A and B bounded normal operators on a Hilbert space $\mathcal{H}$ with spectral measures P_A and P_B, respectively. Then there exists $c > 0$, such that for any two Borel subsets S_1 and S_2 of $\mathbb{C}$ separated by a distance

$$\delta = \inf\{|x - y| : x \in S_1, \ y \in S_2\},$$

the projections $E = P_A(S_1)$, $F = P_B(S_2)$, satisfy the norm estimate

$$\|EF\| \leq \frac{c}{\delta}\|A - B\|.$$

The norm $\|EF\|$ represents the angle between the subspaces $\text{ran}(E)$ and $\text{ran}(F)$. Such estimates are useful in numerical computations. Even in finite dimensional Hilbert spaces, the Sylvester-Rosenblum Theorem leads to eigenvalue estimates for matrix norms independent of dimension.

Theorem 8.2 ([13, Theorem 5.1a]). *Let A and B be two normal $(n \times n)$ matrices with eigenvalues $\alpha_1, \ldots, \alpha_n$ and $\beta_1, \ldots, \beta_n$ respectively, counting multiplicity. With the same constant c mentioned above, if $\|A - B\| \leq \epsilon/c$, then there exists a permutation π of the index set $\{1, \ldots, n\}$ such that*

$$|\alpha_i - \beta_{\pi i}| < \epsilon$$

for $i = 1, \ldots, n$.

The Sylvester-Rosenblum Theorem also comes with a representation of the solution X of the equation $AX - XB = Y$ if A and B are bounded linear operators for which $\sigma(A) \cap \sigma(B) = \emptyset$. Suppose that the contour Γ is the union of closed contours in the plane, with total winding numbers 1 around $\sigma(A)$ and 0 around $\sigma(B)$. Then

$$X = \frac{1}{2\pi i} \int_\Gamma (\zeta I - A)^{-1} Y (\zeta I - B)^{-1}. \tag{8.1}$$

Other representations of the solution are possible by utilising the spectral properties of the operators A and B, see [14, Section 9].

In this chapter, we are concerned with solutions X of the operator equation $AX - XB = Y$ when A is an unbounded selfadjoint or normal operator acting in a Hilbert space $\mathcal{H}$ and B is a closed unbounded operator. If the spectra $\sigma(A)$ and $\sigma(B)$ are a positive distance apart, then we hope to construct the solution X of $AX - XB = Y$ by the formula

$$X = \int_{\sigma(A)} dP_A(\zeta) Y (\zeta I - B)^{-1} \tag{8.2}$$

in place of (8.1) with respect to the spectral measure P_A of A. The operator valued measure P_A acts on the values of the operator valued function $\zeta \longmapsto Y(\zeta I - B)^{-1}$. As in the case of scattering theory considered in Chapter 5, for $h \in \mathcal{H}$, the vector $Xh \in \mathcal{H}$ often has the *decoupled* representation

$$Xh = J \int_{\sigma(A)} P_A(d\zeta) \otimes (Y(\zeta I - B)^{-1}h),$$

where the $\mathcal{H}$-valued function $\zeta \longmapsto Y(\zeta I - B)^{-1}h$, $\sigma \in \sigma(A)$, is P_A-integrable in the space $\mathcal{L}(\mathcal{H})\widehat{\otimes}_\tau \mathcal{H}$ and $J : \mathcal{L}(\mathcal{H})\widehat{\otimes}_\tau \mathcal{H} \to \mathcal{H}$ is the continuous linear extension of the composition map $T \otimes h \mapsto Th$, $T \in \mathcal{L}(\mathcal{H})$, $h \in \mathcal{H}$.

If the operator B is itself a bounded linear operator, then the simpler representation (8.1) may be employed with the contour Γ winding once around $\sigma(B)$ and zero times around $\sigma(A)$.

Because we shall be dealing with unbounded operators A and B, we have to be careful about domains when interpreting the equation $AX - XB = Y$. We follow the treatment in [3, Section 2]. Applications of equation (8.2) to perturbation theory and the spectral shift function may also be found in [3].

8.1 Operator equations

Definition 8.1. Let $\mathcal{H}$ and $\mathcal{K}$ be Hilbert spaces. Suppose that $A : \mathcal{D}(A) \to \mathcal{K}$ and $B : \mathcal{D}(B) \to \mathcal{H}$ are closed and densely defined linear operators with domains $\mathcal{D}(A) \subset \mathcal{K}$ and $\mathcal{D}(B) \subset \mathcal{H}$. Given $Y \in \mathcal{L}(\mathcal{H},\mathcal{K})$, a continuous linear operator $X \in \mathcal{L}(\mathcal{H},\mathcal{K})$ is said to be a *weak solution* of the equation

$$AX - XB = Y \qquad (8.3)$$

if for every $h \in \mathcal{D}(B)$ and $k \in \mathcal{D}(A^*)$, the equality

$$(Xh, A^*k) - (XBh, k) = (Yh, k)$$

holds with respect to the inner product $(\cdot, \cdot)$ of $\mathcal{K}$.

The domain $\mathcal{D}(A^*)$ of the adjoint A^* of A is the set of all elements k of $\mathcal{K}$ such that the linear map $h \longmapsto (Ah, k)$, $h \in \mathcal{D}(A)$, is the restriction to $\mathcal{D}(A)$ of $h \longmapsto (h, y)$, $h \in \mathcal{H}$, for an element $y \in \mathcal{K}$ and then $y = A^*k$.

A *strong solution* $X \in \mathcal{L}(\mathcal{H},\mathcal{K})$ of (8.3) has the property that

$$\mathrm{ran}(X \restriction \mathcal{D}(B)) \subset \mathcal{D}(A)$$

and

$$AXh - XBh = Yh, \quad h \in \mathcal{D}(B).$$

The existence of strong solutions of the operator equation (8.3) is discussed in [106] under the assumption that A and $-B$ are the generators of C_0-semigroups, a situation that arises in delay or partial differential equations and control theory. Strong solutions of (8.3) may not exist in this setting, even when the spectra $\sigma(A)$ and $\sigma(B)$ are separated by a vertical strip [106, Example 9].

In the case that A and B are both selfadjoint operators, the following result is a consequence of [13, Theorem 4.1], see [3, Theorem 2.7].

Theorem 8.3. *Let $\mathcal{H}$ and $\mathcal{K}$ be Hilbert spaces. Suppose that $A : \mathcal{D}(A) \to \mathcal{K}$ and $B : \mathcal{D}(B) \to \mathcal{H}$ are selfadjoint operators whose spectra $\sigma(A)$ and $\sigma(B)$ are a distance $\delta > 0$ apart. Then equation (8.3) has a unique weak solution*

$$X = \int_{\mathbb{R}} e^{-itA} Y e^{itB} f_\delta(t) \, dt$$

for any function $f_\delta \in L^1(\mathbb{R})$, continuous on $\mathbb{R} \setminus \{0\}$, such that

$$\int_{\mathbb{R}} e^{-isx} f_\delta(s) \, ds = \frac{1}{x} \quad \text{for } |x| > \frac{1}{\delta}.$$

Moreover $\|X\| \leq \frac{\pi}{2\delta} \|Y\|$.

The integral representing the solution X is a Bochner integral for the strong operator topology.

For a selfadjoint operator A in a Hilbert space $\mathcal{K}$ and a closed, densely defined operator B in a Hilbert space $\mathcal{H}$, the domains $\mathcal{D}(B)$ and $\mathcal{D}(A)$ are endowed with the respective graph norms associated with the closed operators B and A. Suppose also, that τ is the topology on the tensor product $\mathcal{L}(\mathcal{X}) \otimes \mathcal{X}$ defined by formula (5.13) with $\mathcal{X} = \mathcal{K}$ and $\mathcal{Y} = \mathcal{H}$ and let $E = \mathcal{L}(\mathcal{K}) \widehat{\otimes}_\tau \mathcal{K}$ be the completion of the tensor product with the norm topology τ. According to Lemma 5.3, the Banach space E is bilinear admissible for the Hilbert space $\mathcal{K}$ in the sense of Section 2.3 and the composition map

$$T \otimes k \longmapsto Tk, \quad T \in \mathcal{L}(\mathcal{K}), \quad k \in \mathcal{K},$$

has a continuous linear extension $J_E : E \to \mathcal{K}$. We may adopt the definition analogous to Definition 2.3.

Definition 8.2. Let $\mathcal{K}$ be a Hilbert space. A function $f : \Omega \to \mathcal{K}$ is said to be *m-integrable in $E = \mathcal{L}(\mathcal{K}) \widehat{\otimes}_\tau \mathcal{K}$* for an operator valued measure $m : \mathcal{S} \to \mathcal{L}(\mathcal{K})$, if for each $x, x', y' \in \mathcal{K}$, the scalar function (f, x') is integrable with respect to the scalar measure (mx, y') and for each $S \in \mathcal{S}$, there exists an element $(m \otimes f)(S)$ of E such that

$$((m \otimes f)(S), x \otimes y' \otimes x') = \int_S (f, x') \, d(mx, y')$$

for every $x, x', y' \in \mathcal{K}$.

If f is m-integrable in E, then $mf(S) \in \mathcal{K}$ is defined for each $S \in \mathcal{S}$ by

$$mf(S) = J_E \big((m \otimes f)(S) \big).$$

We also denote $mf(S)$ by $\int_S dm \, f$ or $\int_S dm(\omega) \, f(\omega)$.

In the present context, the representation of solutions of equation (8.3) via bilinear integration is analogous to Example 5.1 in scattering theory.

Example 8.1. Suppose that A is a bounded selfadjoint operator defined on a Hilbert space $\mathcal{K}$ such that $\sigma(A) \subset (-\infty, -\delta)$ for some $\delta > 0$. Let $-B$ be the generator of a uniformly bounded C_0-semigroup e^{-tB}, $t \geq 0$, on the Hilbert space $\mathcal{H}$.

We can employ (8.1) in this situation to represent the weak solution of equation (8.3), but it is instructive to see how the integral (8.2) converges with the assumptions above.

Let $E = \mathcal{L}(\mathcal{K}) \widehat{\otimes}_\pi \mathcal{K}$ be the projective tensor product of the Hilbert space $\mathcal{K}$ with the space $\mathcal{L}(\mathcal{K})$ of bounded linear operators on $\mathcal{K}$ with the uniform norm. Then $e^{tA} \otimes (Ye^{-tB}h)$ belongs to the tensor product $\mathcal{L}(\mathcal{K}) \otimes \mathcal{K}$ for each $t \geq 0$ and $h \in \mathcal{H}$ and the function $t \longmapsto e^{tA} \otimes (Ye^{-tB}h)$, $t \geq 0$, is continuous in $\mathcal{L}(\mathcal{K}) \widehat{\otimes}_\pi \mathcal{K}$, because A is assumed to be bounded so

$$e^{tA} \otimes (Ye^{-tB}h) = I \otimes (Ye^{-tB}h) + \sum_{n=1}^{\infty} \frac{t^n}{n!} (A^n \otimes (Ye^{-tB}h))$$

converges in $\mathcal{L}(\mathcal{K}) \widehat{\otimes}_\pi \mathcal{K}$ uniformly for t in any bounded interval. The inequalities

$$\int_0^\infty \left\| e^{tA} \otimes (Ye^{-tB}h) \right\|_{\mathcal{L}(\mathcal{K}) \widehat{\otimes}_\pi \mathcal{K}} \leq \int_0^\infty \|e^{tA}\| . \|(Ye^{-tB}h)\| \, dt$$

$$\leq \left(\int_0^\infty e^{-\delta t} \|e^{-tB}\| \, dt \right) . \|Y\|_{\mathcal{L}(\mathcal{H},\mathcal{K})} . \|h\|$$

ensure that $\int_0^\infty e^{tA} \otimes (Ye^{-tB}h) \, dt$ converges as a Bochner integral in the projective tensor product in $\mathcal{L}(\mathcal{K}) \widehat{\otimes}_\pi \mathcal{K}$ and

$$\int_{\sigma(A)} P_A(d\zeta) \otimes (Y(\zeta I - B)^{-1}h) = \int_{\sigma(A)} P_A(d\zeta) \otimes \left(Y \int_0^\infty e^{\zeta t} e^{-tB} h \, dt \right)$$

$$= \int_0^\infty \left(\int_{\sigma(A)} e^{\zeta t} P_A(d\zeta) \right) \otimes (Ye^{-tB}h) \, dt$$

$$= \int_0^\infty e^{tA} \otimes (Ye^{-tB}h) \, dt$$

belongs to $\mathcal{L}(\mathcal{K}) \widehat{\otimes}_\pi \mathcal{K}$ too. Then

$$\int_{\sigma(A)} P_A(d\zeta) (Y(\zeta I - B)^{-1}h) = J_E \int_{\sigma(A)} P_A(d\zeta) \otimes (Y(\zeta I - B)^{-1}h)$$

defines a continuous linear operator

$$\int_{\sigma(A)} P_A(d\zeta) Y(\zeta I - B)^{-1} : h \longmapsto \int_{\sigma(A)} P_A(d\zeta) (Y(\zeta I - B)^{-1}h), \quad h \in \mathcal{H},$$

belonging to $\mathcal{L}(\mathcal{H}, \mathcal{K})$ with norm bounded by

$$\frac{\sup_{t \geq 0} \|e^{-tB}\|}{\delta} \|Y\|_{\mathcal{L}(\mathcal{H}, \mathcal{K})}.$$

In order to deal with unbounded operators, we replace the projective tensor product topology π by the topology τ defined by formula (5.13).

Lemma 8.1. *Let $\mathcal{H}$ and $\mathcal{K}$ be Hilbert spaces. Suppose that $A : \mathcal{D}(A) \to \mathcal{K}$ is a selfadjoint operator with spectral measure P_A and $B : \mathcal{D}(B) \to \mathcal{H}$ is a densely defined, closed linear operator such that $\sigma(A) \cap \sigma(B) = \emptyset$.*

Let $Y \in \mathcal{L}(\mathcal{H}, \mathcal{K})$. For each $h \in \mathcal{H}$, the $\mathcal{K}$-valued function

$$\Phi_h : \zeta \longmapsto Y(\zeta I - B)^{-1}h, \quad \zeta \in \sigma(A), \tag{8.4}$$

is P_A-integrable in $\mathcal{L}(\mathcal{K}) \widehat{\otimes}_\pi \mathcal{K}$ on every compact subset of $\sigma(A)$.

Furthermore, there exist $\mathcal{L}(\mathcal{H}, \mathcal{K})$-valued $\mathcal{B}(\sigma(A))$-simple functions

$$s_n : \sigma(A) \to \mathcal{L}(\mathcal{H}, \mathcal{K}), \quad n = 1, 2, \ldots,$$

such that for each $h \in \mathcal{H}$, $s_n(\omega)h \to \Phi_h(\omega)$ in $\mathcal{K}$ as $n \to \infty$ for P_A-almost all $\omega \in \sigma(A)$ and for each compact subset of K of $\sigma(A)$,

$$\sup_{S \in \mathcal{B}(K)} \|(P_A \otimes \Phi_h)(S) - (P_A \otimes (s_n h))(S)\|_{\mathcal{L}(\mathcal{K}) \widehat{\otimes}_\pi \mathcal{K}} \to 0$$

as $n \to \infty$.

Proof. For a closed and densely defined operator T, the resolvent $(\lambda I - T)^{-1}$ is defined for all complex numbers λ belonging to the resolvent set $\rho(T) = \mathbb{C} \setminus \sigma(T)$. Suppose that $\rho(T)$ is nonempty. Then the resolvent equation

$$(\lambda I - T)^{-1} - (\mu I - T)^{-1} = (\mu - \lambda)(\lambda I - T)^{-1}(\mu I - T)^{-1}$$

for $\lambda, \mu \in \rho(T)$ ensures that $\lambda \longmapsto (\lambda I - T)^{-1}$, $\lambda \in \rho(T)$, is a holomorphic operator valued function for the uniform operator topology [76, Equation I (5.6)]. It follows that for each $h \in \mathcal{H}$, the function

$$\lambda \longmapsto (\lambda I - A)^{-1} \otimes (Y(\lambda I - B)^{-1}h)$$

is continuous in the projective tensor product $\mathcal{L}(\mathcal{K}) \widehat{\otimes}_\pi \mathcal{K}$ for the uniform norm on $\mathcal{L}(\mathcal{K})$. For a compact subset K of $\sigma(A)$, let $A_K = P_A(K)A$ be the part of A on K. Then for a contour Γ_K with winding number 1 around K and zero around the closed set $\sigma(B)$, the integral

$$\int_{\Gamma_K} \|(\lambda I - A_K)^{-1}\| . \|Y(\lambda I - B)^{-1}h\|_{\mathcal{K}} |d\lambda|$$

is bounded by

$$(|\Gamma_K| \cdot \sup_{\lambda \in \Gamma_K} \|(\lambda I - A_K)^{-1}\| \cdot \|(\lambda I - B)^{-1}\|) \cdot \|Y\|_{\mathcal{L}(\mathcal{H}, \mathcal{K})} \cdot \|h\|_{\mathcal{H}},$$

so the function $\int_{\Gamma_K} (\lambda I - A_K)^{-1} \otimes (Y(\lambda I - B)^{-1}h) \, d\lambda$ converges as a Bochner integral in $\mathcal{L}(\mathcal{K}) \widehat{\otimes}_\pi \mathcal{K}$. For every Borel subset S of the set K and $x, x', y' \in \mathcal{K}$, an application of Cauchy's integral formula yields

$$\int_S (P_A x, x')(d\zeta)(Y(\zeta I - B)^{-1} h, y')$$

$$= \frac{1}{2\pi i} \int_S (P_A x, x')(d\zeta) \int_{\Gamma_K} \frac{(Y(\lambda I - B)^{-1} h, y')}{\lambda - \zeta} \, d\lambda$$

$$= \frac{1}{2\pi i} \int_{\Gamma_K} ((\lambda I - P_A(S) A_K)^{-1} x, x')(Y(\lambda I - B)^{-1} h, y') \, d\lambda,$$

so according to Definition 8.2 (replacing the topology τ by the stronger projective topology π), the function $\zeta \longmapsto Y(\zeta I - B)^{-1} h$, $\zeta \in \sigma(A)$, is P_A-integrable in $\mathcal{L}(\mathcal{K}) \widehat{\otimes}_\pi \mathcal{K}$ on the set K and

$$\int_S dP_A(\zeta) \otimes (Y(\zeta I - B)^{-1} h)$$

$$= \frac{1}{2\pi i} \int_{\Gamma_K} (\lambda I - P(S) A_K)^{-1} \otimes (Y(\lambda I - B)^{-1} h) \, d\lambda \qquad (8.5)$$

as an element of the projective tensor product $\mathcal{L}(\mathcal{K}) \widehat{\otimes}_\pi \mathcal{K}$ for each Borel subset S of K.

Because the operator valued function $\lambda \longmapsto (\lambda I - B)^{-1}$, $\lambda \in \sigma(A)$, is uniformly continuous on the compact set K, for each $\epsilon > 0$, there exists an $\mathcal{L}(\mathcal{H})$-valued $\mathcal{B}(\sigma(A))$-simple function φ_ϵ such that

$$\sup_{\lambda \in K} \|(\lambda I - B)^{-1} - \varphi_\epsilon(\lambda)\|_{\mathcal{L}(\mathcal{H})} < \epsilon,$$

so that

$$\sup_{S \in \mathcal{B}(K)} \int_{\Gamma_K} \|(\lambda I - P(S) A_K)^{-1}\| \cdot \|Y(\lambda I - B)^{-1} h - Y\varphi_\epsilon(\lambda) h\|_{\mathcal{K}} \, |d\lambda| \to 0$$

as $\epsilon \to 0+$ for each $h \in \mathcal{H}$. According to the identity (8.5), it follows that

$$\sup_{S \in \mathcal{B}(K)} \|(P_A \otimes \Phi_h)(S) - (P_A \otimes (Y\varphi_\epsilon h))(S)\|_{\mathcal{L}(\mathcal{K}) \widehat{\otimes}_\pi \mathcal{K}} \to 0$$

as $\epsilon \to 0+$. Because the spectral measure P_A is inner regular on compact sets, the simple functions s_n, $n = 1, 2, \ldots$, can be pieced together from the simple functions $\varphi_{1/n}$, $n = 1, 2, \ldots$, on each compact set K. $\qquad \square$

If both operators A and B are selfadjoint, then Theorem 8.3 ensures that a weak solution X of equation (8.3) exists and gives a norm estimate for X. If just one operator is selfadjoint, the following result is applicable.

Theorem 8.4. *Let $\mathcal{H}$ and $\mathcal{K}$ be Hilbert spaces. Suppose that $A : \mathcal{D}(A) \to \mathcal{K}$ is a selfadjoint operator with spectral measure P_A and $B : \mathcal{D}(B) \to \mathcal{H}$ is a densely defined, closed linear operator such that $\sigma(A) \cap \sigma(B) = \emptyset$. Let $Y \in \mathcal{L}(\mathcal{H}, \mathcal{K})$.*

(i) *Equation (8.3) has a strong solution if and only if there exists an operator valued measure $M : \mathcal{B}(\sigma(A)) \to \mathcal{L}(\mathcal{H}, \mathcal{K})$ such that*

$$M(K)h = \int_K dP_A(\zeta)(Y(\zeta I - B)^{-1}h), \quad h \in \mathcal{H}, \qquad (8.6)$$

for each compact subset K of $\sigma(A)$. The operator valued measure M exists if and only if

$$\sup_K \left\| \int_K dP_A(\zeta)(Y(\zeta I - B)^{-1}h) \right\|_{\mathcal{L}(\mathcal{H},\mathcal{K})} < \infty \qquad (8.7)$$

for every $h \in \mathcal{H}$. Then $X = M(\sigma(A))$ is the unique strong solution of equation (8.3).

(ii) *If for each $h \in \mathcal{H}$ the function Φ_h given by formula (8.4) is P_A-integrable in $E = \mathcal{L}(\mathcal{K}) \widehat{\otimes}_\tau \mathcal{K}$ on $\sigma(A)$, then the map $h \longmapsto J_E \int_{\sigma(A)} dP_A \otimes \Phi_h$, $h \in \mathcal{H}$, defines a continuous linear operator $\int_{\sigma(A)} dP_A(\zeta) Y(\zeta I - B)^{-1} \in \mathcal{L}(\mathcal{H}, \mathcal{K})$ and the operator*

$$X = \int_{\sigma(A)} dP_A(\zeta) Y(\zeta I - B)^{-1}$$

is the unique strong solution of equation (8.3).
Let $h \in \mathcal{H}$. The function Φ_h is P_A-integrable in $E = \mathcal{L}(\mathcal{K}) \widehat{\otimes}_\tau \mathcal{K}$ on $\sigma(A)$ if and only if

$$\sup_K \left\| \int_K dP_A(\zeta) \otimes (Y(\zeta I - B)^{-1}h) \right\|_{\mathcal{L}(\mathcal{K}) \widehat{\otimes}_\tau \mathcal{K}} < \infty. \qquad (8.8)$$

Proof. (i) According to Lemma 8.1, the right-hand side of equation (8.6) is the image by the composition map of an element of the projective tensor product $\mathcal{L}(\mathcal{K}) \widehat{\otimes}_\pi \mathcal{K}$.

Then for $h \in \mathcal{D}(B)$, we have

$$A P_A(K) \int_{\sigma(A)} P(d\zeta) \Phi_h(\zeta) - P_A(K) \int_{\sigma(A)} P(d\zeta) \Phi_{Bh}(\zeta) = P_A(K) Y h$$

because $P_A(K) \int_{\sigma(A)} P(d\zeta)\Phi_u(\zeta) = \int_K P_A(d\zeta)P(K)\Phi_u(\zeta)$ for all $u \in \mathcal{H}$ and by formula (8.1), the operator $X_K = \int_K dP_A(\zeta)P(K)Y(\zeta I - B)^{-1}$ is the unique solution of the equation

$$(P_A(K)A)X_K h - X_K Bh = P_A(K)Yh, \quad h \in \mathcal{D}(B).$$

The case of unbounded B is mentioned in [3, Lemma 2.5]. Because $P_A(K)X_K = X_K$ and A and $P_A(K)$ commute, we have $AX_K h - X_K Bh = P_A(K)Yh$ for all $h \in \mathcal{D}(B)$.

Suppose that operator valued measure $M : \mathcal{B}(\sigma(A)) \to \mathcal{L}(\mathcal{H}, \mathcal{K})$ exists. Then $X_K u = P_A(K)M(\sigma(A)u$ converges in $\mathcal{K}$ as $K \uparrow \sigma(A)$ for each $u \in \mathcal{H}$, so $X = \lim_K X_K$ belongs to $\mathcal{L}(\mathcal{H}, \mathcal{K})$ by the uniform boundedness principle. Suppose that $h \in \mathcal{D}(B)$. Then $\lim_K AX_K h = XBh + Yh$, so Xh belongs to the closure of A restricted to the subspace

$$\{P_A(K)u : u \in \mathcal{K}, \ K \subset \sigma(A) \text{ compact}\}.$$

Hence $Xh \in \mathcal{D}(A)$ and X is therefore a strong solution of equation (8.3).

Given a strong solution X of equation (8.3), let $M(B) = P_A(B)X$ for $B \in \mathcal{B}(\sigma(A))$. Then M is an operator valued measure and $X_K = M(K)$ satisfies equation (8.6) for each compact subset K of $\sigma(A)$.

Any operator valued measure is inner regular for the strong operator topology, so the bound (8.7) is equivalent to the existence of M.

(ii) Now suppose that for each $h \in \mathcal{H}$ the function Φ_h is P_A-integrable in $E = \mathcal{L}(\mathcal{K})\widehat{\otimes}_\tau \mathcal{K}$ on $\sigma(A)$.

Now $X_K u = P_A(K) \int_{\sigma(A)} P(d\zeta)\Phi_u(\zeta)$ converges in $\mathcal{K}$ as $K \uparrow \sigma(A)$ for each $u \in \mathcal{H}$, so $X = \lim_K X_K$ belongs to $\mathcal{L}(\mathcal{H}, \mathcal{K})$ by the uniform boundedness principle. Suppose that $h \in \mathcal{D}(B)$. An argument similar to that in (i) shows that

$$AX_K h - X_K Bh = P_A(K)Yh.$$

Then $\lim_K AX_K h = XBh + Yh$, so Xh belongs to the closure of A restricted to the subspace

$$\{P_A(K)u : u \in \mathcal{K}, \ K \subset \sigma(A) \text{ compact}\}.$$

Hence $Xh \in \mathcal{D}(A)$ and X is therefore a strong solution of equation (8.3).

Conversely, suppose that the bound (8.8) holds for every $h \in \mathcal{H}$. There exists an increasing sequence of compact subsets K_j, $j = 1, 2, \ldots$, of $\sigma(A)$, such that $\|P_A((\sigma(A) \setminus K_j) \cap S)\| < 1/j$ for every $j = 1, 2, \ldots$ and $S \in \sigma(A)$, because the spectral measure P_A is a regular operator valued Borel measure. Let $\Omega_j = K_j \setminus (\cup_{i<j} K_i)$. Then $\sigma(A) \setminus \cup_j \Omega_j$ is P_A-null and $\Omega_1, \Omega_2, \ldots$ are pairwise disjoint.

For each $y' \in \mathcal{K}$, $S \in \mathcal{B}(\sigma(A))$ and $j = 1, 2, \ldots$,

$$\int_{\Omega_j \cap S} (Y(\zeta I - B)^{-1}h) \otimes (P_A(d\zeta)y') \in \mathcal{K}\widehat{\otimes}_\pi \mathcal{K}.$$

If the bound (8.8) holds, then

$$C_h = \sup_{n,S,\|y'\|\leq 1} \left\| \int_{(\cup_{j=1}^n \Omega_j) \cap S} (Y(\zeta I - B)^{-1}h) \otimes (P_A(d\zeta)y') \right\|_{\mathcal{K}\widehat{\otimes}_\pi \mathcal{K}} < \infty.$$

The projective tensor product $\mathcal{K}\widehat{\otimes}_\pi \mathcal{K}$ is associated with the trace class operators on $\mathcal{K}$ via the embedding $u : \mathcal{K}\widehat{\otimes}_\pi \mathcal{K} \to \mathcal{L}(\mathcal{K})$ defined by $u(x\otimes y)k = (k, y)x$. Then

$$u\left(\int_{(\cup_{j=1}^n \Omega_j) \cap S} (Y(\zeta I - B)^{-1}h) \otimes (P_A(d\zeta)y') \right) k$$

$$= \sum_{j=1}^n \int_{\Omega_j \cap S} (Y(\zeta I - B)^{-1}h)(k, P_A(d\zeta)y')$$

for $x, y, k \in \mathcal{K}$ and the bound

$$\sum_{j=1}^n \int_{\Omega_j \cap S} |(Y(\zeta I - B)^{-1}h, x')|.|(k, P_A y')|(d\zeta) \leq 4C_h \|x'\|.\|y'\|.\|k\|$$

holds for each $x', y', k \in \mathcal{K}$ and $S \in \mathcal{B}(\sigma(A))$ by Proposition 1.1. It follows from the weak sequential completeness of the Hilbert space $\mathcal{K}$ and the Orlicz-pettis Theorem that the sum

$$\sum_{j=1}^\infty \int_{\Omega_j \cap S} (Y(\zeta I - B)^{-1}h)(k, P_A(d\zeta)y')$$

converges unconditionally in $\mathcal{K}$ for each $S \in \mathcal{B}(\sigma(A))$ and

$$k \longmapsto \sum_{j=1}^\infty \int_{\Omega_j \cap S} (Y(\zeta I - B)^{-1}h)(k, P_A(d\zeta)y'), \quad k \in \mathcal{K},$$

is a bounded linear operator whose norm is bounded by $4C_h\|y'\|$. According to the noncommutative Fatou lemma Proposition 3.8,

$$\int_S (Y(\zeta I - B)^{-1}h) \otimes (P_A(d\zeta)y') = \sum_{j=1}^\infty \int_{\Omega_j \cap S} (Y(\zeta I - B)^{-1}h) \otimes (P_A(d\zeta)y')$$

belongs to $\mathcal{K}\widehat{\otimes}_\pi \mathcal{K}$ and

$$\left\| \int_S (Y(\zeta I - B)^{-1}h) \otimes (P_A(d\zeta)y') \right\|_{\mathcal{K}\widehat{\otimes}_\pi \mathcal{K}} \leq 4C_h\|y'\|$$

for each $S \in \mathcal{B}(\sigma(A))$. Hence the function Φ_h is P_A-integrable in $\mathcal{L}(\mathcal{K})\widehat{\otimes}_\tau\mathcal{K}$ on $\sigma(A)$ and $\left\|\int_{\sigma(A)} dP_A \otimes \Phi_h\right\|_{\mathcal{L}(\mathcal{K})\widehat{\otimes}_\tau\mathcal{K}} \leq 4Ch$. The uniform boundedness principle and the Vitali-Hahn-Saks Theorem ensures that the formula $M(S) = \int_S P_A(d\zeta)(Y(\zeta I - B)^{-1})$ defines an $\mathcal{L}(\mathcal{H},\mathcal{K})$-valued measure M for the strong operator topology, so that (i) applies. $\qquad\square$

Remark 8.1. The operator valued measure $M : \mathcal{B}(\sigma(A)) \to \mathcal{L}(\mathcal{H},\mathcal{K})$ is called a *strong operator valued Stieltjes integral* in [3, 4]. According to Lemma 8.1, for each compact subset K of $\sigma(A)$, the operator $M(K) \in \mathcal{L}(\mathcal{H})$ can be written as a Stieltjes integral

$$M(K)h = \lim_{n\to\infty} \int_K P_A(d\zeta)Y s_n(\zeta)h$$

for $\mathcal{B}(\sigma(A))$-simple function $s_n : \sigma(A) \to \mathcal{L}(\mathcal{H})$, $n = 1, 2, \ldots$, which may be chosen to be step functions based on finite intervals, restricted to the spectrum $\sigma(A)$ of A.

Example 8.2. The solution X in Theorem 8.3 is actually a strong solution. If A and B are selfadjoint, and $d(\sigma(A), \sigma(B)) = \delta > 0$ then

$$\int_S dP_A(\zeta) \otimes (Y(\zeta I - B)^{-1}h) = \int_\mathbb{R} (P_A(S)e^{itA}) \otimes (Ye^{-itB}h)f_\delta(t)\, dt$$

belongs to $\mathcal{L}(\mathcal{K})\widehat{\otimes}_\tau\mathcal{K}$ for each $S \in \mathcal{B}(\sigma(A))$ and $h \in \mathcal{H}$. To see this, let $k \in \mathcal{K}$. Then the integral

$$\int_\mathbb{R} (Ye^{-itB}h) \otimes (P_A(S)e^{-itA}k)f_\delta(t)\, dt$$

converges in $\mathcal{K}\widehat{\otimes}_\pi\mathcal{K}$ because $t \longmapsto (Ye^{-itB}h) \otimes (P_A(S)e^{-itA}k)$, $t \in \mathbb{R}$, is continuous in $\mathcal{K}\widehat{\otimes}_\pi\mathcal{K}$ and $f_\delta \in L^1(\mathbb{R})$, so

$$\int_\mathbb{R} \|Ye^{-itB}h) \otimes (P_A(S)e^{-itA}k)f_\delta(t)\|_{\mathcal{K}\widehat{\otimes}_\pi\mathcal{K}}\, dt$$

$$\leq \int_\mathbb{R} \|Ye^{-itB}h\|.\|e^{-itA}k\|.|f_\delta(t)|\, dt$$

$$\leq \|Y\|_{\mathcal{L}(\mathcal{H},\mathcal{K})}\|h\|_\mathcal{H}\|k\|_\mathcal{K}\|f_\delta\|_1$$

and $\|\int_S dP_A(\zeta) \otimes (Y(\zeta I - B)^{-1}h)\|_{\mathcal{L}(\mathcal{K})\widehat{\otimes}_\tau\mathcal{K}} \leq \|Y\|_{\mathcal{L}(\mathcal{H},\mathcal{K})}\|h\|_\mathcal{H}\|f_\delta\|_1$. Then an appeal to Theorem 8.4 (ii), the operator

$$X = \int_{\sigma(A)} dP_A(\zeta)Y(\zeta I - B)^{-1} = \int_\mathbb{R} e^{-itA}Ye^{itB}f_\delta(t)\, dt$$

is the unique strong solution of equation (8.3).

Example 8.3. If A is selfadjoint, B is densely defined and closed, $\sup \sigma(A) \leq 0$ and there exists $0 < \omega < \pi/2$ and a sector

$$S_{\omega-} = \{-z : z \in \mathbb{C} \setminus \{0\}, \arg|z| < \omega\} \cup \{0\}$$

that is contained in $\rho(B)$, then according to [106, Theorem 15]

$$\int_S dP_A(\zeta) \otimes (Y(\zeta I - B)^{-1}) \in \mathcal{L}(\mathcal{K}) \widehat{\otimes}_\pi \mathcal{K}, \quad S \in \mathcal{B}(\sigma(A)).$$

8.2   Double operator integrals

As mentioned in Example 8.2 above, if $A : \mathcal{D}(\mathcal{H}) \to \mathcal{H}$ and $B : \mathcal{D}(\mathcal{H}) \to \mathcal{H}$ are selfadjoint operators, $d(\sigma(A), \sigma(B)) = \delta > 0$ and $Y \in \mathcal{L}(\mathcal{H})$, then for each $h \in \mathcal{H}$ the function $\zeta \mapsto Y(\zeta I - B)^{-1}h$, $\zeta \in \sigma(A)$, is P_A-integrable in $\mathcal{L}(\mathcal{H}) \widehat{\otimes}_\tau \mathcal{H}$ and $X = \int_{\sigma(A)} dP_A(\zeta) Y(\zeta I - B)^{-1}$ is the unique strong solution of equation (8.3). Because B is selfadjoint, we can rewrite the solution X as an iterated integral

$$X = \int_{\sigma(A)} dP_A(\zeta) Y \left(\int_{\sigma(B)} \frac{dP_B(\mu)}{\zeta - \mu} \right)$$

with respect to the spectral measures P_A, P_B associated with the A and B.

An application of the Fubini strategy sees the expression

$$X = \int_{\sigma(A) \times \sigma(B)} \frac{dP_A(\zeta) Y dP_B(\mu)}{\zeta - \mu} \tag{8.9}$$

as a representation of the strong solution of the operator equation

$$AX - XB = Y$$

in the case that both A and B are selfadjoint operators.

Integrals like (8.9) have been studied extensively in the case that $Y \in \mathcal{L}(\mathcal{H})$ is a Hilbert-Schmidt operator [15] and, more generally, when Y belongs to the Schatten ideal $\mathcal{C}_p(\mathcal{H})$ in $\mathcal{L}(\mathcal{H})$ for some $1 \leq p < \infty$, where they are called *double operator integrals*. In this section, the operator ideal $\mathcal{C}_p(\mathcal{H})$ consists of compact linear operators on $\mathcal{H}$ whose singular values belong to ℓ^p.

For a bounded linear operator T on a Hilbert space $\mathcal{H}$, the expression

$$\mathcal{I}_\varphi(T) = \int_{\Lambda \times M} \varphi(\lambda, \mu) F(d\lambda) T E(d\mu)$$

is a double operator integral if F is an $\mathcal{L}(\mathcal{H})$-valued spectral measure on the measurable space $(M, \mathcal{F})$ and E is an $\mathcal{L}(\mathcal{H})$-valued spectral measure on the measurable space $(\Lambda, \mathcal{E})$. The function $\varphi : \Lambda \times M \to \mathbb{C}$ is taken to be uniformly bounded on $\Lambda \times M$. In formula (8.9), $\varphi(\lambda, \mu) = (\lambda - \mu)^{-1}$, so that $|\varphi(\lambda, \mu)|$ is bounded by $1/\delta$ for $(\lambda, \mu) \in \sigma(A) \times \sigma(B)$ when the spectra $\sigma(A)$ and $\sigma(B)$ are a distance δ apart.

The map $T \longmapsto \mathcal{I}_\varphi(T)$, $T \in \mathcal{C}_2(\mathcal{H})$, is continuous into the space $\mathcal{C}_2(\mathcal{H})$ of Hilbert-Schmidt operators and

$$\|\mathcal{I}_\varphi\|_{\mathcal{C}_2(\mathcal{H})} = \|\varphi\|_{L^\infty(\Lambda \times M)},$$

so that the map $(E \otimes F)_{\mathcal{C}_2(\mathcal{H})} : U \longmapsto \mathcal{I}_{\chi_U}$, $U \in \mathcal{E} \otimes \mathcal{F}$, is actually a countably additive spectral measure acting on $\mathcal{C}_2(\mathcal{H})$ and the equality

$$\mathcal{I}_\varphi = \int_{\Lambda \times M} \varphi \, d(E \otimes F)_{\mathcal{C}_2(\mathcal{H})}$$

holds for all bounded measurable functions $\varphi : \Lambda \times M \to \mathbb{C}$ [15, Section 3.1].

The situation is more complicated if the space $\mathcal{C}_2(\mathcal{H})$ of Hilbert-Schmidt operators (with the Hilbert-Schmidt norm) is replaced by the Schatten ideal $\mathfrak{S} = \mathcal{C}_p(\mathcal{H})$ in $\mathcal{L}(\mathcal{H})$ for some $1 \leq p < \infty$ not equal to 2, or as in the case of formula (8.9), by $\mathfrak{S} = \mathcal{L}(\mathcal{H})$ itself, because the map $U \times V \mapsto \mathcal{I}_{\chi_{U \times V}}$, $U \in \mathcal{E}$, $V \in \mathcal{F}$, only defines a *finitely additive* set function $(E \otimes F)_{\mathfrak{S}}$ acting on elements $T \in \mathfrak{S}$ so that

$$(E \otimes F)_{\mathfrak{S}}(U \times V)T = E(U)TF(V), \quad U \in \mathcal{E}, \ V \in \mathcal{F}.$$

For a bounded function $\varphi : \Lambda \times M \to \mathbb{C}$, the double operator integral $\mathcal{I}_\varphi$ may be viewed as a continuous generalisation of a classical Schur multiplier

$$T_\mu : x \longmapsto \sum_{i,j} \mu_{ij} \alpha_{ij} e_{ij}, \quad x = \sum_{i,j} \alpha_{ij} e_{ij}, \tag{8.10}$$

for an infinite matrix $\mu = \{\mu_{ij}\} \in \mathcal{M}$, with respect to the matrix units e_{ij} corresponding to an orthonormal basis $\{h_j\}$ of $\mathcal{H}$. If P_j denotes the orthogonal projection onto the linear space $\text{span}\{h_j\}$ for each $j = 1, 2, \ldots$, then

$$T_\mu = \sum_{i,j} \mu_{ij} (P_i \otimes P_j)_{\mathcal{M}}$$

for the operators $(P_i \otimes P_j)_{\mathcal{M}} : x \longmapsto P_i x P_j$ acting on the infinite matrix $x \in \mathcal{M}$ for $i, j = 1, 2 \ldots$.

To be more precise, let $\mathfrak{S}$ be a symmetrically normed ideal in $\mathcal{L}(\mathcal{H})$. The linear map $\mathcal{J}_{\mathfrak{S}} : \mathcal{L}(\mathcal{H}) \otimes \mathcal{L}(\mathcal{H}) \to \mathcal{L}(\mathfrak{S})$ is defined by $\mathcal{J}_{\mathfrak{S}}(A \otimes B)T = ATB$ for $T \in \mathfrak{S}$ and $A, B \in \mathcal{L}(\mathcal{H})$. In the language of [15, Section 4], the element $\mathcal{J}_{\mathfrak{S}}(A \otimes B)$ of $\mathcal{L}(\mathfrak{S})$ is the *transformer* on $\mathfrak{S}$ associated with $A \otimes B$.

Definition 8.3. Let $(\Lambda, \mathcal{E})$ and $(M, \mathcal{F})$ be measurable spaces and $\mathcal{H}$ a separable Hilbert space. Let $m : \mathcal{E} \to \mathcal{L}_s(\mathcal{H})$ be an operator valued measure for the strong operator topology and $n : \mathcal{F} \to \mathcal{H}$ be a $\mathcal{H}$-valued measure. Suppose that τ is the topology on the tensor product $\mathcal{L}(\mathcal{H}) \otimes \mathcal{H}$ defined by formula (5.13).

An $(\mathcal{E} \otimes \mathcal{F})$-measurable function $\varphi : \Lambda \times M \to \mathbb{C}$ is said to be $(m \otimes n)$-*integrable in* $E = \mathcal{L}(\mathcal{H}) \widehat{\otimes}_\tau \mathcal{H}$ if for every $x, x', y' \in \mathcal{H}$, the function φ is integrable with respect to the scalar measure $(mx, x') \otimes (n, y')$ and for every $A \in \mathcal{E} \otimes \mathcal{F}$, there exists $\varphi.(m \otimes n)(A) \in \mathcal{L}(\mathcal{H}) \widehat{\otimes}_\tau \mathcal{H}$ such that

$$\langle \varphi.(m \otimes n)(A), x \otimes x' \otimes y' \rangle = \int_A \varphi \, d((mx, x') \otimes (n, y'))$$

for every $x, x', y' \in \mathcal{H}$.

If φ is $(m \otimes n)$-integrable in $\mathcal{L}(\mathcal{H}) \widehat{\otimes}_\tau \mathcal{H}$ and $J_E : \mathcal{L}(\mathcal{H}) \widehat{\otimes}_\tau \mathcal{H} \to \mathcal{H}$ is the multiplication map, then

$$\int_A \varphi \, d(mn) = J_E(\varphi.(m \otimes n)(A)), \quad A \in \mathcal{E} \otimes \mathcal{F}.$$

The following observation is useful for treating double operator integrals. Let τ_w be the relative topology on $\mathcal{L}(\mathcal{H}) \widehat{\otimes}_\tau \mathcal{H}$ defined by $\mathcal{L}(\mathcal{H}, \mathcal{H} \widehat{\otimes}_\pi \mathcal{H})$ as in Lemma 5.1 above.

Proposition 8.1. *Let* $(\Lambda, \mathcal{S})$ *and* $(M, \mathcal{T})$ *be measurable spaces and let* $\mathcal{H}$, $m : \mathcal{S} \to \mathcal{L}_s(\mathcal{H})$ *and* $n : \mathcal{T} \to \mathcal{H}$ *be as in Definition 8.3. If* $U \in \mathcal{C}_1(\mathcal{H})$, *then there exists a unique vector measure*

$$m \otimes (Un) : \mathcal{S} \otimes \mathcal{T} \to \mathcal{L}(\mathcal{H}) \widehat{\otimes}_\tau \mathcal{H}$$

σ-*additive for the topology* τ_w *such that*

$$(m \otimes (Un))(S \times T) = m(S) \otimes (Un(T)) \in \mathcal{L}(\mathcal{H}) \otimes \mathcal{H}$$

for each $S \in \mathcal{S}$ *and* $T \in \mathcal{T}$. *Consequently, every bounded* $(\mathcal{S} \otimes \mathcal{T})$-*measurable function* $\varphi : \Lambda \times M \to \mathbb{C}$ *is* $(m \otimes (Un))$-*integrable in* $E = \mathcal{L}(\mathcal{H}) \widehat{\otimes}_\tau \mathcal{H}$ *and*

$$\int_A \varphi \, d(m(Un)) = J_E(\varphi.(m \otimes (Un))(A)), \quad A \in \mathcal{S} \otimes \mathcal{T}.$$

Proof. If U is a trace class operator on $\mathcal{H}$, then there exist orthonormal sets $\{\phi_j\}_j$, $\{\psi_j\}_j$ and a summable sequence $\{\lambda_j\}_j$ of scalars such that $Uh = \sum_{j=1}^\infty \lambda_j \phi_j(h, \psi_j)$ for every $h \in H$. For each $j = 1, 2, \ldots$, the total variation of the product measure

$$(mh, k) \otimes (n, \psi_j) : S \times T \longmapsto (m(S)h, k) \otimes (n(T), \psi_j), \quad S \in \mathcal{S}, \ T \in \mathcal{T},$$

is bounded by $\|m\|(\Lambda).\|n\|(M).\|h\|.\|k\|$ for every $h, k \in \mathcal{H}$. Here $\|m\|$ and $\|n\|$ denote the semivariation of m and n, respectively.

Appealing to formula (5.13), the norm $\||(n, \psi_j)|(M)(m(S) \otimes \phi_j)\|_\tau$ of the finitely additive set function $|(n, \psi_j)|(M)(m \otimes \phi_j)$ evaluated at the set $S \in \mathcal{S}$ is

$$\||(n, \psi_j)|(M)(m(S) \otimes \phi_j)\|_\tau = |(n, \psi_j)|(M) \sup_{\|k\| \le 1} \|\phi_j \otimes (m(S)^*k)\|_\pi$$

$$\le \|n\|(M)\|m\|(\Lambda).$$

It follows that the finitely additive set function

$$S \times T \longmapsto (n(S), \psi_j)(m(T) \otimes \phi_j), \quad S \in \mathcal{S}, \ T \in \mathcal{T},$$

admits a unique τ_w-countably additive extension $M_j : \mathcal{S} \otimes \mathcal{T} \to \mathcal{L}_s(\mathcal{H}) \otimes \mathcal{H}$ whose semivariation with respect to the norm (5.13) is bounded by $\|m\|(\Lambda).\|n\|(M)$ and $m \otimes (Un) = \sum_j \lambda_j M_j$ converges in $\mathcal{L}(\mathcal{H}) \widehat{\otimes}_\tau \mathcal{H}$ uniformly on $\mathcal{S} \otimes \mathcal{T}$. $\square$

Corollary 8.1. *Let* $\{\Lambda, \mathcal{S})$ *and* $(M, \mathcal{T})$ *be measurable spaces and* $\mathcal{H}$ *a separable Hilbert space. Let* $m : \mathcal{S} \to \mathcal{L}_s(\mathcal{H})$ *and* $n : \mathcal{T} \to \mathcal{L}_s(\mathcal{H})$ *be operator valued measures for the strong operator topology. Then there exists a unique operator valued measure* $\mathcal{J}_{\mathcal{C}_1(\mathcal{H})}(m \otimes n) : \mathcal{S} \otimes \mathcal{T} \to \mathcal{L}_s(\mathcal{C}_1(\mathcal{H}), \mathcal{L}_s(\mathcal{H}))$ *such that*

$$\mathcal{J}_{\mathcal{C}_1(\mathcal{H})}(m \otimes n)(S \times T) = \mathcal{J}_{\mathcal{C}_1(\mathcal{H})}(m(S) \otimes n(T)), \quad S \in \mathcal{S}, \ T \in \mathcal{T}.$$

Proof. It is easy to check that for $A \in \mathcal{S} \otimes \mathcal{T}$ and $T \in \mathcal{C}_1(\mathcal{H})$, the formula

$$\left(\left[\mathcal{J}_{\mathcal{C}_1(\mathcal{H})}(m \otimes n)(A)\right] T\right) h = J_E((m \otimes (T(nh)))(A)), \quad h \in \mathcal{H},$$

defines a linear operator $\left[\mathcal{J}_{\mathcal{C}_1(\mathcal{H})}(m \otimes n)(A)\right] T$ on $\mathcal{H}$ whose operator norm is bounded by $\|m\|(\Lambda).\|n\|(M)\|T\|_{\mathcal{C}_1(\mathcal{H})}$ and $A \longmapsto \left[\mathcal{J}_{\mathcal{C}_1(\mathcal{H})}(m \otimes n)(A)\right] T$, $A \in \mathcal{S} \otimes \mathcal{T}$, is countably additive in $\mathcal{L}(\mathcal{H})$ for the strong operator topology for each $T \in \mathcal{C}_1(\mathcal{H})$. $\square$

The following notation gives an interpretation of formula (8.9) in the case that Y belongs to the ideal $\mathfrak{S} = \mathcal{C}_p(\mathcal{H})$, $1 \le p < \infty$ or $\mathfrak{S} = \mathcal{L}(\mathcal{H})$. The collection $\mathcal{C}_p(\mathcal{H})$ of trace class operators is a linear subspace of $\mathfrak{S}$ in each case.

Let $(m \otimes n)_\mathfrak{S}$ be the finitely additive set function defined by

$$(m \otimes n)_\mathfrak{S}(E \times F) = \mathcal{J}_\mathfrak{S}(m(E) \otimes n(F)), \quad E \in \mathcal{E}, \ F \in \mathcal{F},$$

that is, $(m \otimes n)_\mathfrak{S} : \mathcal{A} \to \mathcal{L}(\mathfrak{S})$ is finitely additive on the algebra $\mathcal{A}$ of all finite unions of product sets $E \times F$ for $E \in \mathcal{E}$, $F \in \mathcal{F}$.

Suppose that the function $\varphi : \Lambda \times M \to \mathbb{C}$ is integrable with respect to the $\mathcal{L}_s(\mathcal{C}_1(\mathcal{H}), \mathcal{L}_s(\mathcal{H}))$-valued measure $\mathcal{J}_{\mathcal{C}_1(\mathcal{H})}(m \otimes n)$. If for $E \in \mathcal{E}$ and $F \in \mathcal{F}$, the linear map

$$u \longmapsto \left(\int_{E \times F} \varphi \, d[\mathcal{J}_{\mathcal{C}_1(\mathcal{H})}(m \otimes n)] \right) u, \quad u \in \mathcal{C}_1(\mathcal{H}),$$

is the restriction to $\mathcal{C}_1(\mathcal{H})$ of a continuous linear map $T_\varphi \in \mathcal{L}(\mathfrak{S})$, then we write $\int_{E \times F} \varphi \, d(m \otimes n)_{\mathfrak{S}}$ for T_φ and we say that φ is $(m \otimes n)_{\mathfrak{S}}$-*integrable* if $\int_{E \times F} \varphi \, d(m \otimes n)_{\mathfrak{S}} \in \mathcal{L}(\mathfrak{S})$ for every $E \in \mathcal{E}$ and $F \in \mathcal{F}$.

To check that the operator $\int_{E \times F} \varphi \, d(m \otimes n)_{\mathfrak{S}} \in \mathcal{L}(\mathfrak{S})$ is uniquely defined, observe that $\mathcal{C}_1(\mathcal{H})$ is norm dense in $\mathcal{C}_p(\mathcal{H})$ for $1 < p < \infty$. In the case $\mathfrak{S} = \mathcal{L}(\mathcal{H})$, the closure in the ultraweak topology $\sigma(\mathcal{L}(\mathcal{H}), \mathcal{C}_1(\mathcal{H}))$ can be taken, so that

$$\int_{E \times F} \varphi \, d(m \otimes n)_{\mathcal{L}(\mathcal{H})} = \left(\int_{E \times F} \varphi \, d(m \otimes n)_{\mathcal{C}_1(\mathcal{H})} \right)'.$$

Although $(m \otimes n)_{\mathfrak{S}}$ is only a *finitely additive* set function, the $\mathcal{L}(\mathfrak{S})$-valued set function

$$E \times F \longmapsto \int_{E \times F} \varphi \, d(m \otimes n)_{\mathfrak{S}}, \quad E \in \mathcal{E}, \ F \in \mathcal{F},$$

of an $(m \otimes n)_{\mathfrak{S}}$-integrable function φ defines a finitely additive $\mathcal{L}(\mathfrak{S})$-valued set function on the algebra generated by all product sets $E \times F$ for $E \in \mathcal{E}$ and $F \in \mathcal{F}$.

Corollary 8.1 tells us that for an $(m \otimes n)_{\mathcal{C}_1(\mathcal{H})}$-integrable function $\varphi : \Lambda \times M \to \mathbb{C}$, the $\mathcal{L}(\mathcal{H})$-valued set function

$$A \longmapsto \left(\int_A \varphi \, d(m \otimes n)_{\mathcal{C}_1(\mathcal{H})} \right) T, \quad A \in \mathcal{E} \otimes \mathcal{F},$$

is countably additive in the strong operator topology for each $T \in \mathcal{C}_1(\mathcal{H})$. The following result describes the situation for other operator ideals $\mathfrak{S}$.

Proposition 8.2. *Suppose that $\varphi : \Lambda \times M \to \mathbb{C}$ is an $(m \otimes n)_{\mathfrak{S}}$-integrable function. For each $T \in \mathfrak{S}$, the set function*

$$E \times F \longmapsto \left(\int_{E \times F} \varphi \, d(m \otimes n)_{\mathfrak{S}} \right) T, \quad E \in \mathcal{E}, \ F \in \mathcal{F},$$

is separately σ-additive in the strong operator topology of $\mathcal{L}(\mathcal{H})$, that is,

$$\left(\int_{(\cup_{j=1}^\infty E_j) \times F} \varphi \, d(m \otimes n)_{\mathfrak{S}} \right) T = \sum_{j=1}^\infty \left(\int_{E_j \times F} \varphi \, d(m \otimes n)_{\mathfrak{S}} \right) T, \quad F \in \mathcal{F},$$

$$\left(\int_{E \times (\cup_{j=1}^\infty F_j)} \varphi \, d(m \otimes n)_{\mathfrak{S}} \right) T = \sum_{j=1}^\infty \left(\int_{E \times F_j} \varphi \, d(m \otimes n)_{\mathfrak{S}} \right) T, \quad E \in \mathcal{E},$$

for all pairwise disjoint $E_j \in \mathcal{E}$, $j = 1, 2, \ldots$, and all pairwise disjoint $F_j \in \mathcal{F}$, $j = 1, 2, \ldots$.

The following result was proved by M. Birman and M. Solomyak [15, Section 3.1].

Theorem 8.5. *Let $(\Lambda, \mathcal{E})$ and $(M, \mathcal{F})$ be measurable spaces and $\mathcal{H}$ a separable Hilbert space. Let $P : \mathcal{E} \to \mathcal{L}_s(\mathcal{H})$ and $Q : \mathcal{F} \to \mathcal{L}_s(\mathcal{H})$ be spectral measures.*

Then there exists a unique spectral measure $(P\overline{\otimes}Q)_{\mathcal{C}_2(\mathcal{H})} : \mathcal{E} \otimes \mathcal{F} \to \mathcal{L}(\mathcal{C}_2(\mathcal{H}))$ such that

$$(P\overline{\otimes}Q)_{\mathcal{C}_2(\mathcal{H})}(A) = (P \otimes Q)_{\mathcal{C}_2(\mathcal{H})}(A)$$

for every set A belonging to the algebra $\mathcal{A}$ of all finite unions of product sets $E \times F$ for $E \in \mathcal{E}$, $F \in \mathcal{F}$, and

$$\int_A \varphi \, d(P \otimes Q)_{\mathcal{C}_2(\mathcal{H})} = \int_A \varphi \, d(P\overline{\otimes}Q)_{\mathcal{C}_2(\mathcal{H})} \in \mathcal{L}(\mathcal{C}_2(\mathcal{H})), \quad A \in \mathcal{E} \otimes \mathcal{F},$$

for every bounded $(\mathcal{E} \otimes \mathcal{F})$-measurable function $\varphi : \Lambda \times M \to \mathbb{C}$. Moreover,

$$\|(P\overline{\otimes}Q)_{\mathcal{C}_2(\mathcal{H})}(\varphi)\|_{\mathcal{L}(\mathcal{C}_2(\mathcal{H}))} = \|\varphi\|_\infty.$$

For spectral measures P and Q, the formula

$$\left(\int_{E \times F} \varphi \, d(P \otimes Q)_{\mathfrak{S}} \right) T = \left(\int_{\Lambda \times M} \varphi \, d(P \otimes Q)_{\mathfrak{S}} \right) P(E) T Q(F)$$

holds for each $E \in \mathcal{E}$, $F \in \mathcal{F}$ and $T \in \mathfrak{S}$, so it is only necessary to verify that $\int_{\Lambda \times M} \varphi \, d(P \otimes Q)_{\mathfrak{S}} \in \mathcal{L}(\mathfrak{S})$ in order to show that φ is $(P \otimes Q)_{\mathfrak{S}}$-integrable.

The following observation gives an interpretation of formula (8.9) as a double operator integral. The *Fourier transform* of $f \in L^1(\mathbb{R})$ is the function $\hat{f} : \mathbb{R} \to \mathbb{C}$ defined by $\hat{f}(\xi) = \int_{\mathbb{R}} e^{-i\xi x} f(x) \, dx$ for $\xi \in \mathbb{R}$.

Theorem 8.6. *Let $\mathcal{H}$ be a separable Hilbert space. Let $P : \mathcal{B}(\mathbb{R}) \to \mathcal{L}_s(\mathcal{H})$ and $Q : \mathcal{B}(\mathbb{R}) \to \mathcal{L}_s(\mathcal{H})$ be spectral measures on $\mathbb{R}$. Let $\mathfrak{S} = \mathcal{C}_p(\mathcal{H})$ for some $1 \le p < \infty$ or $\mathfrak{S} = \mathcal{L}(\mathcal{H})$. Suppose that $f \in L^1(\mathbb{R})$ and $\varphi(\lambda, \mu) = \hat{f}(\lambda - \mu)$ for all $\lambda, \mu \in \mathbb{R}$. Then $\int_{\mathbb{R} \times \mathbb{R}} \varphi \, d(P \otimes Q)_{\mathfrak{S}} \in \mathcal{L}(\mathfrak{S})$ and*

$$\left\| \int_{\mathbb{R} \times \mathbb{R}} \varphi \, d(P \otimes Q)_{\mathfrak{S}} \right\|_{\mathcal{L}(\mathfrak{S})} \le \|f\|_1.$$

Proof. For each $T \in \mathcal{C}_1(\mathcal{H})$, the set function $E \times F \longmapsto P(E) T Q(F)$, $E, F \in \mathcal{B}(\mathbb{R})$, is the restriction to all measurable rectangles of an $\mathcal{L}(\mathcal{H})$-valued measure σ-additive for the strong operator topology and the integral

$$\int_{\mathbb{R} \times \mathbb{R}} \varphi \, d(PTQ) = \int_{\mathbb{R} \times \mathbb{R}} \left(\int_{\mathbb{R}} e^{-it(\lambda - \mu)t} f(t) \, dt \right) d(PTQ)(\lambda, \mu)$$

$$= \int_{\mathbb{R}} \left(\int_{\mathbb{R} \times \mathbb{R}} e^{-it(\lambda - \mu)t} \, d(PTQ)(\lambda, \mu) \right) f(t) \, dt$$

$$= \int_{\mathbb{R}} e^{-itA} T e^{itB} f(t) \, dt \tag{8.11}$$

converges as a Bochner integral in the strong operator topology to an element of the operator ideal $\mathcal{C}_1(\mathcal{H})$ of trace class operators. The interchange of integrals is verified scalarly.

It follows that φ is a $(P \otimes Q)_{\mathcal{C}_1(\mathcal{H})}$-integrable function and

$$\left\| \int_{\mathbb{R} \times \mathbb{R}} \varphi \, d(P \otimes Q)_{\mathcal{C}_1(\mathcal{H})} \right\|_{\mathcal{L}(\mathcal{C}_1(\mathcal{H}))} \leq \|f\|_1.$$

The corresponding bound for $\mathfrak{S} = \mathcal{C}_p(\mathcal{H})$ for $1 \leq p < \infty$ and $\mathfrak{S} = \mathcal{L}(\mathcal{H})$ follows by duality and interpolation, or directly from formula (8.11). $\square$

The following corollary is a consequence of Theorem 8.3.

Corollary 8.2. *Let $\mathcal{H}$ be a separable Hilbert space and let A, B be self-adjoint operators with spectral measures $P_A : \mathcal{B}(\sigma(A)) \to \mathcal{L}_s(\mathcal{H})$ and $P_B : \mathcal{B}(\sigma(B)) \to \mathcal{L}_s(\mathcal{H})$, respectively. Let $\mathfrak{S} = \mathcal{C}_p(\mathcal{H})$ for some $1 \leq p < \infty$ or $\mathfrak{S} = \mathcal{L}(\mathcal{H})$. If the spectra of A and B are separated by a distance $d(\sigma(A), \sigma(B)) = \delta > 0$, then $\int_{\sigma(A) \times \sigma(B)} (\lambda - \mu)^{-1} (P_A \otimes P_B)_{\mathfrak{S}}(d\lambda, d\mu) \in \mathcal{L}(\mathfrak{S})$ and*

$$\left\| \int_{\sigma(A) \times \sigma(B)} \frac{(P_A \otimes P_B)_{\mathfrak{S}}(d\lambda, d\mu)}{\lambda - \mu} \right\|_{\mathcal{L}(\mathfrak{S})} \leq \frac{\pi}{2\delta}.$$

In particular, equation (8.3) has a unique strong solution for $Y \in \mathfrak{S}$ given by the double operator integral

$$X = \int_{\sigma(A) \times \sigma(B)} \frac{dP_A(\lambda) Y \, dP_B(\mu)}{\lambda - \mu} := \left(\int_{\sigma(A) \times \sigma(B)} \frac{(P_A \otimes P_B)_{\mathfrak{S}}(d\lambda, d\mu)}{\lambda - \mu} \right) Y,$$

so that $\|X\|_{\mathfrak{S}} \leq \frac{\pi}{2\delta} \|Y\|_{\mathfrak{S}}$.

Although the Heaviside function $\chi_{(0,\infty)}$ is not the Fourier transform of an L^1-function, the following result of I. Gohberg and M. Krein [52, Section III.6] holds, in case $P = Q$.

Theorem 8.7. *Let $\mathcal{H}$ be a separable Hilbert space. Let $P : \mathcal{B}(\mathbb{R}) \to \mathcal{L}_s(\mathcal{H})$ and $Q : \mathcal{B}(\mathbb{R}) \to \mathcal{L}_s(\mathcal{H})$ be spectral measures on $\mathbb{R}$. Then*

$$\int_{\mathbb{R} \times \mathbb{R}} \chi_{\{\lambda > \mu\}} \, d(P \otimes Q)_{\mathcal{C}_p(\mathcal{H})} \in \mathcal{L}(\mathcal{C}_p(\mathcal{H}))$$

for every $1 < p < \infty$.

The following recent result of F. Sukochev and D. Potapov [113] settled a long outstanding conjecture of M. Krein for the index p in the range $1 < p < \infty$.

Theorem 8.8. *Let $\mathcal{H}$ be a separable Hilbert space. Let $P : \mathcal{B}(\mathbb{R}) \to \mathcal{L}_s(\mathcal{H})$ and $Q : \mathcal{B}(\mathbb{R}) \to \mathcal{L}_s(\mathcal{H})$ be spectral measures on $\mathbb{R}$. Suppose that $f : \mathbb{R} \to \mathbb{R}$ is a continuous function for which the difference quotient*

$$\varphi_f(\lambda, \mu) = \begin{cases} \frac{f(\lambda) - f(\mu)}{\lambda - \mu}, & \lambda \neq \mu, \\ 0, & \lambda = \mu, \end{cases}$$

is uniformly bounded. Then for every $1 < p < \infty$,

$$\int_{\mathbb{R} \times \mathbb{R}} \varphi_f \, d(P \otimes Q)_{\mathcal{C}_p(\mathcal{H})} \in \mathcal{L}(\mathcal{C}_p(\mathcal{H}))$$

and there exists $C_p > 0$ such that

$$\left\| \int_{\mathbb{R} \times \mathbb{R}} \varphi_f \, d(P \otimes Q)_{\mathcal{C}_p(\mathcal{H})} \right\|_{\mathcal{C}_p(\mathcal{H})} \leq C_p \|\varphi_f\|_\infty.$$

Such a function f is said to be *uniformly Lipschitz* on $\mathbb{R}$ and $\|f\|_{\mathrm{Lip}_1} := \|\varphi_f\|_\infty$.

Corollary 8.3. *Suppose that $f : \mathbb{R} \to \mathbb{R}$ is a uniformly Lipschitz function. Then for every $1 < p < \infty$, there exists $C_p > 0$ such that*

$$\|f(A) - f(B)\|_{\mathcal{C}_p(\mathcal{H})} \leq C_p \|f\|_{\mathrm{Lip}_1} \|A - B\|_{\mathcal{C}_p(\mathcal{H})}$$

for any selfadjoint operators A and B on a separable Hilbert space $\mathcal{H}$.

Proof. Let P_A and P_B be the spectral meaures of A and B, respectively, and suppose that $\|A - B\|_{\mathcal{C}_p(\mathcal{H})} < \infty$. Then according to [15, Theorem 8.1] (see also [101, Corollary 7.2]), the equality

$$f(A) - f(B) = \left(\int_{\mathbb{R} \times \mathbb{R}} \varphi_f \, d(P_A \otimes P_B)_{\mathcal{C}_p(\mathcal{H})} \right) (A - B)$$

holds and the norm estimate follows from Theorem 8.8. $\square$

8.3 Traces of double operator integrals

In this section, let $(\Lambda, \mathcal{E})$ and $(M, \mathcal{F})$ be given measurable spaces, $\mathcal{H}$ a separable Hilbert space and $P : \mathcal{E} \to \mathcal{L}_s(\mathcal{H})$, $Q : \mathcal{F} \to \mathcal{L}_s(\mathcal{H})$ spectral measures. Let $\mathfrak{S} = \mathcal{C}_p(\mathcal{H})$ for some $1 \leq p < \infty$ or $\mathfrak{S} = \mathcal{L}(\mathcal{H})$. The

Banach space $L^1(P)$ of P-integrable functions is isomorphic to the C*-algebra $L^\infty(P)$ of P-essentially bounded functions. The analagous result for $(P \otimes Q)_\mathfrak{S}$-integrable functions follows.

Proposition 8.3. *For an $(\mathcal{E} \otimes \mathcal{F})$-measurable function $\varphi : \Lambda \times M \to \mathbb{C}$, let $[\varphi]$ be the equivalence class of all functions equal to φ $(P \otimes Q)$-almost everywhere. Let $L^1((P \otimes Q)_\mathfrak{S}) = \{[\varphi] : \varphi \text{ is } (P \otimes Q)_\mathfrak{S}\text{-integrable}\}$ with the pointwise operations of addition and scalar multiplication with the norm*

$$\|[\varphi]\|_\mathfrak{S} = \left\| \int_{\Lambda \times \Lambda} \varphi \, d(P \otimes Q)_\mathfrak{S} \right\|_{\mathcal{L}(\mathfrak{S})}.$$

*Then $\|[\varphi]\|_\infty \leq \|[\varphi]\|_\mathfrak{S}$ and $L^1((P \otimes Q)_\mathfrak{S})$ is a commutative Banach *-algebra under pointwise multiplication.*

If $\mathfrak{S} = \mathcal{C}_2(\mathcal{H})$, then $L^1((P \otimes Q)_\mathfrak{S}) = L^\infty(P \otimes Q)$ is a commutative C-algebra. Furthermore, the Banach *-algebras*

$$L^1((P \otimes Q)_{\mathcal{C}_1(\mathcal{H})}) = L^1((P \otimes Q)_{\mathcal{C}_\infty(\mathcal{H})}) = L^1((P \otimes Q)_{\mathcal{L}(\mathcal{H})}),$$

are isometric, where $\mathcal{C}_\infty(\mathcal{H})$ is the uniformly closed subspace of $\mathcal{L}(\mathcal{H})$ consisting of compact operators on $\mathcal{H}$.

Remark 8.2. The analogy of double operator integrals with multiplier theory in harmonic analysis is fleshed out in [101, Example 2.13], as follows. If Λ is a locally compact abelian group with Fourier transform $\mathcal{F}$, the spectral measure Q is defined by multiplication by characteristic functions on $L^2(\Lambda)$ and $P = \mathcal{F}^{-1}Q\mathcal{F}$ is the spectral measure of the "momentum operator" on Λ, then for $1 < p < \infty$, the space $\mathcal{M}_p(\Lambda)$ of Fourier multipliers on $L^p(\Lambda)$ coincides with the commutative Banach *-algebra $L^1(P_p)$ for the finitely additive set function $P_p : \mathcal{A} \to \mathcal{L}(L^p(\Lambda))$ defined as in [101, Example 2.13] by the spectral measure P acting on $L^2(\Lambda)$. For example, when $\Lambda = \mathbb{R}$, the operator $\int_\mathbb{R} \operatorname{sgn} dP_p \in \mathcal{L}(L^p(\mathbb{R}))$ is the Hilbert transform for $1 < p < \infty$.

It is only in the case $p = 2$, that $L^1(P_2) = L^\infty(P)$. One might argue that multiplier theory in commutative harmonic analysis is devoted to the study of the commutative Banach *-algebra $L^1(P_p)$ for $1 < p < \infty$. The analysis of the commutative Banach *-algebra $L^1((E \otimes F)_\mathfrak{S})$ for a general spectral measures E and F and symmetric operator ideal $\mathfrak{S}$ has many applications to scattering theory and quantum physics [15].

The commutative Banach *-algebra $L^1((P \otimes Q)_{\mathcal{L}(\mathcal{H})})$ is characterised by a result of V. Peller [104].

Theorem 8.9. *Let $\varphi : \Lambda \times M \to \mathbb{C}$ be a uniformly bounded function. Then $[\varphi] \in L^1((P \otimes Q)_{\mathcal{L}(\mathcal{H})})$ if and only if there exists a finite measure space*

$(T, \mathcal{S}, \nu)$ *and measurable functions* $\alpha : \Lambda \times T \to \mathbb{C}$ *and* $\beta : M \times T \to \mathbb{C}$ *such that* $\int_T \|\alpha(\cdot, t)\|_{L^\infty(P)} \|\beta(\cdot, t)\|_{L^\infty(Q)} \, d\nu(t) < \infty$ *and*

$$\varphi(\lambda, \mu) = \int_T \alpha(\lambda, t) \beta(\mu, t) \, d\nu(t), \quad \lambda \in \Lambda, \ \mu \in M. \tag{8.12}$$

Then norm of $[\varphi] \in L^1((P \otimes Q)_{\mathcal{L}(\mathcal{H})})$ *with the representation (8.12) satisfies*

$$K_G^{-1} \int_T \|\alpha(\cdot, t)\|_{L^\infty(P)} \|\beta(\cdot, t)\|_{L^\infty(Q)} \, d\nu(t) \leq \|[\varphi]\|_{L^1((P \otimes Q)_{\mathcal{L}(\mathcal{H})})}$$

$$\leq \left\| \left(\int_T |\alpha(\cdot, t)|^2 \, d\nu(t) \right)^{\frac{1}{2}} \right\|_{L^\infty(P)} \left\| \left(\int_T |\beta(\cdot, t)|^2 \, d\nu(t) \right)^{\frac{1}{2}} \right\|_{L^\infty(Q)} \tag{8.13}$$

for Grothendieck's constant K_G. *Moreover,* $\|[\varphi]\|_{L^1((P \otimes Q)_{\mathcal{L}(\mathcal{H})})}$ *is the infimum of all numbers on the right hand side of the inequality (8.13) for which there exists a finite measure* ν *such that the representation (8.12) holds for* φ.

Formula (8.12) is to be interpreted in the sense that φ is a special representative of the equivalence class $[\varphi] \in L^1((P \otimes Q)_{\mathcal{L}(\mathcal{H})})$. It is worthwhile to make a few remarks on the significance of formula (8.12) in order to motivate its proof below.

If the functions α and β in the representation (8.12) have the property that $t \longmapsto \alpha(\cdot, t)$, $t \in T$, and $t \longmapsto \beta(\cdot, t)$, $t \in T$, are strongly ν-measurable in $L^\infty(P)$ and $L^\infty(Q)$, respectively, then the function $t \longmapsto \alpha(\cdot, t) \otimes \beta(\cdot, t)$, $t \in T$, is strongly measurable in the projective tensor product $L^\infty(P) \widehat{\otimes}_\pi L^\infty(Q)$, and

$$\int_T \|\alpha(\cdot, t)\|_{L^\infty(P)} \|\beta(\cdot, t)\|_{L^\infty(Q)} \, d\nu(t) < \infty,$$

hence the function $t \longmapsto \alpha(\cdot, t) \otimes \beta(\cdot, t)$, $t \in T$, is Bochner integrable in $L^\infty(P) \widehat{\otimes}_\pi L^\infty(Q)$, that is, $[\varphi] \in L^\infty(P) \widehat{\otimes}_\pi L^\infty(Q)$. However, it is only assumed α is $(\mathcal{E} \otimes \mathcal{S})$-measurable and β is $(\mathcal{F} \otimes \mathcal{S})$-measurable, so this conclusion is unavailable.

Let $\nu_P : \mathcal{E} \to [0, \infty)$ be a finite measure such that $\nu_P(E) \leq \|P\|(E)$ for $E \in \mathcal{E}$ and $\lim_{\nu_P(E) \to 0} \|Ph\|(E) = 0$ for all $h \in \mathcal{H}$ with $\|h\| \leq 1$. Such a measure exists by the Bartle-Dunford-Schwartz Theorem 1.2, or more simply, $\nu_P = \sum_{n=1}^\infty 2^{-n} (Pe_n, e_n)$ for some orthonormal basis $\{e_n\}_n$ of $\mathcal{H}$. Let $\nu_Q : \mathcal{F} \to [0, \infty)$ be a finite measure corresponding to Q. Then $L^\infty(P) = L^\infty(\nu_P)$ and $L^\infty(Q) = L^\infty(\nu_Q)$.

There is a bijective correspondence between elements $[k]$ of the projective tensor product $L^\infty(\nu_P) \widehat{\otimes}_\pi L^\infty(\nu_Q) \subset L^\infty(\nu_P \otimes \nu_Q)$ and *nuclear operators* $T_k : L^1(\nu_Q) \to L^\infty(\nu_P)$ such that for each $f \in L^1(\nu_Q)$,

$$(T_k f)(\lambda) = \int_M k(\lambda, \mu) f(\mu) \, d\nu_Q(\mu)$$

for ν_P-almost all $\lambda \in \Lambda$, in the sense that for functions with

$$\sum_{j=1}^\infty \|\phi_j\|_{L^\infty(\nu_P)} \|\psi_j\|_{L^\infty(\nu_Q)} < \infty$$

the kernel $[k] = \sum_{j=1}^\infty \phi_j \otimes \psi_j$ correspond to the nuclear operator

$$(T_k f) = \sum_{j=1}^\infty \phi_j \int_M f \psi_j \, d\nu_Q, \quad f \in L^1(\nu_Q).$$

Nuclear operators between Banach space are discussed in [107].

In the case that $\mathcal{H} = \ell^2$ and $P = Q$ are projections onto the standard basis vectors, then $\int_{\mathbb{N} \times \mathbb{N}} \varphi \, d(P \otimes Q)_{\mathcal{L}(\ell^2)}$ is the classical Schur multiplier operator (8.10) and Grothendieck's inequality ensures that $L^1((P \otimes Q)_{\mathcal{L}(\ell^2)}) = \ell^\infty \widehat{\otimes}_\pi \ell^\infty$, see Proposition 1.3 below and [112, Theorem 3.2]. In this case, the measure ν in formula (8.12) is the counting measure on $\mathbb{N}$ and there is no difficulty with strong ν-measurability in an L^∞-space.

The passage from the discrete case to the case of general spectral measures P and Q sees the nuclear operators from $L^1(\nu_Q)$ to $L^\infty(\nu_P)$ replaced by *1-integral operators* from $L^1(\nu_Q)$ to $L^\infty(\nu_P)$, which leads to the Peller representation (8.12).

8.3.1 *Schur multipliers and Grothendieck's inequality*

Grothendieck's inequality (1.7) has already been employed to prove that for any measure μ, an $L^2(\mu)$-valued measure has bounded $\mathcal{K}$-semivariation in $L^2(\mu, \mathcal{K})$ for any Hilbert space $\mathcal{K}$.

If E is any $\mathcal{L}(\mathcal{H})$-valued spectral measure and $h \in \mathcal{H}$, the identity

$$\sum_{n=1}^\infty \|E(f_n)h\|_{\mathcal{H}}^2 = \left(E\left(\sum_{n=1}^\infty |f_n|^2 \right) h, h \right)$$

ensures that the $\mathcal{H}$-valued measure Eh has bounded ℓ^2-semivariation in $\ell^2(\mathcal{H})$—the Hilbert space tensor product $\mathcal{H} \widehat{\otimes} \ell^2 = \oplus_{j=1}^\infty \mathcal{H}$ with norm $\|u\|_{\ell^2(\mathcal{H})}^2 = \sum_{j=1}^\infty \|u_j\|_{\mathcal{H}}^2$. It follows from Theorem 2.2 that for any essentially bounded functions $f : \Lambda \to \ell^2$ and $g : M \to \ell^2$ and $h \in \mathcal{H}$, the

ℓ^2-valued function f is (Ph)-integrable in $\ell^2(\mathcal{H})$ and the ℓ^2-valued function g is (Qh)-integrable in $\ell^2(\mathcal{H})$. Then there exist operator valued measures $f \otimes P : \mathcal{E} \to \mathcal{L}(\mathcal{H}, \ell^2(\mathcal{H}))$ and $g \otimes Q : \mathcal{F} \to \mathcal{L}(\mathcal{H}, \ell^2(\mathcal{H}))$ such that

$$(f \otimes P)(E)h = (f \otimes (Ph))(E), \quad E \in \mathcal{E}, \ h \in \mathcal{H} \text{ and}$$
$$(g \otimes Q)(F)h = (g \otimes (Qh))(F), \quad F \in \mathcal{F}, \ h \in \mathcal{H}.$$

There is a simple sufficient condition for $\varphi \in L^1((P \otimes Q)_{\mathcal{L}(\mathcal{H})})$. Observe first that the linear map $J : \ell^2(\mathcal{H}) \otimes \ell^2(\mathcal{H}) \to \mathcal{H} \widehat{\otimes}_\pi \mathcal{H}$ defined by

$$J((\{\phi_n\}_n) \otimes (\{\psi_m\}_m)) = \sum_{j=1}^{\infty} \phi_j \otimes \psi_j$$

has a continuous linear extension to a contraction $\overline{J} : \mathcal{C}_1(\ell^2(\mathcal{H})) \to \mathcal{C}_1(\mathcal{H})$ corresponding to taking the trace in the discrete index. The formula

$$(((f \otimes P) \otimes (g \otimes Q))_{\mathcal{C}_1(\mathcal{H})}(E \times F))(h \otimes k) := (((f \otimes (Ph))(E)) \otimes (g \otimes (Qk))(F))$$

for $h, k \in \mathcal{H}$, $E \in \mathcal{E}$ and $F \in \mathcal{F}$, defines a finitely additive set function

$$((f \otimes P) \otimes (g \otimes Q))_{\mathcal{C}_1(\mathcal{H})}$$

with values in $\mathcal{L}(\mathcal{C}_1(\mathcal{H}), \mathcal{C}_1(\ell^2(\mathcal{H})))$, because $\mathcal{C}_1(\mathcal{K})$ can be identified with $\mathcal{K} \widehat{\otimes}_\pi \mathcal{K}$ for any Hilbert space $\mathcal{K}$. Moreover,

$$\|((f \otimes P) \otimes (g \otimes Q))_{\mathcal{C}_1(\mathcal{H})}(\Lambda \times M)\|_{\mathcal{L}(\mathcal{C}_1(\mathcal{H}), \mathcal{C}_1(\ell^2(\mathcal{H})))} \leq \|f\|_\infty \cdot \|g\|_\infty.$$

Then the operator

$$\int_{\Lambda \times M} (f, \overline{g}) \, d(P \otimes Q)_{\mathcal{C}_1(\mathcal{H})} = \overline{J}[((f \otimes P) \otimes (g \otimes Q))_{\mathcal{C}_1(\mathcal{H})}(\Lambda \times M)]$$

is an element of $\mathcal{L}(\mathcal{C}_1(\mathcal{H}))$, that is, $\varphi = (f, \overline{g})$ belongs to $L^1((P \otimes Q)_{\mathcal{C}_1(\mathcal{H})}) = L^1((P \otimes Q)_{\mathcal{L}(\mathcal{H})})$ and we have the representation

$$\varphi(\lambda, \mu) = \sum_{n=1}^{\infty} f_n(\lambda) g_n(\mu), \tag{8.14}$$

where $P\text{-ess.sup}\{\sum_{n=1}^{\infty} |f_n|^2\} < \infty$ and $Q\text{-ess.sup}\{\sum_{n=1}^{\infty} |g_n|^2\} < \infty$. Moreover, the bound

$$\left\| \int_{\Lambda \times M} \varphi \, d(P \otimes Q)_{\mathcal{L}(\mathcal{H})} \right\|_{\mathcal{L}(\mathcal{L}(\mathcal{H}))} \leq \left\| \sum_{n=1}^{\infty} |f_n|^2 \right\|_{L^\infty(P)} \cdot \left\| \sum_{n=1}^{\infty} |g_n|^2 \right\|_{L^\infty(Q)} \tag{8.15}$$

holds. Alternatively, for each $T \in \mathcal{L}(\mathcal{H})$, the linear operator

$$\left(\int_{\Lambda \times M} \varphi \, d(P \otimes Q)_{\mathcal{L}(\mathcal{H})} \right) T \in \mathcal{L}(\mathcal{H})$$

can be realised as the operator associated with the bounded sesquilinear form

$$(h, k) \longmapsto \sum_{n=1}^{\infty} (TQ(g_n)h, P(\overline{f_n})k),$$

see [15, Theorem 4.1].

A remarkable consequence of Grothendick's inequality is that for $\varphi \in L^1((P \otimes Q)_{\mathcal{L}(\mathcal{H})})$, Peller's representation (8.12) is *necessary* $(\nu_P \otimes \nu_Q)$-almost everywhere. The analysis of G. Pisier [112] leads the way.

8.3.2   *Schur multipliers on measure spaces*

We first note that for any choice of finite measures ν_P, ν_Q equivalent to P and Q respectively, the Banach algebra $L^1((P \otimes Q)_{\mathcal{L}(\mathcal{H})})$ is isometrically isomorphic to the set of multipliers of the projective tensor product $L^2(\nu_P) \widehat{\otimes}_\pi L^2(\nu_Q)$, that is, $[\varphi] \in L^1((P \otimes Q)_{\mathcal{L}(\mathcal{H})})$ if and only if for every $[h] \in L^2(\nu_P) \widehat{\otimes}_\pi L^2(\nu_Q)$, the function $\varphi.h$ is equal $(\nu_P \otimes \nu_Q)$-a.e. to an element of $L^2(\nu_P) \widehat{\otimes}_\pi L^2(\nu_Q)$ and $\|[\varphi]\|_{L^1((P \otimes Q)_{\mathcal{L}(\mathcal{H})}}$ is equal to the norm of the linear map $[h] \longmapsto [\varphi.h]$, $[h] \in L^2(\nu_P) \widehat{\otimes}_\pi L^2(\nu_Q)$, on $L^2(\nu_P) \widehat{\otimes}_\pi L^2(\nu_Q)$.

If ν'_P and ν'_Q are another pair of such equivalent measures, then the operator of multiplication by $\sqrt{d\nu'_P/d\nu_P}$ is a unitary map from $L^2(\nu_P)$ to $L^2(\nu'_P)$ and similarly for ν_Q, so that multiplication by $\sqrt{d\nu'_P/d\nu_P} \otimes \sqrt{d\nu'_Q/d\nu_Q}$ is an isometric isomorphism from $L^2(\nu_P) \widehat{\otimes}_\pi L^2(\nu_Q)$ onto $L^2(\nu'_P) \widehat{\otimes}_\pi L^2(\nu'_Q)$.

Proposition 8.4. *Let ν_P, ν_Q be finite measures equivalent to the spectral measures P, Q respectively. Then $L^1((P \otimes Q)_{\mathcal{L}(\mathcal{H})})$ is isometrically isomorphic to the set of multipliers of the projective tensor product $L^2(\nu_P) \widehat{\otimes}_\pi L^2(\nu_Q)$ and the identity*

$$\|[\varphi]\|_{L^1((P \otimes Q)_{\mathcal{L}(\mathcal{H})}} = \sup_{\|h\|_{\mathcal{H}} \le 1, \|g\|_{\mathcal{H}} \le 1} \|[\varphi]\|_{L^2((Ph,h)) \widehat{\otimes}_\pi L^2((Qg,g))} \qquad (8.16)$$

holds.

Proof. Let $\{h_n\}_n$ be a sequence of vectors in $\mathcal{H}$ with $\sum_n \|h_n\|^2 < \infty$ such that $\{P(E)h_n : n = 1, 2, \dots\}$ is an orthogonal set of vectors in $\mathcal{H}$ for each $E \in \mathcal{E}$. Such a sequence of vectors can always be manufactured by taking any vectors $\xi_n \in \mathcal{H}$ with $\sum_n \|\xi_n\|^2 < \infty$ and for a measure ν_P equivalent to P, the sets Λ_n where $d(P\xi_n, \xi_n)/d\nu_P > 0$. Then $h_n = P(\Lambda_n \setminus \bigcup_{m<n} \Lambda_n)\xi_n$, $n = 1, 2, \dots$, will do the job. Let $\{g_n\}_n$ be the corresponding vectors for Q.

As noted above, the norm of $L^2(\nu_P)\widehat{\otimes}_\pi L^2(\nu_Q)$ is invariant under a change of equivalent measures, so we may as well assume that

$$\nu_P = \sum_{n=1}^{\infty}(Ph_n, h_n) \text{ and } \nu_Q = \sum_{n=1}^{\infty}(Qg_n, g_n)$$

so that the mappings $\chi_E \to \sum_{n=1}^{\infty} P(E)h_n$, $E \in \mathcal{E}$, and $\chi_F \to \sum_{n=1}^{\infty} Q(F)h_n$, $F \in \mathcal{F}$, define a unitary equivalences U_P, U_Q between $L^2(\nu_P)$ and $L^2(\nu_Q)$ and $\mathcal{H}$, respectively.

The map $T_k : L^2(\nu_Q) \to L^2(\nu_P)$ whose integral kernel k belongs to $L^2(\nu_P)\widehat{\otimes}_\pi L^2(\nu_Q)$ is trace class. Let $\tilde{T}_k \in \mathcal{C}_1(\mathcal{H})$ be the corresponding trace class operator on $\mathcal{H}$. Then

$$\sum_{n,m=1}^{\infty} (\tilde{T}_k Q(F)g_m, P(E)h_n) = \int_{E\times F} k\, d(\nu_P \otimes \nu_Q).$$

Let $[\varphi] \in L^1((P\otimes Q)_{\mathcal{L}(\mathcal{H})}) = L^1((P\otimes Q)_{\mathcal{C}_1(\mathcal{H})})$. Then $\tilde{T}_k \in \mathcal{C}_2(\mathcal{H})$ and

$$\left(\left(\int_{\Lambda\times M} \varphi\, d(P\otimes Q)_{\mathcal{C}_1(\mathcal{H})}\right)\tilde{T}_k\left(\sum_{m=1}^{\infty} Q(F)g_m\right), \left(\sum_{n=1}^{\infty} P(E)h_n\right)\right)$$

$$= \mathrm{tr}\left(\left(\int_{\Lambda\times M} \varphi\, d(P\otimes Q)_{\mathcal{C}_2(\mathcal{H})}\right)\tilde{T}_k\left(\sum_{m=1}^{\infty} Q(F)g_m\right) \otimes \left(\sum_{n=1}^{\infty} P(E)h_n\right)^*\right)$$

$$= \sum_{n,m=1}^{\infty} \int_{E\times F} \varphi\, d((P\tilde{T}_k Q)g_m, h_n)$$

$$= \int_{E\times F} \varphi.k\, d(\nu_P \otimes \nu_Q).$$

It follows that $\varphi.k$ is the kernel of the trace class operator $T_{\varphi.k} : L^2(\nu_Q) \to L^2(\nu_P)$ such that

$$U_P T_{\varphi.k} U_Q^* = \left(\int_{\Lambda\times M} \varphi\, d(P\otimes Q)_{\mathcal{C}_1(\mathcal{H})}\right)\tilde{T}_k \in \mathcal{C}_1(\mathcal{H}),$$

the equality $\|\varphi.k\|_{L^2(\nu_P)\widehat{\otimes}_\pi L^2(\nu_Q)} = \left\|\left(\int_{\Lambda\times M} \varphi\, d(P\otimes Q)_{\mathcal{C}_1(\mathcal{H})}\right)\tilde{T}_k\right\|_{\mathcal{C}_1(\mathcal{H})}$ holds and

$$\|[\varphi]\|_{L^1((P\otimes Q)_{\mathcal{L}(\mathcal{H})})} = \sup\{\|\varphi.k\|_{L^2(\nu_P)\widehat{\otimes}_\pi L^2(\nu_Q)} : \|k\|_{L^2(\nu_P)\widehat{\otimes}_\pi L^2(\nu_Q)} \le 1\}. \tag{8.17}$$

According to the identities above,

$$\left\|\int_{\Lambda\times M} \varphi\, d(P\tilde{T}_k Q)\right\|_{\mathcal{C}_1(\mathcal{H})} = \|\varphi.k\|_{L^2(\nu_P)\widehat{\otimes}_\pi L^2(\nu_Q)}$$

for all $k \in L^2(\nu_P) \otimes L^2(\nu_Q)$, so if

$$\|\varphi.k\|_{L^2(\nu_P)\hat{\otimes}_\pi L^2(\nu_Q)} \leq C$$

for all $k \in L^2(\nu_P) \otimes L^2(\nu_Q)$ satisfying $\|k\|_{L^2(\nu_P)\hat{\otimes}_\pi L^2(\nu_Q)} \leq 1$, then $[\varphi] \in L^1((P \otimes Q)_{\mathcal{L}(\mathcal{H})})$, the identity (8.17) holds and

$$\|[\varphi]\|_{L^1((P\otimes Q)_{\mathcal{L}(\mathcal{H})})} = \sup_{\sup_{\|u\|_2 \leq 1, \|v\|_2 \leq 1}} \|\varphi.(u \otimes v)\|_{L^2(\nu_P)\hat{\otimes}_\pi L^2(\nu_Q)}.$$

The equality (8.16) follow from the identities

$$\int_\Lambda |\psi_1|^2 |\psi_2|^2 \, d\nu_P = \left\| \sum_{n=1}^\infty P(\psi_1.\psi_2)h_n \right\|_{\mathcal{H}}^2$$
$$= (P(|\psi_1|^2)(U_P\psi_2), (U_P\psi_2))$$

for $\psi_1 \in L^\infty(\nu_P)$, $\psi_2 \in L^2(\nu_P)$ and the unitary equivalence U_P defined above. The analogous identities hold for the spectral measure Q. $\qquad\square$

Proof of Theorem 8.9. We proceed by reduction to the ℓ^2-case considered in Proposition 1.3.

Let ν_P, ν_Q be finite measures equivalent to P, Q, respectively. Because both $L^2(\nu_P)$ and $L^2(\nu_Q)$ are isomorphic to the separable Hilbert space $\mathcal{H}$, for the purpose of obtaining the representation (8.12), we may suppose that the underlying σ-algebras are countably generated.

Let $\mathcal{P} = \{\mathcal{P}_n\}_n$ be a Lusin ν_P-filtration and let $\mathcal{Q} = \{\mathcal{Q}_n\}_n$ be a Lusin ν_Q-filtration. Suppose that $n = 1, 2, \ldots$ and $\{A_i\}_{i=1}^\infty$ is the nth partition associated with $\mathcal{P}$ and $\{B_i\}_{j=1}^\infty$ is the nth partition associated with $\mathcal{Q}$. The corresponding projection operators $P_n : L^1(\nu_P) \to \ell^1$ and $Q_n : L^1(\nu_Q) \to \ell^1$ are defined by

$$P_n : f \longmapsto \left\{ \int_{A_i} f \, d\nu_P \right\}_{i=1}^\infty, \ f \in L^1(\nu_P),$$

$$Q_n : g \longmapsto \left\{ \int_{B_j} f \, d\nu_Q \right\}_{j=1}^\infty, \ g \in L^1(\nu_Q).$$

The conditional expectation $(\mathbb{E}_n \otimes \mathbb{F}_n)(f) = \mathbb{E}(f|\mathcal{P}_n \otimes \mathcal{Q}_n)$ is defined for any measurable function $f : \Lambda \times M \to \mathbb{C}$ that is integrable over any set $A_i \times B_j$, $i, j \in \mathbb{N}$.

It is easy to verify that $P_n^* T_{\varphi_n} Q_n = T_{(\mathbb{E}_n \otimes \mathbb{F}_n)\varphi}$ for the matrix

$$\varphi_n = \left\{ \frac{\int_{A_i \times B_j} \varphi \, d(\nu_P \otimes \nu_Q)}{\nu_P(A_i)\nu_Q(B_j)} \right\}_{i,j=1}^\infty$$

and the operator $T_{\varphi_n} : \ell^1 \to \ell^\infty$ with kernel φ_n.

Moreover, for every finite rank operator $U : L^\infty(\nu_P) \to L^1(\nu_Q)$, the bound

$$|\mathrm{tr}(P_n^* T_{\varphi_n} Q_n U)| = |\mathrm{tr}(T_{\varphi_n} Q_n U P_n^*)| \le \|\varphi_n\|_{\ell^\infty \widehat{\otimes}_\pi \ell^\infty} \|Q_n U P_n^*\|.$$

Suppose that there exists $M > 0$ such that $\|\varphi_n\|_{\ell^\infty \widehat{\otimes}_\pi \ell^\infty} \le M$ for all $n = 1, 2, \ldots$.

Then $\mathrm{tr}(P_n^* T_{\varphi_n} Q_n U) = \mathrm{tr}(T_{(\mathbb{E}_n \otimes \mathbb{F}_n)\varphi} U) = \mathrm{tr}(T_\varphi \mathbb{E}_n U \mathbb{F}_n)$, and taking $n \to \infty$, the martingale convergence theorem shows that the bound

$$|\mathrm{tr}(T_\varphi U)| \le M \|U\|_{\mathcal{L}(L^\infty, L^1)}$$

holds for every finite rank operator $U : L^\infty(\nu_P) \to L^1(\nu_Q)$. It follows from [36, Theorem 6.16] that T_φ belongs to the Banach ideal $\mathcal{I}_1(L^1(\nu_Q), L^\infty(\nu_P))$ of 1-integral operators from $L^1(\nu_Q)$ to $L^\infty(\nu_P)$—see Example 1.4 above. Because $L^\infty(\nu_P)$ is a dual space, [36, Corollary 5.4] ensures that T_φ enjoys the factorisation

$$
\begin{array}{ccc}
L^1(\nu_Q) & \xrightarrow{T_\varphi} & L^\infty(\nu_P) \\
T_1 \downarrow & & \uparrow T_2 \\
L^\infty(\nu) & \xrightarrow{j} & L^1(\nu)
\end{array}
$$

for some bounded linear operators T_1 and T_2 and finite measure space $(T, \mathcal{S}, \nu)$. The given factorisation also follows by the original 1954 Grothendieck argument with the choice $E = L^1(\nu_Q)$, $F = L^1(\nu_P)$ in [123, Section IV.9.2].

Every bounded linear operator u from $L^1(\eta_1)$ to $L^\infty(\eta_2)$ is an integral operator with a bounded kernel because $f \otimes g \mapsto \langle uf, g \rangle$ defines a continuous linear functional on $L^1(\eta_1)\widehat{\otimes}_\pi L^1(\eta_2) \equiv L^1(\eta_1 \otimes \eta_2)$ (see [63, Lemma 2.2] for a compactness argument), so there exist bounded measurable functions $\alpha : \Lambda \times T \to \mathbb{C}$ and $\beta : M \times T \to \mathbb{C}$ such that

$$(T_1 f)(t) = \int_M \beta(\mu, t) f(\lambda) d\nu_Q(\lambda), \quad f \in L^1(\nu_Q),$$

$$(T_2 g)(\lambda) = \int_T \alpha(\lambda, t) g(t) d\nu(t), \quad g \in L^1(\nu).$$

The representation (8.12) and the associated bounds follow if we can take $M = K_G \|[\varphi]\|_{L^1((P \otimes Q)_{\mathcal{L}(\mathcal{H})})}$.

We know from the bounds (1.3) that

$$\|\varphi_n\|_{\ell^\infty \widehat{\otimes}_\pi \ell^\infty} \le K_G \|M_{\varphi_n}\|_{\mathcal{L}(\mathcal{L}(\ell^2))} = K_G \|\varphi_n\|_{\ell^\infty \widehat{\otimes}_{\gamma_2} \ell^\infty}.$$

The norm γ_2 defined on $\ell^\infty \otimes \ell^\infty$ is the norm of factorisation through a Hilbert space. For any bounded linear operator $u : X \to Y$ between Banach spaces X and Y, $\gamma_2(u) = \inf\{\|u_1\|, \|u_2\|\}$ where the infimum runs over all Hilbert spaces $\mathcal{H}$ and all possible factorisations

$$u : X \xrightarrow{u_2} \mathcal{H} \xrightarrow{u_1} Y$$

of u through $\mathcal{H}$ with $u = u_1 \circ u_2$. Taking $X = L^1(\nu_Q)$ and $Y = L^\infty(\nu_P)$, the bound (8.15) says that

$$\|\varphi\|_{L^1((P \otimes Q)_{\mathcal{L}(\mathcal{H})})} \leq \|\varphi\|_{L^\infty(\nu_P)\widehat{\otimes}_{\gamma_2} L^\infty(\nu_Q)}$$

with respect to the completion $L^\infty(\nu_P)\widehat{\otimes}_{\gamma_2} L^\infty(\nu_Q)$ of $L^\infty(\nu_P) \otimes L^\infty(\nu_Q)$ in the norm $\phi \mapsto \gamma_2(T_\phi)$, $\phi \in L^\infty(\nu_P) \otimes L^\infty(\nu_Q)$.

The norm estimates

$$\|\varphi_n\|_{\ell^\infty \widehat{\otimes}_{\gamma_2} \ell^\infty} = \|(\mathbb{E}_n \otimes \mathbb{F}_n)\varphi\|_{L^\infty(\nu_P)\widehat{\otimes}_{\gamma_2} L^\infty(\nu_Q)} \leq \|\varphi\|_{L^\infty(\nu_P)\widehat{\otimes}_{\gamma_2} L^\infty(\nu_Q)}$$

follow from the definition of γ_2 and the contractivity of the conditional expectation operators $\mathbb{E}_n, \mathbb{F}_n$.

According to Proposition 8.4, the norm of the linear operator

$$M_\varphi : \mathcal{C}_1(L^2(\nu_Q), L^2(\nu_P)) \to \mathcal{C}_1(L^2(\nu_Q), L^2(\nu_P))$$

associated with multiplication by φ on $L^2(\nu_P)\widehat{\otimes}_\pi L^2(\nu_Q)$ is equal to

$$\|[\varphi]\|_{L^1((P\otimes Q)_{\mathcal{L}(\mathcal{H})})} = \|[\varphi]\|_{L^1((P\otimes Q)_{\mathcal{C}_1(\mathcal{H})})}.$$

The equality $\|\varphi\|_{L^\infty(\nu_P)\widehat{\otimes}_{\gamma_2} L^\infty(\nu_Q)} = \|M_\varphi\|_{\mathcal{L}(\mathcal{L}(L^2(\nu_Q),L^2(\nu_P)))}$ is proved in [129, Theorem 3.3] using complete boundedness techniques, but this can be established in a more elementary way by noting that if $[\varphi] \in L^1((P \otimes Q)_{\mathcal{C}_1(\mathcal{H})})$, then the Martingale Convergence Theorem ensures that $M_{(\mathbb{E}_n \otimes \mathbb{F}_n)\varphi} \to M_\varphi$ in the strong operator topology of

$$\mathcal{L}\big(\mathcal{C}_1(L^2(\nu_Q), L^2(\nu_P)), \mathcal{C}_1(L^2(\nu_Q), L^2(\nu_P))\big)$$

as $n \to \infty$ and also

$$\|(\mathbb{E}_n \otimes \mathbb{F}_n)\varphi\|_{L^\infty(\nu_P)\widehat{\otimes}_{\gamma_2} L^\infty(\nu_Q)} \longrightarrow \|\varphi\|_{L^\infty(\nu_P)\widehat{\otimes}_{\gamma_2} L^\infty(\nu_Q)}$$

as $n \to \infty$. Then $\|M_\varphi\| = \sup_n \|M_{(\mathbb{E}_n \otimes \mathbb{F}_n)\varphi)}\|$ by duality. The equality

$$\|(\mathbb{E}_n \otimes \mathbb{F}_n)\varphi\|_{L^\infty(\nu_P)\widehat{\otimes}_{\gamma_2} L^\infty(\nu_Q)} = \|M_{(\mathbb{E}_n \otimes \mathbb{F}_n)\varphi)}\|_{\mathcal{L}(\mathcal{L}(L^2(\nu_Q),L^2(\nu_P)))}$$

follows for each $n = 1, 2, \ldots$ from Proposition 1.2 by replacing $e_i \otimes e_j$ in (iii) by $\chi_{A_i \times B_j}$ for $i, j = 1, 2, \ldots$. The final assertion of Theorem 8.9 follows from the equality $\|\varphi\|_{L^\infty(\nu_P)\widehat{\otimes}_{\gamma_2} L^\infty(\nu_Q)} = \|M_\varphi\|_{\mathcal{L}(\mathcal{L}(L^2(\nu_Q),L^2(\nu_P)))}$. $\qquad\square$

Remark 8.3. a) The original proof of V. Peller [104], [63, Theorem 2.2] factorises the finite rank operator $U : L^\infty(\nu_P) \to L^1(\nu_Q)$ instead, so the constant K_G^2 appears in place of K_G in the bound associated with (8.12).

b) Let $L^1(\nu_P) \tilde\otimes L^1(\nu_Q)$ be the closure of the linear space of all $k \in L^1(\nu_P) \otimes L^1(\nu_Q)$ in the uniform norm of the space of operators $T_k \in \mathcal{L}(L^\infty(\nu_Q), L^1(\nu_P))$ corresponding to the compact linear operators from $L^\infty(\nu_Q)$ to $L^1(\nu_P)$. By [123, Section IV.9.2], the function $\alpha \otimes \beta$ in formula (8.12) is ν-integrable in the space of 1-integral operators

$$\mathcal{I}_1(L^1(\nu_Q), L^\infty(\nu_P)) \equiv (L^1(\nu_P) \tilde\otimes L^1(\nu_Q))'$$

and $\varphi = \int_T \alpha \otimes \beta \, d\nu$.

c) The proof above shows that operator $T_\varphi : L^1(\nu_Q) \to L^\infty(\nu_P)$ is (strictly) 1-integral in the sense of [36] and [123, Section IV.9.2] if and only if $[\varphi] \in L^1((P \otimes Q)_{\mathcal{L}(\mathcal{H})})$. The reason that $\|[\varphi]\|_{L^\infty(\nu_P) \hat\otimes_\pi L^\infty(\nu_Q)} = \infty$ for some $[\varphi] \in L^1((P \otimes Q)_{\mathcal{L}(\mathcal{H})})$, that is, the function $\alpha \otimes \beta$ associated with the representation (8.12) fails to be ν-integrable in $L^\infty(\nu_P) \hat\otimes_\pi L^\infty(\nu_Q)$ so that $T_\varphi : L^1(\nu_Q) \to L^\infty(\nu_P)$ thereby fails to be a *nuclear* operator, is that the vector measure $E \longmapsto u\chi_E$ associated with a continuous linear map u from L^1 to L^∞ has a weak*-density, but not necessarily a *strongly measurable* density in L^∞.

For any $u \in \mathcal{C}_1(\mathcal{H})$ and $\varphi \in L^1((P \otimes P)_{\mathcal{L}(\mathcal{H})})$, the operator

$$M_\varphi u = \left(\int_{\Lambda \times \Lambda} \varphi \, d(P \otimes P)_{\mathcal{C}_1(\mathcal{H})} \right) u$$

is trace class. Moreover, the expression $E \longmapsto \mathrm{tr}(uP(E))$, $E \in \mathcal{E}$, is a complex measure μ_u on the σ-algebra $\mathcal{E}$ such that $|\mu_u| \ll \nu_P$. As indicated in [15, Section 9.1], the identity

$$\mathrm{tr}(M_\varphi u) = \int_\Lambda \varphi(\lambda, \lambda) \, d\mu_u(\lambda) \tag{8.18}$$

holds. In the case that $u : \mathcal{H} \to \mathcal{H}$ is a finite rank operator, together with the polarisation, the bound (8.16) shows that the operator $T_\varphi : L^2(\mu_u) \to L^2(\mu_u)$ with integral kernel φ is trace class and

$$\|T_\varphi\|_{\mathcal{C}_1(L^2(\mu_u))} \leq 16 \|[\varphi]\|_{L^1(P \otimes Q)_{\mathcal{L}(\mathcal{H})}} \|u\|_{\mathcal{C}_1(\mathcal{H})}.$$

The same bound holds for all $u \in \mathcal{C}_1(\mathcal{H})$. The identity

$$|\psi|^2 . \nu_P = (P(U_P\psi), (U_P\psi)), \quad \psi \in L^2(\nu_P),$$

ensures that $\mathrm{tr}(M_{\phi_1 \otimes \phi_2} u) = \mathrm{tr}(T_{\phi_1 \otimes \phi_2})$ for $T_{\phi_1 \otimes \phi_2} \in \mathcal{C}_1(L^2(\mu_u))$ with $u \in \mathcal{C}_1(\mathcal{H})$ and ϕ_1, ϕ_2 bounded on Λ. Then the equality

$$\mathrm{tr}(M_\varphi u) = \mathrm{tr}(T_\varphi)$$

holds because both sides are continuous for $\varphi \in L^\infty(\nu_P)\widehat{\otimes}_{\gamma_2}L^\infty(\nu_P)$.

The representation (8.14) converges in $L^\infty(\nu_P)\widehat{\otimes}_{\gamma_2}L^\infty(\nu_P)$ and there exists a set Λ_0 of full ν_P-measure such that

$$\varphi(\lambda, \mu) = \sum_{n=1}^{\infty} f_n(\lambda)g_n(\mu)$$

for all $\lambda, \mu \in \Lambda_0$ where the right-hand sum converges absolutely. The expression above constitutes a *distinguished* element of the equivalence class $[\varphi]$. Consequently, formula (8.18) is valid because

$$\begin{aligned}
\operatorname{tr}(T_\varphi) &= \sum_{n=1}^{\infty} \operatorname{tr}(T_{f_n \otimes g_n}) \\
&= \sum_{n=1}^{\infty} \int_\Lambda f_n(\lambda)g_n(\lambda)\, d\mu_u(\lambda) \\
&= \int_\Lambda \varphi(\lambda, \lambda)\, d\mu_u(\lambda).
\end{aligned}$$

From the point of view of our study of traces in Chapter 3, for any Lusin μ_P-filtration $\mathcal{F} = \langle \mathcal{E}_k \rangle_k$ of Λ, for each $k = 1, 2, \ldots$, the conditional expectation operators $\mathbb{E}_k : f \longmapsto \mathbb{E}(f|\mathcal{E}_k)$ with respect to the σ-algebra $\mathcal{E}_k$ and the finite measure ν_P have the property that

$$\sum_{n=1}^{\infty} |f_n.g_n - \mathbb{E}_k(f_n).\mathbb{E}_k(g_n)|$$

$$\leq \sum_{n=1}^{\infty} |(f_n - \mathbb{E}_k(f_n)).g_n| + \sum_{n=1}^{\infty} |\mathbb{E}_k(f_n).(g_n - \mathbb{E}_k(g_n))|$$

$$\leq \left(\sum_{n=1}^{\infty} |f_n - \mathbb{E}_k(f_n)|^2\right)^{\frac{1}{2}} \cdot \left(\sum_{n=1}^{\infty} |g_n|^2\right)^{\frac{1}{2}}$$

$$+ \left(\sum_{n=1}^{\infty} |\mathbb{E}_k(f_n)|^2\right)^{\frac{1}{2}} \cdot \left(\sum_{n=1}^{\infty} |g_n - \mathbb{E}_k(g_n)|^2\right)^{\frac{1}{2}}$$

$$\to 0 \quad \nu_P\text{-almost everywhere as } k \to \infty,$$

by the Martingale Convergence Theorem. Consequently, setting

$$\tilde{\varphi} = \lim_{k \to \infty} (\mathbb{E}_k \otimes \mathbb{E}_k)\varphi$$

wherever the limit exists, the equality $\tilde{\varphi}(\lambda, \lambda) = \varphi(\lambda, \lambda)$ holds for ν_P-almost all $\lambda \in \Lambda$, so that formula (8.18) is also a consequence of [19, Theorem 3.1].

Remark 8.4. There is a representative function φ of the equivalence class $[\varphi]$ that is continuous for the so-called ω-topology of [78, Proposition 9.1], so formula (8.18) is also a consequence of Theorem 8.9. In fact, Peller's representation (8.12) can be deduced directly from Proposition 1.3 by employing the ω-continuity of φ rather than the Martingale Convergence Theorem, see [78, Remark on p. 139].

8.4 The spectral shift function

The following perturbation formula of Birman and Solomyak [15, theorem 8.1] was mentioned in the proof of Corollary 8.3. The operator ideal $\mathfrak{S}$ is taken to be $\mathcal{C}_p(\mathcal{H})$ for $1 \leq p < \infty$ or $\mathcal{L}(\mathcal{H})$ for a given Hilbert space $\mathcal{H}$.

Theorem 8.10. *Let $\mathcal{H}$ be a separable Hilbert space and let A and B be selfadjoint operators with the same domain such that $A - B \in \mathfrak{S}$. Let $P_A : \mathcal{B}(\mathbb{R}) \to \mathcal{L}_s(\mathcal{H})$ and $P_B : \mathcal{B}(\mathbb{R}) \to \mathcal{L}_s(\mathcal{H})$ be the spectral measures on $\mathbb{R}$ associated with A and B, respectively. Suppose that $f : \mathbb{R} \to \mathbb{R}$ is a continuous function for which the difference quotient*

$$\varphi_f(\lambda, \mu) = \begin{cases} \frac{f(\lambda) - f(\mu)}{\lambda - \mu}, & \lambda \neq \mu, \\ 0, & \lambda = \mu, \end{cases}$$

is uniformly bounded and $\varphi_f \in L^1((P_A \otimes Q_B)_\mathfrak{S})$. Then

$$\int_{\mathbb{R} \times \mathbb{R}} \varphi_f \, d(P_A \otimes P_B)_\mathfrak{S} \in \mathcal{L}(\mathfrak{S})$$

and

$$f(A) - f(B) = \left(\int_{\mathbb{R} \times \mathbb{R}} \varphi_f \, d(P_A \otimes P_B)_\mathfrak{S} \right) (A - B).$$

If $\mathfrak{S} = \mathcal{C}_1(\mathcal{H})$, then we would like to calculate the trace of $f(A) - f(B)$. The method of the preceding section is unavailable with different spectral measures P_A, P_B so we can try to invoke the Daletskii-Krein formula [15, Equation (9.10)]. For a sufficiently smooth function f, this takes the form

$$f(A) - f(B) = \int_0^1 \left(\int_{\mathbb{R} \times \mathbb{R}} \varphi_f(\lambda, \mu) \, d(P_{A(t)} \otimes P_{A(t)})_{\mathcal{C}_1(\mathcal{H})} \right) (A - B) \, dt$$

with $A(t) = B + t(A - B)$, $0 \leq t \leq 1$, and $\varphi_f(\lambda, \lambda) = f'(\lambda)$, $\lambda \in \mathbb{R}$. At each point $0 \leq t \leq 1$, the same spectral measure $P_{A(t)}$ is involved, so from formula (8.18), we can expect that

$$\mathrm{tr}(f(A) - f(B)) = \int_{\mathbb{R}} f'(\lambda) \, d\Xi(\lambda)$$

for the complex measure $\Xi : E \longmapsto \int_0^1 \text{tr}(V P_{A(t)}(E)) \, dt$, $E \in \mathcal{B}(\mathbb{R})$, with $V = (A - B) \in \mathcal{C}_1(\mathcal{H})$. It turns out that Ξ is absolutely continuous with respect to Lebesgue measure on $\mathbb{R}$ from which the formula

$$\text{tr}(f(A) - f(B)) = \int_{\mathbb{R}} f'(\lambda) \xi(\lambda) \, d\lambda \qquad (8.19)$$

is obtained. The function $\xi : \mathbb{R} \to \mathbb{C}$ is *Krein's spectral shift function*.

We now turn to establishing the validity of formula (8.19) for a restricted class of functions f. Better results are known, for example from [113], but our purpose is to describe applications of singular bilinear integrals such as double operator integrals to problems in the perturbation theory of linear operators. The approach of K. Boyadzhiev [17] best suits our purpose.

Setting $V = A - B \in \mathcal{C}_1(\mathcal{H})$, we first note that $e^{isA(t)} - e^{isB} \in \mathcal{C}_1(\mathcal{H})$ for each $s \in \mathbb{R}$ and $0 \le t \le 1$ because the perturbation series

$$e^{isA(t)} = e^{isB}$$

$$+ \sum_{n=1}^{\infty} (is)^n \int_0^t \cdots \int_0^{s_2} e^{isB(s-s_n)} V \cdots e^{isB(s_2-s_1)} V e^{isBs_1} \, ds_1 \cdots ds_n$$

converges in the norm of $\mathcal{C}_1(\mathcal{H})$ and $t \longmapsto e^{isA(t)} - e^{isB}$ is norm differentiable in $\mathcal{C}_1(\mathcal{H})$. Moreover,

$$\|e^{isA(t)} - e^{isB}\|_{\mathcal{C}_1(\mathcal{H})} \le (e^{|s| \|V\|_{\mathcal{C}_1(\mathcal{H})}} - 1). \qquad (8.20)$$

The following result is straightforward but it depends on some measure theoretic facts. It establishes that Ξ is a complex measure.

Lemma 8.2. *The function $t \longmapsto P_{A(t)}(E)h$, $t \in [0,1]$, is strongly measurable in $\mathcal{H}$ for each $h \in \mathcal{H}$ and $E \in \mathcal{B}(\mathbb{R})$. There exists an operator valued measure $M : \mathcal{B}([0,1]) \otimes \mathcal{B}(\mathbb{R}) \to \mathcal{L}_s(\mathcal{H})$, σ-additive for the strong operator topology, such that the equality*

$$(M(X \times Y)h, h) = \int_X (P_{A(t)}(Y)h, h) \, dt, \quad X \in \mathcal{B}([0,1]), \ Y \in \mathcal{B}(\mathbb{R}),$$

holds for each $h \in \mathcal{H}$. For each $V \in \mathcal{C}_1(\mathcal{H})$, the set function $E \longmapsto \text{tr}(VM(E))$, $E \in \mathcal{B}([0,1]) \otimes \mathcal{B}(\mathbb{R})$, is a complex measure and we have

$$\text{tr}(VM([0,1] \times Y)) = \int_0^1 \text{tr}(V P_{A(t)}(Y)) \, dt. \qquad (8.21)$$

Proof. If $f = \hat{\mu}$ is the Fourier transform of a finite measure μ, then

$$(P_{A(t)}h)(f) = \int_{\mathbb{R}} e^{-i\xi A(t)} h \, d\mu(\xi)$$

as a Bochner integral and by dominated convergence $t \longmapsto (P_{A(t)}h)(f)$, $0 \leq t \leq 1$, is continuous in $\mathcal{H}$ for each $h \in \mathcal{H}$. By a monotone class argument, $t \longmapsto (P_{A(t)}h)(f)$, $0 \leq t \leq 1$, is strongly measurable for all bounded Borel measurable functions f.

For each $h \in \mathcal{H}$, the set function (Mh, h) is nonnegative and finitely additive and the algebra $\mathcal{A}$ generated by product sets $X \times Y$ for $X \in \mathcal{B}([0,1])$ and $Y \in \mathcal{B}(\mathbb{R})$, so $|(M(A)h, h)| \leq \|h\|_1^2$, $A \in \mathcal{A}$. The set function $(Mh, h) : \mathcal{A} \to [0, \|h\|_2]$ is separately countably additive with respect to Borel sets, so it is inner regular with respect to compact product sets and so, countably additive (note that countable additivity may fail without inner-regularity).

Denoting the extended measure by the same symbol, $|(M(E)h, h)| \leq \|h\|_1^2$ for all $E \in \mathcal{B}([0,1]) \otimes \mathcal{B}(\mathbb{R})$. The $\mathcal{H}$-valued measure Mh is weakly countable additive by polarity and so norm countably additive by the Orlicz-Pettis Theorem.

For each $V \in \mathcal{C}_1(\mathcal{H})$ and orthonormal basis $\{h_j\}_j$ of $\mathcal{H}$, the bound

$$\sum_{j=1}^{\infty} |(VM(E)h_j, h_j)| \leq 4\|V\|_{\mathcal{C}_1(\mathcal{H})}, \quad E \in \mathcal{B}([0,1]) \otimes \mathcal{B}(\mathbb{R})$$

holds and

$$\text{tr}(VM([0,1] \times Y)) = \sum_{j=1}^{\infty} (VM([0,1] \times Y)h_j, h_j)$$

$$= \int_0^1 \sum_{j=1}^{\infty} (VP_{A(t)}(Y)h_j, h_j) \, dt$$

by the Beppo Levi convergence theorem, because

$$\sum_{j=1}^{\infty} |(VP_{A(t)}(Y)h_j, h_j)| \leq 4\|V\|_{\mathcal{C}_1(\mathcal{H})}, \quad 0 \leq t \leq 1,$$

so equation (8.21) holds. $\qquad\qquad\qquad\qquad\qquad\qquad\qquad\qquad\qquad\qquad$ □

An application of Fubini's Theorem for disintegrations of measures [16, Section 10.6] shows that

$$\int_{[0,1] \times \mathbb{R}} e^{-i\lambda s} \, d(Mh, h)(t, \lambda) = \int_{\mathbb{R}} e^{-i\lambda s} \left(\int_0^1 (P_{A(t)}h, h) \, dt \right) (d\lambda)$$

$$= \int_0^1 \int_{\mathbb{R}} e^{-i\lambda s} (P_{A(t)}h, h)(d\lambda) \, dt$$

$$= \int_0^1 (e^{-isA(t)}h, h) \, dt$$

for each $h \in \mathcal{H}$. The identity

$$\int_{\mathbb{R}} e^{-i\lambda s} \left(\int_0^1 (V P_{A(t)} h, h) \, dt \right) (d\lambda) = \int_0^1 (V e^{-isA(t)} h, h) \, dt$$

follows for each $h \in \mathcal{H}$ by polarisation. Because

$$\Xi(E) = \int_0^1 \mathrm{tr}(V P_{A(t)}(E)) \, dt = \sum_{j=1}^{\infty} \int_0^1 (V P_{A(t)}(E) h_j, h_j) \, dt, \quad E \in \mathcal{B}(\mathbb{R}),$$

for any orthonormal basis $\{h_j\}_j$ of $\mathcal{H}$, the Fourier transform of the measure Ξ is

$$\int_{\mathbb{R}} e^{-i\lambda s} \, d\Xi(\lambda) = \int_0^1 \mathrm{tr}(V e^{-isA(t)}) \, dt$$

$$= i \int_0^1 s^{-1} \frac{d}{dt} \mathrm{tr}(e^{-isA(t)}) \, dt$$

$$= i \frac{\mathrm{tr}(e^{-isA} - e^{-isB})}{s}.$$

We need to establish that the inverse Fourier transform $\check{\Phi}$ of the uniformly bounded, continuous function

$$\Phi : s \longmapsto i \frac{\mathrm{tr}(e^{-isA} - e^{-isB})}{s}, \quad s \in \mathbb{R} \setminus \{0\}, \quad \Phi(0) = \mathrm{tr}(V),$$

belongs to $L^1(\mathbb{R})$. Then $\xi = \check{\Phi}$ is the spectral shift function. Clearly, the value of Φ at 0 is irrelevant.

It suffices to show that there exists $\xi \in L^1(\mathbb{R})$ such that

$$\mu(\Phi) = 2\pi \int_{\mathbb{R}} \xi(t) \overline{\check{\mu}(t)} \, dt = \int_{\mathbb{R}} \xi(t) \hat{\mu}(t) \, dt$$

with $\check{\mu}(t) = (2\pi)^{-1} \int_{\mathbb{R}} e^{ist} \, d\mu(s)$ and $\hat{\mu}(t) = \int_{\mathbb{R}} e^{-ist} \, d\mu(s)$, $t \in \mathbb{R}$, for every finite positive measure μ, because then $\check{\Phi} = \xi$ as elements of the space $\mathcal{S}'$ of Schwartz distributions on $\mathbb{R}$ [121, Definition 7.11]. So, we consider the class of functions $f : \mathbb{R} \to \mathbb{R}$ for which $f' = \hat{\mu}$ and $f(0) = 0$ and consequently $\mathrm{tr}(f(A) - f(B)) = (2\pi)^{-1} \mu(\Phi)$.

Theorem 8.11. *Let $\mathcal{H}$ be a separable Hilbert space and let A and B be selfadjoint operators with the same domain such that $A - B \in \mathcal{C}_1(\mathcal{H})$. Then there exists a function $\xi \in L^1(\mathbb{R})$ such that*

$$\mathrm{tr}(f(A) - f(B)) = \int_{\mathbb{R}} f'(\lambda) \xi(\lambda) \, d\lambda \tag{8.22}$$

for every function $f : \mathbb{R} \to \mathbb{C}$ for which there exists a finite positive Borel measure μ on $\mathbb{R}$ such that

$$f(x) = i \int_{\mathbb{R}} \frac{e^{-isx} - 1}{s} \, d\mu(s), \quad x \in \mathbb{R}.$$

Furthermore, ξ possesses the following properties.

a) $\operatorname{tr}(A - B) = \int_{\mathbb{R}} \xi(\lambda) \, d\lambda$.
b) $\|\xi\|_1 \leq \|A - B\|_{\mathcal{C}_1(\mathcal{H})}$.
c) *If $B \leq A$, then $\xi \geq 0$ a.e.*
d) *ξ is zero a.e. outside the interval $(\inf(\sigma(A) \cup \sigma(B)), \sup(\sigma(A) \cup \sigma(B)))$.*

Proof. The following proof is adapted from [17].

The estimate

$$\|f(A) - f(B)\|_{\mathcal{C}_1(\mathcal{H})} \leq \mu(\mathbb{R})\|A - B\|_{\mathcal{C}_1(\mathcal{H})} \tag{8.23}$$

follows from the bound (8.20) and the calculation

$$f(A) - f(B) = \frac{i}{2\pi} \int_{\mathbb{R}} \frac{e^{-isA} - e^{-isB}}{s} \, d\mu(s)$$

obtained from an application of Fubini's theorem with respect to $P_A \otimes \mu$ and $P_B \otimes \mu$ on $\mathbb{R} \times [\epsilon, \infty)$ for $\epsilon > 0$. Then

$$\operatorname{tr}(f(A) - f(B)) = \frac{1}{2\pi} \int_{\mathbb{R}} \Phi \, d\mu.$$

An expression for the spectral shift function ξ may be obtained from Fatou's Theorem [120, Theorem 11.24]. Suppose that ν is a finite measure on $\mathbb{R}$

$$\phi_\nu(z) = \frac{1}{2\pi i} \int_{\mathbb{R}} \frac{d\nu(\lambda)}{\lambda - z}, \quad z \in \mathbb{C} \setminus \mathbb{R},$$

is the Cauchy transform of ν. Then ν is absolutely continuous if

$$\hat{\nu}(\xi) = \int_{\mathbb{R}} e^{-i\xi x}(\phi_\nu(x + i0+) - \phi_\nu(x + i0-)) \, dx, \quad \xi \in \mathbb{R}.$$

The function $x \longmapsto \phi_\nu(x + i0+) - \phi_\nu(x + i0-)$ defined for almost all $x \in \mathbb{R}$ is then the density of ν with respect to Lebesgue measure. For $\nu = \Xi$, if the representation

$$\Phi(s) = i \frac{\operatorname{tr}(e^{-isA} - e^{-isB})}{s}$$

$$= \frac{1}{2\pi i} \int_{\mathbb{R}} e^{-isx} \left(\lim_{\epsilon \to 0+} \int_0^1 \operatorname{tr}(V(A + tV - x - i\epsilon)^{-1} - V(A + tV - x + i\epsilon)^{-1}) \, dt \right) dx$$

were valid, we would expect that $\xi = \check{\Phi}$ has the representation

$$\xi(s) = \frac{1}{2\pi i} \lim_{\epsilon \to 0+} \int_{\mathbb{R}} e^{isx - \epsilon|x|} \frac{\text{tr}(e^{-ixA} - e^{-ixB})}{x} dx, \quad s \in \mathbb{R},$$

$$= \lim_{\epsilon \to 0+} \frac{1}{\pi} \text{tr} \left[\arctan\left(\frac{A - sI}{\epsilon}\right) - \arctan\left(\frac{B - sI}{\epsilon}\right) \right], \qquad (8.24)$$

where the arctan function may be expressed as

$$\arctan t = \frac{1}{2i} \int_{\mathbb{R}} \frac{e^{ist} - 1}{s} e^{-|s|} ds, \quad t \in \mathbb{R}. \qquad (8.25)$$

For the function defined by

$$h(x, y) = \frac{1}{\pi} \text{tr} \left[\arctan\left(\frac{A - xI}{y}\right) - \arctan\left(\frac{B - xI}{y}\right) \right]$$

we have the bounds

$$\pi|h(x, y)| \leq \left\| \arctan\left(\frac{A - xI}{y}\right) - \arctan\left(\frac{B - xI}{y}\right) \right\|_{\mathcal{C}_1(\mathcal{H})}$$

$$\leq \frac{1}{y} \|A - B\|_{\mathcal{C}_1(\mathcal{H})},$$

from the bound (8.23) and the representation (8.25). Rewriting

$$h(x, y) = \frac{1}{2\pi i} \int_{\mathbb{R}} e^{-ixs - y|s|} \text{tr} \left[\frac{e^{isA} - e^{isB}}{s} \right] ds$$

using (8.25), it follows that $h(x, y)$ is harmonic in the upper half-plane $\{(x, y) : x \in \mathbb{R}, \ y > 0\}$.

We first look at the case that $A - B = \alpha(\cdot, w)w$ for $\alpha > 0$ and $w \in \mathcal{H}$, $\|w\| = 1$, so that A is a rank one perturbation of the bounded selfadjoint operator B.

If we set

$$X = 2 \arctan \frac{A - x}{y}, Y = 2 \arctan \frac{B - x}{y},$$

then $2\pi h = \text{tr}(X - Y)$. The formula $\text{tr} \log(e^{iX} e^{-iY}) = i\text{tr}(X - Y)$ follows from the Baker-Campbell-Hausdorff formula for large $y > 0$, see [17, Lemma 1.1]. Let $T_A = e^{-iX}$, $T_B = e^{-iY}$. Then for $z = x + iy$, spectral theory gives

$$T_A = (A - \bar{z}I)(A - zI)^{-1} = I + 2iy(A - zI)^{-1},$$
$$T_B = (B - \bar{z}I)(B - zI)^{-1} = I + 2iy(B - zI)^{-1}.$$

Our aim is to compute $\text{tr} \log(U)$ for the unitary operator $U = T_A^* T_B$. Because

$$U - I = T_A^* T_B - T_B^* T_B$$

$$= (T_A^* - T_B^*) T_B$$

$$= -i2y[(A - \bar{z}I)^{-1} - (B - \bar{z}I)^{-1}] T_B,$$

we obtain

$$U = I + i2y(A - \bar{z}I)^{-1}(A - B)(B - zI)^{-1}.$$

Substituting $A - B = \alpha(\cdot, w)w$ gives

$$U = I + i2y\alpha(\cdot, (B - \bar{z}I)^{-1}w)(A - \bar{z}I)^{-1}w.$$

The vector $(A - \bar{z}I)^{-1}w$ is an eigenvector for the unitary operator U with eigenvalue

$$1 + i2y\alpha((A - \bar{z}I)^{-1}w, (B - \bar{z}I)^{-1}w)$$

which can be expressed as $e^{i2\pi\theta(x,y)}$ for a continuous function θ in the upper half plane such that $0 < \theta < 1$. Consequently, for large $y > 0$,

$$i2\pi\theta = \operatorname{tr}\log(U) = i\operatorname{tr}(X - Y) = i2\pi h.$$

Then θ is harmonic for large $y > 0$ so it is harmonic on the upper half plane and it is equal to h there, so $0 < h < 1$.

By Fatou's Theorem, the boundary values $\xi(x) = \lim_{y \to 0+} h(x, y)$ are defined for almost all $x \in \mathbb{R}$ and satisfy

$$\lim_{y \to \infty} \pi y h(x, y) = \int_{\mathbb{R}} \xi(t)\, dt = \|\xi\|_1 \leq \|A - B\|_{C_1(\mathcal{H})}$$

for every $x \in \mathbb{R}$, so in the case that $A - B$ has rank one, formula (8.24) is valid.

For an arbitrary selfadjoint perturbation $V = \sum_{j=1}^{\infty} \alpha_j(\cdot, w_j)w_j$ with $\sum_{j=1}^{\infty} |\alpha_j| = \|A - B\|_{C_1(\mathcal{H})} < \infty$, the function $\xi_n \in L^1(\mathbb{R})$ may be defined in a similar fashion for $A_n = B + \sum_{j=1}^{n} \alpha_j(\cdot, w_j)w_j$, $n = 1, 2, \ldots$, so that $\xi_n \to \xi$ in $L^1(\mathbb{R})$ as $n \to \infty$ from which it verified that $\xi = \check{\Phi}$. $\square$

The representation $\xi = \check{\Phi}$ obtained above may be viewed as the Fourier transform approach. In the case of a rank one perturbation $V = \alpha(\cdot, w)w$, the Cauchy transform approach is developed by B. Simon [126] with the formula

$$\operatorname{tr}((A - zI)^{-1} - (B - zI)^{-1}) = -\int_{\mathbb{R}} \frac{\xi(\lambda)}{(\lambda - z)^2}\, d\lambda$$

for $z \in \mathbb{C} \setminus [a, \infty)$ for some $a \in \mathbb{R}$, established in [126, Theorem 1.9] by computing a contour integral. Here the boundary value $\xi(x) = \lim_{y \to 0+} h(x, y)$ is expressed as

$$\xi(x) = \frac{1}{\pi}\operatorname{Arg}(1 + \alpha F(\lambda + i0+))$$

for almost all $x \in \mathbb{R}$ with respect to the Cauchy transform

$$F(z) = \int_{\mathbb{R}} \frac{d(P_B w, w)(\lambda)}{\lambda - z}, \quad z \in \mathbb{C} \setminus (-\infty, a).$$

The Cauchy transform approach is generalised to type II von Neumann algebras in [10].

Many different proofs of Krein's formula (8.22) are available for a wide class of functions f, especially in a form that translates into the setting of noncommutative integration [10, 105, 114]. As remarked in [15, p. 163], an ingredient additional to double operator integrals (such as complex function theory) is needed to show that the measure Ξ is absolutely continuous with respect to Lebesgue measure on $\mathbb{R}$. Krein's original argument uses perturbation determinants from which follows the representation $\mathrm{Det}(S(\lambda)) = e^{-2\pi i \xi(\lambda)}$ for the scattering matrix $S(\lambda)$ for A and B [136, Chapter 8].

Bibliography

[1] D. Adams and L. Hedberg, Function spaces and potential theory, Grundlehren der Mathematischen Wissenschaften, **314**, Springer-Verlag, Berlin, 1996.

[2] S. Albeverio, Z. Brzeźniak and L. Dąbrowski, Fundamental solution of the heat and Schrödinger equations with point interaction, *J. Funct. Anal.* **130** (1995), 220–254.

[3] S. Albeverio, K. Makarov and A. Motovilov, Graph subspaces and the spectral shift function, *Canad. J. Math.* **55** (2003), 449–503.

[4] S. Albeverio and A. Motovilov, Operator Stieltjes integrals with respect to a spectral measure and solutions of some operator equations, *Trans. Moscow Math. Soc.* (2011) 45–77.

[5] W. Amrein, *Hilbert Space Methods in Quantum Mechanics*, EPFL Press, 2009.

[6] W. Amrein, A. Boutet de Monvel and V. Georgescu, C_0-*Groups, Commutator Methods and Spectral Theory of N-Body Hamiltonians* (Progress in Mathematics **135**, Basel: Birkhäuser), 1996.

[7] W.O. Amrein, V. Georgescu and J. Jauch, *Stationary state scattering theory*, Helv. Phys. Acta **44** (1971) 407–434.

[8] W. Amrein, J. Jauch and K. Sinha, *Scattering Theory in Quantum Mechanics: Physical Principles and Mathematical Methods* (Reading: W.A. Benjamin), 1977.

[9] J. Arthur, An introduction to the trace formula, *Harmonic Analysis, the Trace Formula, and Shimura Varieties*, Clay Math. Proc. **4**, Amer. Math. Soc., Providence, RI, 2005, 1–263

[10] N. Azamov, P. Dodds and F. Sukochev, The Krein Spectral Shift Function in Semifinite von Neumann Algebras, *Integr. Equ. Oper. Theory* **55** (2006), 347–362.

[11] R. Bartle, A general bilinear vector integral, *Studia Math.* **15** (1956), 337-351.

[12] A. Berthier, *Spectral Theory and Wave Operators for the Schrödinger Equation.* (Research Notes in Mathematics), Pitman, 1982.

[13] R. Bhatia, C. Davis and A. McIntosh, Perturbation of spectral subspaces

and solution of linear operator equations, *Linear Algebra Appl.* **52/53** (1983), 45–67.

[14] R. Bhatia and P. Rosenthal, How and why to solve the operator equation $AX - XB = Y$, *Bull. London Math. Soc.* **29** (1997), 1–21.

[15] M. Birman and M. Solomyak, Double operator integrals in a Hilbert space, *Integr. Equ. Oper. Theory* **47** (2003) 131–168.

[16] V. Bogachev, *Measure Theory*, Springer-Verlag, Berlin, 2007.

[17] K. Boyadzhiev, Krein's trace formula and the spectral shift function, *Int. J. Math. Math. Sci.* **25** (2001), 239–252.

[18] C. Brislawn, Kernels of trace class operators, *Proc. Amer. Math. Soc.* **104** (1988), 1181–1190.

[19] _____, Traceable integral kernels on countably generated measure spaces, *Pacific J. Math.* **150** (1991), 229–240.

[20] A. Bukhvalov, A. Gutman, V. Korotkov, A. Kusraev, S. Kutateladze and B. Makarov, *Vector lattices and integral operators*, Mathematics and its Applications, **358**, Kluwer Academic Publishers Group, Dordrecht, 1996.

[21] A. Carey and F. Sukochev, Dixmier traces and some applications in non-commutative geometry, *Russian Math. Surveys* **61** (2006), 1039–1099.

[22] T. Carleman, Über die Fourierkoeffizienten einer stetigen Funktion, *Acta Math.* **41** (1916), 377–384.

[23] M. Castro, V. Menegatto and A. Péron, Integral operators generated by Mercer-like kernels on topological spaces, *Colloq. Math.* **126** (2012), 125–138.

[24] _____, Traceability of positive integral operators in the absence of a metric, *Banach J. Math. Anal.* **6** (2012), 98–112.

[25] R. Chivukula and A. Sastry, Product vector measures via Bartle integrals, *J. Math. Anal. Appl.* **96** (1983), 180–194.

[26] S. Chobanyan, V. Tarieladze and V. Vakhania, *Probability Distributions on Banach Spaces*, Mathematics and its Applications (Soviet Series), **14** (transl. W. Woyczynski) D. Reidel Publishing Co., Dordrecht, 1987.

[27] K.-L. Chung and J. Doob, Fields, optionality and measurability. *Amer. J. Math.* (1964) **87**, 397–424.

[28] K.-L. Chung and J. Walsh, *Markov Processes, Brownian Motion, and Time Symmetry*, 2nd Ed., Grundlehren der Mathematischen Wissenschaften **249**, Springer, New York, 2005.

[29] A. Connes and M. Marcolli, *Noncommutative Geometry, Quantum Fields and Motives*, http://www.alainconnes.org/docs/bookwebfinal.pdf

[30] M. Cwikel, Weak type estimates for singular values and the number of bound states of Schrödinger operators, *Ann. Math.* **106** (1977), 93–100.

[31] J. Delgado, Trace formulas for nuclear operators in spaces of Bochner integrable functions, *Monatsh. Math.* **172** (2013), 259–275.

[32] _____, The trace of nuclear operators on $L^p(\mu)$ for σ-finite Borel measures on second countable spaces, *Integ. Equ. Oper. Theory* **68** (2010), 61–74.

[33] _____, A trace formula for nuclear operators on L^p, *Oper. Theory Adv. Appl.* **205** (2009), 181–193.

[34] C. Dellacherie and P.-A. Meyer, *Probabilités et potentiel*, Hermann, Paris,

1975, Chapitres I á IV, Édition entièrement refondue, Publications de l'Institut de Mathématique de l'Université de Strasbourg, No. XV, Actualités Scientifiques et Industrielles, No. 1372.

[35] J. Dereziński and C. Gérard, *Scattering Theory of Classical and Quantum N-particle Systems* (Springer), 1997.

[36] J. Diestel, H. Jarchow and A. Tonge, *Absolutely Summing Operators*, Cambridge Studies in Advanced Mathematics **43**, Cambridge University Press, Cambridge, 1995.

[37] G. Di Nunno and Yu. A. Rozanov, On measurable modification of stochastic functions, *Teor. Veroyatnost. i Primenen.* **46** (2001), 175–180, transl. *Theory Probab. Appl.* **46** (2002), 122–127.

[38] J. Diestel and J.J. Uhl, Jr., *Vector Measures*, Math. Surveys No. **15**, Amer. Math. Soc., Providence, 1977.

[39] I. Dobrakov, On integration in Banach spaces I, *Czech. Math. J.* **20** (1970), 511–36.

[40] ———, On integration in Banach spaces II, *Czech. Math. J.* **20** (1970), 680–95.

[41] ———, On representation of linear operators on $C_0(T, \mathbf{X})$, *Czech. Math. J.* **21** (1971), 13–30.

[42] I. Dobrakov and T. Panchapagesan, A generalized Pettis measurability criterion and integration of vector functions, *Studia Math.* **164** (2004), 205–229.

[43] N. Dunford and J. Schwartz, *Linear Operators*, Part I, Interscience, New York, 1958.

[44] J. Ferreira, V. Menegatto and C. Oliveira, On the nuclearity of integral operators, *Positivity* **13** (2009), 519–541.

[45] D. Fremlin, *Measure Theory. Vol. 2*, (Broad Foundations) Torres Fremlin, Colchester, 2003.

[46] ———, *Measure Theory. Vol. 4*, (Topological measure spaces. Part I, II, Corrected second printing of the 2003 original) Torres Fremlin, Colchester, 2006.

[47] F. Freniche and J. Garcia-Vázquez, The Bartle bilinear integration and Carleman operators, *J. Math. Anal. Appl.* **240** (1999), 324–339.

[48] M. Fukushima, Y. Oshima and M. Takedai, *Dirichlet Forms and Symmetric Markov Processes*, de Gruyter Studies in Mathematics **19**, Walter de Gruyter & Co., Berlin, 2011.

[49] L.M. Garcia-Raffi and B. Jefferies, An application of bilinear integration to quantum scattering, *J. Math. Anal. Appl.* **415** (2014) 394–421.

[50] D.J.H. Garling, Brownian motion and UMD-spaces, *Probability and Banach Spaces* (Zaragoza, 1985), 36–49, *Lecture Notes in Math.* **1221**, Springer-Verlag, Berlin, 1986.

[51] I. Gohberg and M. Krein, *Introduction to the Theory of Linear Nonselfadjoint Operators*, Translations of Mathematical Monographs **18**, Amer. Math. Soc., Providence, R.I., 1969.

[52] I. Gohberg and M. Krein, *Theory and Applications of Volterra Operators*

in Hilbert Space, Izdat. Nauka, Moscow 1967, transl. Amer. Math. Soc., Providence (1970).

[53] I. Gohberg, S. Goldberg and N. Krupnik, *Traces and Determinants of Linear Operators*, Operator Theory: Advances and Applications **116**, Birkhäuser Verlag, Basel, 2000.

[54] L. Grafakos, *Classical Fourier Analysis*, Graduate Texts in Mathematics **249**, 2nd Ed., Springer, New York, 2008.

[55] N. Gretsky and J. Uhl, Jr., Carleman and Korotkov operators on Banach spaces, *Acta Sci. Math. (Szeged)* (1981), 207–218.

[56] R. Griego and R. Hersh, Random evolutions, Markov chains and systems of partial differential equations, *Proc. Nat. Acad. Sci. USA* **62** (1969), 305–308.

[57] L. Gross, Logarithmic Sobolev inequalities and contractivity properties of semigroups, *Dirichlet Forms* (Varenna, 1992), Lecture Notes in Math. **1563**, Springer, Berlin, 1993, 54–88.

[58] J.W. Hagood, The operator-valued Feynman-Kac formula with noncommutative operators, *J. Funct. Anal.* **38** (1980), 99–117.

[59] P. Halmos, *Measure Theory*, Van Nostrand, New York, 1950.

[60] P. Halmos and V. Sunder, *Bounded Integral Operators on L^2 Spaces*, Ergebnisse der Mathematik **96**, Springer-Verlag, Berlin-New York, 1978.

[61] R. Hersh, Random evolutions: A survey of results and problems, *Rocky Mountain J. Math.* **4** (1974), 443–477.

[62] _____, The birth of random evolutions, *Math. Intelligencer* **25** (2003), 53–60.

[63] F. Hiai and H. Kosaki, *Means of Hilbert Space Operators*, Lecture Notes in Mathematics **1820**, Springer-Verlag, Berlin, 2003.

[64] B. Jefferies, *Evolution Processes and the Feynman-Kac Formula*, Kluwer Academic Publishers, Dordrecht/Boston/London, 1996.

[65] _____, Some recent applications of bilinear integration, *Vector Measures, Integration and Related Topics*, 255–269, Oper. Theory Adv. Appl. **201**, Birkhäuser Verlag, Basel, 2010.

[66] _____, Lattice trace operators, *Journal of Operators* **2014**, Article ID 629502.

[67] _____, The CLR inequality for dominated semigroups, *Math. Phys. Anal. Geom.* **17** (2014), 115–137, DOI 10.1007/s11040-014-9145-6.

[68] _____, Measurable processes and the Feynman-Kac formula, *Indag. Math. (N.S.)* **27** (2016), 296–306.

[69] B. Jefferies and S. Okada, Pettis integrals and singular integral operators, *Illinois J. Math.*, **38** (1994), 250–272.

[70] _____ Bilinear integration in tensor products, *Rocky Mountain J. Math.* **28** (1998), 517–545.

[71] _____, Dominated semigroups of operators and evolution processes, *Hokkaido Math. J.* **33** (2004), 127–151.

[72] B. Jefferies, S. Okada and L. Rodrigues-Piazza, L^p-valued measures without finite X-semivariation for $2 < p < \infty$ *Quaest. Math.* **30** (2007), 437–449.

[73] B. Jefferies and P. Rothnie, Bilinear integration with positive vector mea-

sures, *J. Aust. Math. Soc.* **75** (2003), 279–93.

[74] S. Kaden and J. Potthoff, Progressive stochastic processes and an application to the Itô integral, *Stochastic Anal. Appl.* **22** (2004), 843–865.

[75] I. Karatzas and S. Shreve, *Brownian Motion and Stochastic Calculus*, Graduate Texts in Mathematics **113**, 2nd Ed., Springer-Verlag, New York, 1991.

[76] T. Kato, *Perturbation Theory for Linear Operators*, Springer-Verlag, New York, 1980.

[77] S. Karlin and M. Pinsky, *An Introduction to Stochastic Modeling*, Academic Press, New York/Oxford, 2011.

[78] E. Kissin and V. Shulman, Operator multipliers, *Pacific J. Math.* **227** (2006), 109–141.

[79] A. Yu. Kitaev, A. Shen and M. Vyalyi, *Classical and quantum computation*, Graduate Studies in Mathematics, **47**, transl. L. Senechal, Amer. Math. Soc., Providence, RI, 2002.

[80] I. Kluvánek, The extension and closure of vector measure, in *Vector and operator valued measures and applications* (Proc. Sympos., Alta, Utah, 1972), 175–190, Academic Press, New York, 1973.

[81] _____, Représentations intégrales d'évolutions perturbées. (French. English summary) *C. R. Acad. Sci. Paris Sér. A-B* **288** (1979), no. 23, A1065–A1067.

[82] _____, Applications of Vector Measures, *Contemporary Mathematics* **2** (1980), Amer. Math. Soc., Providence, Rhode Island, 101–133.

[83] _____, Operator valued measures and perturbations of semi-groups , *Arch. Rat. Mech. & Anal.* **81** (1983), 161–180.

[84] _____, Integration and the Feynman-Kac formula, *Studia Mathematica* **86** (1987), 36–37.

[85] _____, *Integration structures*, Australian Nat. Univ., Canberra, Proc. Centre for Mathematical Analysis **18**, 1988.

[86] I. Kluvánek and G. Knowles, *Vector Measures and Control Systems*, North Holland, Amsterdam, 1976.

[87] G. Köthe, *Topological Vector Spaces I*, Springer-Verlag, Berlin, 1969.

[88] _____, *Topological Vector Spaces II*, Springer-Verlag, Berlin, 1979.

[89] P. Li and S-T. Yau, On the Schrödinger equation and the eigenvalue problem, *Comm. Math. Phys.* **88** (1983), 309–318.

[90] E.H. Lieb, Bounds on the eigenvalues of the Laplace and Schrödinger operators, *Bull. Amer. Math. Soc.* **82** (1976), 751–753.

[91] D. Levin and M. Solomyak: The Rozenblum-Lieb-Cwikel inequality for Markov generators, *J. Anal. Math.* **71** (1997), 173–193.

[92] D. Lewis, An isomorphic characterization of the Schmidt class, *Compos. Math.* **30** (1975), 293–297.

[93] J. Lindenstrauss and L. Tzafriri, *Classical Banach Spaces I*, Sequence Spaces, Springer-Verlag, Berlin, New York, 1977.

[94] G.L. Litvinov, Nuclear operators, *Encyclopedia of Mathematics* (ed. M. Hazewinkel), Springer, 2001.

[95] P. Masani, Orthogonally scattered measures, *Advances in Math.* **2** (1968), 61–117.

[96] P. Meyer-Neiberg, *Banach Lattices*, Springer-Verlag, Berlin, 1991.

[97] S. Molchanov and B. Vainberg, On general Cwikel-Lieb-Rozenblum and Lieb-Thirring inequalities, *Around the research of Vladimir Maz'ya. III*, Int. Math. Ser. (N. Y.) **13**, Springer, New York, 2010, 201–246.

[98] K. Musial, Pettis integral, *Handbook of Measure Theory, Vol. I, II*, 531–586, North-Holland, Amsterdam, 2002.

[99] S. Okada, W. Ricker and E. Sánchez Pérez, *Optimal Domain and Integral Extension of Operators. Acting in Function Spaces*, Operator Theory: Advances and Applications **180**, Birkhäuser Verlag, Basel, 2008.

[100] M. Ondreját and J. Seidler, On existence of progressively measurable modifications, *Electron. Commun. Probab.* **18** (2013), 1–6.

[101] B. de Pagter, W. Witvliet and F. Sukochev, Double operator integrals, *J. Funct. Anal.* **92** (2002), 52–111.

[102] R. Pallu de La Barrière, Integration of vector functions with respect to vector measures, *Studia Univ. Babeş-Bolyai Math.* **43** (1998), 55–93.

[103] T. Panchapagesan, On the distinguishing features of the Dobrakov integral, *Divulg. Mat.* **3** (1995), 79–114.

[104] V. Peller, Hankel operators in the theory of perturbations of unitary and selfadjoint operators. (Russian) *Funktsional. Anal. i Prilozhen.* **19** (1985), 37–51, Eng. Transl. *Functional Anal. Appl.* **19** (1985), 111–123.

[105] ———, Hankel operators in the perturbation theory of unbounded selfadjoint operators, in C. Sadosky (ed.), *Analysis and partial differential equations. A collection of papers dedicated to M. Cotlar.* Lecture Notes in Pure and Applied Mathematics **122**, Marcel Dekker, New York, N.Y., 1990, 529–544.

[106] V.-Q. Phóng, The operator equation $AX - XB = C$ with unbounded operators A and B and related abstract Cauchy problems, *Math. Z.* **208** (1991), 567–588.

[107] A. Pietsch, *Nuclear locally convex spaces*, Ergebnisse der Mathematik und ihrer Grenzgebiete **66**, Springer-Verlag, New York-Heidelberg, 1972.

[108] A. Pietsch, *Eigenvalues and s-Numbers*, Geest & Portig, Leipzig, and Cambridge Univ. Press, 1987.

[109] ———, Traces and shift invariant functionals, *Math. Nachr.* **145** (1990), 7–43.

[110] ———, Traces on operator ideals and related linear forms on sequence ideals (part I), *Indag. Math. (N.S.)* **25** (2014), 341–365.

[111] ———, Traces on operator ideals and related linear forms on sequence ideals (part II), *Integr. Equ. Oper. Theory* **79** (2014), 255–299.

[112] G. Pisier, Grothendieck's theorem, past and present, *Bull. Amer. Math. Soc. (N.S.)* **49** (2012), 237–323.

[113] D. Potapov and F. Sukochev, Operator-Lipschitz functions in Schatten-von Neumann classes, *Acta Math.* **207** (2011), 375–389.

[114] D. Potapov, F. Sukochev, and D. Zanin, Kreins trace theorem revisited, *J. Spectr. Theory* **4** (2014), 1–16.

[115] M. Reed and B. Simon, *Methods of Modern Mathematical Physics I-IV* Academic Press, New York, 1973.

[116] W. Ricker, Separability of the L^1-space of a vector measure, *Glasgow Math. J.* **34** (1992), 1–9.

[117] J. Rosiński and Z. Suchanecki, On the space of vector-valued functions integrable with respect to the white noise, *Colloq. Math.* **43** (1980), 183–201.

[118] G.V. Rozenbljum, The distribution of the discrete spectrum for singular differential operators, *Dokl. Akad. Nauk SSSR* **202** (1972), 1012–1015.

[119] G. Rozenbljum and M. Solomyak, CLR-estimate revisited: Lieb's approach with no path integrals, *Journeé "Équations aux Deriveés Partielles"* (Saint-Jean-de-Monts, 1997), Exp.No.XVI, École Polytech., Palaiseau, 1997, 1–10.

[120] W. Rudin, *Real and Complex Analysis*, 3rd Ed., McGraw-Hill, 1986.

[121] _____, *Functional Analysis*, 2nd Ed., McGraw-Hill New York, 1987.

[122] L. Saloff-Coste, Sobolev inequalities in familiar and unfamiliar settings, *Sobolev Spaces in Mathematics* I, Int. Math. Ser. (N.Y.) **8** Springer, New York, 2009, 299–343.

[123] H. Schaefer, *Topological Vector Spaces*, Graduate Texts in Mathematics **3**, Springer-Verlag, Berlin/Heidelberg/New York, 1980.

[124] _____, *Banach Lattices and Positive Operators*, Springer-Verlag, Grundlehren Math. Wiss. **215**, Berlin/Heidelberg/New York, 1974.

[125] L. Schwartz, *Radon Measures on Arbitrary Topological Spaces and Cylindrical Measures*, Tata Institute Publications, Oxford University Press, Bombay, 1973.

[126] B. Simon, Spectral analysis of rank one perturbations and applications, *Mathematical Quantum Theory. II. Schroedinger Operators* (Vancouver, BC, 1993), pp. 109–149 (J. Feldman, R Froese, and L. M. Rosen, eds.), CRM Proc. Lecture Notes **8**, Amer. Math. Soc., Providence, RI, 1995.

[127] B. Simon, *Trace ideals and their applications*. 2nd ed., Mathematical Surveys and Monographs **120**, Amer. Math. Soc., Providence, RI, 2005.

[128] _____, *Functional Integration and Quantum Physics*, 2nd Ed., Amer. Math. Soc. Chelsea, Providence, 2005.

[129] N. Spronk, Measurable Schur multipliers and completely bounded multipliers of the Fourier algebras, *Proc. London Math. Soc. (3)* **89** (2004), 161–192,

[130] M. Talagrand, Pettis integral and measure theory, *Mem. Amer. Math. Soc.*, **51**, 1984.

[131] I. Todorov and L. Turowska, Schur and operator multipliers, *Banach Algebras 2009*, Banach Center Publ. **91**, 385–410.

[132] J.M.A.M. van Neerven and L. Weis, Stochastic integration of functions with values in a Banach space, *Studia Math.* **166** (2005), 131–170.

[133] J.M.A.M. van Neerven, M.C. Veraar and L. Weis, Stochastic integration in UMD Banach spaces, *Ann. Probab.* **35** (2007), 1438–1478.

[134] J. von Neumann, *Mathematical Foundations of Quantum Mechanics*, Princeton, NJ: Princeton University Press, 1955 [First published in German in 1932: *Mathematische Grundlagen der Quantenmechank*, Berlin: Springer]; http://plato.stanford.edu/entries/qt-nvd/#1

[135] J. Weidman, Integraloperatoren der spurklasse, *Math. Ann.* **163** (1966), 340–345.

[136] D. Yafaev, *Mathematical Scattering Theory: General Theory*, Providence, RI, Amer. Math. Soc., 1992.

[137] ———— *Scattering Theory: Some Old and New Problems* Lecture Notes in Mathematics **1735**, Berlin, Springer, 2000.

[138] A.C. Zaanen, *Riesz Spaces II*, North Holland, Amsterdam, New York, Oxford, 1983.

Index

Printed in the United States
By Bookmasters